Fast Light, Slow Light and Left-Handed Light

Series in Optics and Optoelectronics

Series Editors: **R G W Brown**, Queens University Belfast, UK
E R Pike, Kings College, London, UK

Other titles in the series

Optical Fibre Devices
J-P Goure and I Verrier

Handbook of Moiré Measurement
C A Walker (ed)

Laser-Induced Damage of Optical Materials
R M Wood

Optical Applications of Liquid Crystals
L Vicari (ed)

Diffractional Optics of Millimetre Waves
I V Minin and O V Minin

Diode Lasers
D Sands

Forthcoming titles in the series

High Speed Photonic Devices
N Dagli (ed)

Photonic Crystals
M Charlton and G Parker (eds)

An Introduction to Quantum Optics
Y Shih

High Aperture Focussing of Electromagnetic Waves and Applications in Optical Microscopy
C J R Sheppard and P Torok

An Introduction to Biomedical Optics
R Splinter and B Hooper

Series in Optics and Optoelectronics

Fast Light, Slow Light and Left-Handed Light

P W Milonni

Los Alamos, New Mexico

CRC Press
Taylor & Francis Group
Boca Raton London New York

CRC Press is an imprint of the
Taylor & Francis Group, an **informa** business
A TAYLOR & FRANCIS BOOK

First published 2005 by Taylor & Francis Group

Published 2020 by CRC Press
Taylor & Francis Group
6000 Broken Sound Parkway NW, Suite 300
Boca Raton, FL 33487-2742

First issued in paperback 2020

© 2005 by Taylor & Francis Group, LLC
CRC Press is an imprint of Taylor & Francis Group, an Informa business

No claim to original U.S. Government works

ISBN 13: 978-0-367-57820-6 (pbk)
ISBN 13: 978-0-7503-0926-4 (hbk)

Visit the Taylor & Francis Web site at
http://www.taylorandfrancis.com

and the CRC Press Web site at
http://www.crcpress.com

Library of Congress Cataloging-in-Publication Data

Catalog record is available from the Library of Congress

Taylor & Francis Group
is the Academic Division of T&F Informa plc.

To Enes Novelli Burns, my favourite teacher

My books are water; those of the great geniuses is wine.
Everybody drinks water.

Mark Twain
Notebooks and Journals, Volume III (1883–1891)

Contents

Preface

It has been a century since R W Wood observed anomalous dispersion and Sommerfeld, Brillouin and others developed the theory of the propagation of light in anomalously dispersive media. The problem was to reconcile (1) the possibility that the (measurable) group velocity of light could exceed c with (2) the requirement of relativity theory that no signal can be transmitted superluminally. Sommerfeld and Brillouin concluded that a group velocity is not, in general, the velocity with which a signal, properly defined as a carrier of information, can be transmitted.

The work of Sommerfeld and Brillouin, especially Brillouin's *Wave Propagation and Group Velocity* (1960), is often cited. They focused attention on signal velocity, group velocity, and the velocity of energy propagation; and, according to Brillouin, 'a galaxy of eminent scientists, from Voigt to Einstein, attached great importance to these fundamental definitions'. But apparently this classic work is not widely read, for otherwise the recent demonstrations of superluminal group velocity would not have sparked so much discussion. The news media, with the hyperbole characteristic of the times, have often as not been misleading or wrong but so have the reported comments of some physicists.

The principal development since the publication of Brillouin's monograph is the *experimental* study of 'abnormal' group velocities—group velocities that are superluminal, infinite, negative, or zero. The literature on the subject has grown substantially. One purpose of this book is to review, *vis-à-vis* this development, the most basic ideas about dispersion relations, causality, propagation of light in dispersive media, and the different velocities used to characterize the propagation of light.

Another aspect of the subject is the role of quantum effects. Fermi was among the first to discuss the problem of light propagation in quantum electrodynamics at the most basic level, namely the emission of a photon by an atom and its subsequent absorption by another atom. He obtained the right answer, or part of the right answer, for the time dependence of the excitation probability of the second atom. But his approach, based as it was on a certain approximation, did not provide proof of causal propagation and, consequently, the 'Fermi problem' has been revisited periodically in the past few decades.

Quantum theory 'protects' special relativity from what might otherwise appear to be superluminal communication. Thus, it is impossible to use the 'spooky action at a distance' suggested by quantum correlations of the Einstein–Podolsky–Rosen (EPR) type to devise a superluminal communication scheme. In one suggested scheme, it is the spontaneous emission noise that prevents superluminal communication when one photon of an EPR pair is amplified by stimulated emission. The fact that such schemes must, in general, be impossible led to the no-cloning theorem.

One point that is emphasized here is that any *measurable* advance in time of a 'superluminal' pulse is reduced by noise arising from the field, the medium in which the field propagates, or the detector.

The group velocity of light can also be extremely small. 'Slow light' with group velocities on the order of 10 m s^{-1} was first directly observed in 1999 and shortly thereafter it was demonstrated that pulses of light could even be brought to a full stop, stored, and then regenerated. These developments have been based largely on the quantum interference effects associated with electromagnetically induced transparency. Slow light raises less fundamental questions, perhaps, than 'fast light' but it might have greater potential for applications. One application might be to quantum memories, as the storage and regeneration of light can be done without loss of information as to the quantum state of the original pulse: this information is temporarily imprinted in the slow-light medium. The ability to coherently control light in this way could also find applications eventually in optical communications.

The third major topic addressed in this book is 'left-handed light'—light propagation in media with negative refraction. Here it is not so much the variation of the refractive index with frequency that matters, as in the case of fast light and slow light, but rather the index itself at a given frequency. Left-handedness refers to the fact that, when the refractive index is negative, the electric field vector **E**, the magnetic field vector **H**, and the wavevector **k** of a plane waveform a *left*-handed triad. Nature has apparently not produced media with negative refractive indices; however, so-called *metamaterials* with this property have been created in the laboratory.

The propagation of light in metamaterials is predicted to exhibit various unfamiliar properties. For instance, the Doppler effect is reversed, so that a detector moving towards a source of radiation sees a *smaller* frequency than a stationary observer. Light bends the 'wrong' way when it is incident upon a metamaterial and it is theoretically possible to construct a 'perfect' lens in a narrow spectral range. The many potential applications of metamaterials have spurred a very rapid growth in the number of publications in this area. The last two chapters are an introduction to some of the foundational work on metamaterials and left-handed light.

My recent interest in these areas began with enlightening discussions with R Y Chiao. I also enjoyed talking with other participants in a three-week workshop at the Institute for Theoretical Physics in Santa Barbara in 2002, and discussing related matters on that and other occasions with many excellent physicists including Y Aharonov, J F Babb, S M Barnett, P R Berman, H A Bethe, M S Bigelow, R W Boyd, R J Cook, G D Doolen, J H Eberly, G V Eleftheriades, H Fearn, M Fleischhauer, K Furuya, I R Gabitov, D J Gauthier, S A Glasgow, R J Glauber, D F V James, P L Knight, P G Kwiat, W E Lamb, Jr, U Leonhardt, R Loudon, G J Maclay, L Mandel, M Mojahedi, G Nimtz, K E Oughstun, J Peatross, J B Pendry, E A Power, B Reznik, M O Scully, B Segev, D R Smith, A M Steinberg, L J Wang, H G Winful, E Wolf, and R W Ziolkowski. I have probably left out the names of many other people with whom I had helpful but long-forgotten discussions.

I apologize to the many authors whose work I have not cited. There is a huge literature relating to the topics covered in this book, and I have not cited work that I have not read or understood, let alone publications I have not even seen.

The three major subjects of this book have attracted particular interest in just the past few years. They are related by the fact that they all involve unusual values or variations of the refractive index. I have tried to focus on the basic underlying physics. The many citations to recent work do not represent an attempt to make this book as up-to-date as possible; it does reflect my opinion that this work is of considerable fundamental importance.

I thank Tom Spicer of the Institute of Physics for suggesting this book and for his patience when I failed to finish it by the promised delivery date. Dan Gauthier of Duke University made helpful suggestions for which I am grateful.

Peter W Milonni
Los Alamos, New Mexico

Chapter 1

In the Beginning

1.1 Maxwell's equations and the velocity of light

The variations of the phase velocity or the group velocity of light in different media are of great practical importance. We will be concerned primarily with situations where these variations are unusual and not yet of any practical utility. Our considerations will be based on the laws of electromagnetism:

$$\nabla \cdot E = \rho/\epsilon_0 \tag{1.1}$$

$$\nabla \cdot B = 0 \tag{1.2}$$

$$\nabla \times E = -\frac{\partial B}{\partial t} \tag{1.3}$$

$$\nabla \times B = \mu_0 J + \epsilon_0 \mu_0 \frac{\partial E}{\partial t}. \tag{1.4}$$

These equations are so incredibly important that we begin with a brief discussion of their conceptual foundations, even though this has been done thousands of times before.

The definite pattern formed by iron filings around a bar magnet, or by sawdust around an electrified body, led Faraday to suggest that the space around such objects is filled with lines of force. Electric and magnetic forces, from this point of view, are transmitted by the medium between the objects rather than arising from 'action at a distance'. Maxwell was greatly impressed and influenced by this idea of what he called an electromagnetic *field* [1]:

> Faraday ...saw lines of force traversing all space where the mathematicians saw centres of force attracting at a distance; Faraday saw a medium where they saw nothing but distance; Faraday sought the seat of the phenomena in real actions going on in the medium, they were satisfied that they had found it in a power of action at a distance
> ...

When I had translated what I considered to be Faraday's ideas into a mathematical form, I found that in general the results of the two methods coincided ... but that Faraday's methods resembled those in which we begin with the whole and arrive at the parts by analysis, while the ordinary mathematical methods were founded on the principle of beginning with the parts and building up the whole by synthesis.

It has been said that Maxwell's first great achievement in electromagnetism was to 'translate into mathematical form' the fundamental laws discovered by his predecessors and that his second great achievement was to deduce that these laws were incomplete.

The laws stated in the first three equations required no modification. Equation (1.1) is Gauss's law: the electric flux $\phi_E = \oint E \cdot dS$ over any closed surface is proportional to the electric charge Q inside the surface; the differential form (1.1) follows from the divergence theorem. Gauss's law can be obtained from the formula $E(r) = qr/4\pi\epsilon_0 r^3$ for the electric field of a point charge q but this formula applies only if the charge is at rest in our reference frame, whereas Gauss's law applies always.

Equation (1.2) says there are no magnetic 'charges'. The magnetic flux $\phi_B = \oint B \cdot dS$ over any closed surface is zero.

Equation (1.3) is Faraday's law of induction. In integral form, it states that a changing magnetic flux induces an electromotive force

$$\text{emf} = \oint_C E \cdot dr = -\frac{d\phi_B}{dt} \tag{1.5}$$

where C is a closed circuit (e.g. a wire loop or just a closed path in free space) and ϕ_B is the magnetic flux over any surface enclosed by C. Historians tell us that Oersted's discovery, that an electric current can cause a deflection of a compass needle, led Faraday to believe that magnetism can likewise produce electricity. Evidently he tried for some years to prove this by looking, for instance, for a *steady* current in a copper ring wrapped around a bar magnet. In 1831, he demonstrated that a current is induced in a conducting coil of wire if the current in a second coil increases or decreases, i.e. if a conductor is in relative motion with respect to a magnetic flux. A changing current in a circuit A not only induces a current in a neighbouring circuit B but also, as discovered by Henry, a smaller, opposing current in circuit A. The minus sign in equations (1.3) and (1.5) states that *any* induced current will be in a direction such that its own magnetic field will oppose the change in the magnetic flux. That is, the minus sign enforces Lenz's law.

The magnetic field produced by an electric current is governed by Ampère's law: the line integral of B around a closed loop C is proportional to the electric current I flowing through the area bounded by C,

$$\oint B \cdot dr = \mu_0 I = \mu_0 \int J \cdot dS \tag{1.6}$$

or

$$\nabla \times \boldsymbol{B} = \mu_0 \boldsymbol{J}. \tag{1.7}$$

Maxwell's 'second great achievement' (in electromagnetism) was to replace (1.7) by (1.4), i.e. to add to $\boldsymbol{J}$ in (1.7) the displacement current

$$\boldsymbol{J}_\mathrm{D} = \epsilon_0 \frac{\partial \boldsymbol{E}}{\partial t}. \tag{1.8}$$

How he arrived at this modification with his mechanical models and analogies is a fascinating story that is not necessary or appropriate to recount here. (An excellent, succinct discussion is given by Longair [2].) Let us just remind ourselves that, without the displacement current in (1.4), we would not obtain the equation expressing conservation of electric charge:

$$\nabla \cdot \boldsymbol{J} + \frac{\partial \rho}{\partial t} = 0. \tag{1.9}$$

In static situations, the fields $\boldsymbol{E}$ and $\boldsymbol{B}$ are uncoupled and electricity and magnetism are separate concerns. A time-varying $\boldsymbol{E}$ field, however, can create a $\boldsymbol{B}$ field and *vice versa*. In a region of space with no charges and currents ($\rho = \boldsymbol{J} = 0$), Maxwell's equations imply

$$\nabla^2 \boldsymbol{E} - \epsilon_0 \mu_0 \frac{\partial^2 \boldsymbol{E}}{\partial t^2} = \nabla^2 \boldsymbol{B} - \epsilon_0 \mu_0 \frac{\partial^2 \boldsymbol{B}}{\partial t^2} = 0 \tag{1.10}$$

i.e. electromagnetic waves with propagation velocity

$$c = 1/\sqrt{\epsilon_0 \mu_0}. \tag{1.11}$$

The first evidence that the velocity of light is finite was obtained, as everyone knows, by Roemer (1676), prior to whom the majority opinion was that light travels instantaneously. The orbital plane of Jupiter's moons is close to the plane in which Jupiter and the earth orbit the sun and the moons, as seen from the earth, are periodically eclipsed by Jupiter. Roemer noticed that the time between successive eclipses (about $42\frac{1}{2}$ hours) of the innermost moon was larger when the earth was moving away from Jupiter and smaller when the earth was moving towards Jupiter and he attributed this variation to the finite velocity of light. Thus, when the earth is moving away from Jupiter, each succeeding 'signal' that an eclipse has taken place must travel for a greater time (about 14 s greater) in order to 'catch up' with the earth. Roemer estimated that it takes light about 22 min to travel a distance equal to the diameter of the earth's orbit about the sun. (Using $c = 2.998 \times 10^8$ m s^{-1} and 2.98×10^{11} m for the diameter D of the earth's orbit, we obtain 17 min for the time it takes light to traverse a distance D, compared with Roemer's estimate of 22 min obtained by adding up the time delays in successive eclipses as the earth moved from a point where it is closest to Jupiter to the diametrically opposite point in its orbit.)

4   *In the Beginning*

Wroblewski [3] has written an entertaining article about the incorrect accounts of Roemer's work found in many textbooks. For instance, whereas specific (and different!) values of c are ascribed to Roemer in various texts, Roemer did not actually state a number for the velocity of light: as discussed by Wroblewski, he was interested primarily in arguing that c is finite rather than in figuring out an accurate numerical value for it.

Nearly two centuries later, Fizeau (1849) made the first terrestrial measurement of c using a rotating toothed wheel. Light passing through an opening in the wheel could pass through an opening or be blocked after reflection from a mirror about 8.6 km away. From the wheel radius and angular velocity, and the distance between the openings, Fizeau obtained 3.15×10^8 m s^{-1} for the velocity of light. A year later, Foucault peformed similar experiments using a rotating mirror instead of a toothed wheel and obtained 2.986×10^8 m s^{-1}. He also found by this method that the velocity of light in water is 1.3 times smaller than in air.[1] The rotating-mirror method was used by Michelson, in a long series of experiments, to determine a velocity of light close to $c = 2.998$ m s^{-1}. It is perhaps worth noting that, with photodiodes and the other niceties of modern technology, the rotating-mirror method can easily be employed to measure c to an accuracy of a few per cent or better in undergraduate laboratories.

It is not generally recognized that these experiments actually determine the *group velocity* of light (section 1.5).

Another kind of determination of c is suggested [4] by comparing the force between two charges separated by a distance r,

$$F_c = \frac{1}{4\pi\epsilon_0}\frac{q_1 q_2}{r^2} \equiv k\frac{q_1 q_2}{r^2} \tag{1.12}$$

with the force between two parallel wires of length ℓ and separation r, carrying electric currents i_1 and i_2:

$$F_m = \mu_0 \ell \frac{i_1 i_2}{2\pi r} \equiv K\frac{i_1 i_2}{r}\ell. \tag{1.13}$$

The ratio $k/K = c^2/2$. Along these lines, Maxwell wrote in a letter to Faraday on 19 October 1861 that [5]

[F]rom the determination by Kohlrausch and Weber of the numerical relation between the statical and magnetic effects of electricity, I have determined the *elasticity* of the medium in air, and assuming that it is the same with the luminiferous ether I have determined the velocity of propagation of transverse vibrations.

[1] In the corpuscular theory of light, it was assumed that the particles of light would be attracted by the denser medium and accelerated by it rather than slowed down. Newton's writing on the subject suggests that his adherence to the corpuscular theory was not as strong as that of many of his successors.

The result is 193 088 miles per second (deduced from electrical and magnetic experiments).

Fizeau has determined the velocity of light = 193 118 miles per second by direct experiment.

This coincidence is not merely numerical. I worked out the formulae in the country, before seeing Weber's number, which is in millimetres, and I think we have now strong reason to believe, whether my theory is correct or not, that the luminiferous and the electromagnetic medium are one.

Maxwell took part in similar experiments. Kirchhoff, in theoretical work evidently unknown to Maxwell, had noted earlier (1857) that the velocity he found for the propagation of an electric potential along a telegraph wire, in the limit of zero resistance, was close to the velocity of light [6].

The direct experimental proof of Maxwell's theory by Hertz in 1887 came only after Maxwell's death in 1879. Hertz produced oscillatory sparks between two metal spheres with an induction coil and found that this caused sparks across a second pair of metal spheres some metres away. He showed that the disturbance produced at the transmitter was reflected by conductors, focused by a concave mirror, and refracted by dielectrics. He also measured the wavelength (9.6 m when the frequency of the transmitter was 3×10^7 cycles per second) by producing standing waves and using the spark gap detector to determine the nodes of the field. He thereby deduced that the velocity of the electromagnetic disturbance was 3×10^8 m s^{-1}.

Bates [7] has given a concise summary and nearly 100 references on the modern methods for measuring the velocity of light, which are based on the relation $\nu\lambda = c$ between frequency (ν) and wavelength (λ). These methods were made possible by the development of frequency-stabilized lasers allowing accurate measurements of both ν and λ. In 1983, the redefinition of the metre by the International Committee on Weights and Measurements resulted in the following value for the velocity of light in vacuum:

$$c = 299\,792\,458 \text{ m s}^{-1}. \tag{1.14}$$

1.2 Refractive index

Fast light, slow light, and left-handed light are all associated with unusual values or variations of the refractive index. It will suffice, in the beginning, to consider the refractive index $n(\omega)$ of a dilute gas of atoms. We will follow the semiclassical approach of treating the field classically and the atoms quantum mechanically. Since the phenomena we consider are linear in the electric and magnetic fields and the Heisenberg-picture operators for the field in quantum electrodynamics satisfy formally the same (Maxwell) equations as their classical counterparts, it is easy but not necessary for our purposes to quantize the field [8].

An electric field $E_0 \cos \omega t$ induces an electric dipole moment

$$p = \alpha(\omega)E_0 \cos \omega t \qquad (1.15)$$

in an atom, where $\alpha(\omega)$ is the polarizability. p is actually the quantum-mechanical expectation value of the induced dipole moment which, for our purposes, can be treated as a classical variable. Equation (1.15) is valid as long as the field is not too strong; otherwise p depends nonlinearly on E_0. How 'strong' the field has to be to produce a nonlinear response by the atom depends on the frequency ω. We will assume for now that the field is not strong.

If there are N atoms per unit volume, the polarization density is $P = Np$. Recall that a charge density $\rho_{\text{pol}} = -\nabla \cdot P$ is associated with the polarization density, so that, in the case of a material medium, equation (1.1) is replaced by

$$\nabla \cdot E = \frac{1}{\epsilon_0}(\rho - \nabla \cdot P) \qquad (1.16)$$

or

$$\nabla \cdot D = \rho \qquad (1.17)$$

where

$$D = \epsilon_0 E + P \equiv \epsilon E \qquad (1.18)$$

and ρ is now the 'free' charge density and ϵ_0 is the permittivity of the medium:

$$\epsilon = \epsilon_0 \left(1 + \frac{N\alpha}{\epsilon_0}\right) \equiv \epsilon_0(1 + \chi) \equiv \kappa \epsilon_0 \qquad (1.19)$$

where χ and κ are the electric susceptibility and dielectric constant, respectively, of the medium.

At optical frequencies, the magnetic permeability is essentially unaltered from its free-space value μ_0, so we will take $\mu = \mu_0$ and not bother at this point to introduce the vector H. That is, we take the Maxwell equations in the case of a material medium to be

$$\nabla \cdot D = \rho \qquad (1.20)$$

$$\nabla \cdot B = 0 \qquad (1.21)$$

$$\nabla \times E = -\frac{\partial B}{\partial t} \qquad (1.22)$$

$$\nabla \times B = \mu_0 J + \mu_0 \frac{\partial D}{\partial t}. \qquad (1.23)$$

Consider, for simplicity, a medium with no free charges or currents ($\rho = J = 0$) and permittivity $\epsilon_0(\omega)$ independent of position. Then, with $E = E_0 \exp(-i\omega t)$ and $B = B_0 \exp(-i\omega t)$, equations (1.20)–(1.23) imply

$$\nabla^2 E_0 + n^2(\omega)\frac{\omega^2}{c^2}E_0 = \nabla^2 B_0 + n^2(\omega)\frac{\omega^2}{c^2}B_0 = 0 \qquad (1.24)$$

where[2]

$$n^2(\omega) = \epsilon(\omega)/\epsilon_0 = \kappa(\omega) = 1 + \frac{N}{\epsilon_0}\alpha(\omega). \tag{1.25}$$

For $N\alpha(\omega)/\epsilon_0 \ll 1$, as is typical of a dilute gas, the refractive index is given by

$$n(\omega) \cong 1 + \frac{N}{2\epsilon_0}\alpha(\omega). \tag{1.26}$$

It is a straightforward exercise in perturbation theory to derive the 'Kramers–Heisenberg' formula for the polarizability of an atom in state i. For a one-electron atom, this formula is

$$\alpha_i(\omega) = \frac{e^2}{m}\sum_j \frac{f_{ij}}{\omega_{ji}^2 - \omega^2} \tag{1.27}$$

where e and m are the electron charge and mass, $\omega_{ji} = (E_j - E_i)/\hbar$ is the transition frequency (rad s^{-1}) between eigenstates of energy E_j and E_i, and f_{ij} is the transition oscillator strength. For non-degenerate states, i and j, $f_{ij} = 2m\omega_{ji}|d_{ji}|^2/3\hbar$, where d_{ji} is the electric dipole matrix element between states i and j. The oscillator strengths satisfy the (Thomas–Reiche–Kuhn) sum rule, $\sum_j f_{ij} = 1$. (For a Z-electron atom, $\sum_j f_{ij} = Z$.)

Suppose a plane wave of frequency ω is incident on a half-space in which there is a uniform distribution of N atoms per unit volume. Each atom has a dipole moment induced by the *total* electric field at its position, i.e. the incident field plus the fields radiated by all the other atoms. The total field at any point is the incident field at that point plus the field at that point produced by all the atoms. This total field, at any point inside the medium, has two parts. One part exactly cancels the incident field. The other part propagates with phase velocity $c/n(\omega)$, where

$$n(\omega) = 1 + \frac{1}{2\epsilon_0}\sum_i N_i\alpha_i(\omega) = 1 + \frac{e^2}{2m\epsilon_0}\sum_i \sum_j \frac{N_i f_{ij}}{\omega_{ji}^2 - \omega^2} \tag{1.28}$$

is the refractive index. N_i is the number density of atoms in energy eigenstate i and $\sum_i N_i = N$. In writing (1.28), it is assumed that the gas is sufficiently dilute that local field corrections can be ignored, i.e. $n(\omega) \cong 1$. Note that the expression (1.26) for $n(\omega)$ assumes that all the atoms are in a single (ground) state.

The fact that part of the field radiated by the induced dipoles cancels the incident field, while the remaining part propagates at the phase velocity $c/n(\omega)$, is called the Ewald–Oseen extinction theorem [9]. (A pedagogical treatment may be found in [10].) The summation of the applied field and the dipole fields at points inside or outside the medium results in the Fresnel formulas for transmission and reflection [9].

---

[2] More generally, $n^2(\omega) = \epsilon(\omega)\mu(\omega)/\epsilon_0\mu_0 = [\mu(\omega)/\mu_0][1 + N\alpha(\omega)/\epsilon_0]$.

These results of the *integral* formulation of Maxwell's equations are very pretty. Of course, the same results are obtained in the much more commonly employed *differential* form of Maxwell's equations, although the 'extinction' of the incident field is not explicit as in the integral formulation.

The contribution to the refractive index (1.28) from the $1 \leftrightarrow 2$ transition ($i = 1, j = 2$) is

$$n(\omega)_{12} = 1 + \frac{e^2}{2m\epsilon_0}\left(\frac{N_1 f_{12}}{\omega_{21}^2 - \omega^2} + \frac{N_2 f_{21}}{\omega_{12}^2 - \omega^2}\right) = 1 + \frac{e^2 f_{12}}{2m\epsilon_0}\left(\frac{N_1 - N_2}{\omega_{21}^2 - \omega^2}\right).$$
(1.29)

Thus, the population N_2 of the upper state of a transition makes a contribution to the index that has the opposite sign to the contribution of the lower-state population N_1. This effect was observed by Ladenburg and Kopfermann in their study of the variation of the refractive index with currect in an electric discharge [11].

Effects like collisions and spontaneous emission require us to modify equation (1.29) to include a frictional damping rate γ in the denominator:

$$n(\omega) = 1 + \frac{e^2 f_{12}}{2m\epsilon_0}\left(\frac{N_1 - N_2}{\omega_{21}^2 - \omega^2 - 2i\gamma\omega}\right).$$
(1.30)

This modification ensures that $n(\omega)$ in our theory does not 'blow up' when the field frequency ω equals the transition frequency ω_{21}. When $\omega \cong \omega_{21}$, we have

$$n(\omega) \cong 1 + \frac{e^2 f_{12}}{4m\epsilon_0\omega_{21}} \frac{N_1 - N_2}{\omega_{21} - \omega - i\gamma}.$$
(1.31)

Let us assume that $N_1 - N_2 \approx N_1$, i.e. that most atoms remain with high probability in the lower state of the transition. Then

$$n(\omega) \cong 1 + \frac{Ne^2 f}{4m\epsilon_0\omega_0} \frac{1}{\omega_0 - \omega - i\gamma} = 1 + \frac{K}{\omega_0 - \omega - i\gamma} = n_R(\omega) + in_I(\omega) \quad (1.32)$$

where

$$n_R(\omega) = 1 + K\frac{\omega_0 - \omega}{(\omega_0 - \omega)^2 + \gamma^2}$$
(1.33)

$$n_I(\omega) = K\frac{\gamma}{(\omega_0 - \omega)^2 + \gamma^2}$$
(1.34)

and, for notational simplicity, we have replaced f_{12} by f, N_1 by N, and ω_{21} by ω_0.

Normally the refractive index increases with increasing frequency but near an absorption line $n_R(\omega)$ decreases with increasing frequency (figure 1.1). Such 'anomalous dispersion' in sodium vapour was observed by R W Wood in 1904 [12]. As discussed in section 1.5, the group velocity of light can exceed c in a spectral region of anomalous dispersion and this possibility raised serious concerns in connection with the special theory of relativity.

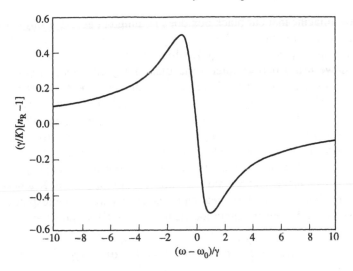

Figure 1.1. $(\gamma/K)[n_R(\omega) - 1]$ *versus* $(\omega - \omega_0)/\gamma$.

1.3 Causality and dispersion relations

There are many systems and devices such that a time-dependent input $F_{in}(t)$ produces an output $F_{out}(t)$ that (a) depends linearly on $F_{in}(t)$ and (b) is time-invariant in the sense that a shift in time of F_{in} produces the same shift in time of F_{out}. These properties are accounted for by writing

$$F_{out}(t) = \int_{-\infty}^{\infty} dt'\, G(t - t') F_{in}(t'). \tag{1.35}$$

Introducing the Fourier transforms $f_{out}(\omega)$, $g(\omega)$, and $f_{in}(\omega)$ by writing

$$G(\tau) = \frac{1}{2\pi} \int_{-\infty}^{\infty} d\omega\, g(\omega) e^{-i\omega t} \tag{1.36}$$

and likewise for F_{out} and F_{in}, we see that (1.35) implies that

$$f_{out}(\omega) = g(\omega) f_{in}(\omega). \tag{1.37}$$

Suppose there is no input to our 'black box' until some time $t = 0$, so that

$$F_{in}(t) = \frac{1}{2\pi} \int_{-\infty}^{\infty} d\omega\, f_{in}(\omega) e^{-i\omega t} = 0 \qquad \text{for all } t < 0 \tag{1.38}$$

meaning that there is complete destructive interference of the Fourier components of F_{in} for $t < 0$. Causality, in the sense that there should be no output before there is any input, requires that $F_{out}(t) = 0$ for $t < 0$.

Now imagine that our black box does nothing but absorb a single frequency component ω of the input. Then its output would be $F_{in}(t)$ minus the Fourier component at frequency ω of $F_{in}(t)$, which *does not vanish for $t < 0$*. In other words, if we had a perfect filter—one that simply absorbs a *single* frequency component of an input signal—there would be an output before there is any input, a violation of causality (figure 1.2). *It must, therefore, be impossible to have a perfect filter, one that absorbs one frequency without affecting any other frequency components of an input signal.* Any realizable filter must evidently produce phase shifts in other Fourier components, in such a way that they interfere destructively for all $t < 0$, so that, in fact, there is no output before any input [13].

The mathematical expression of this requirement is a Kramers–Krönig dispersion relation between the real and imaginary parts of the response function $g(\omega)$. No output before any input means that $G(t - t') = 0$ for $t < t'$, so that

$$g(\omega) = \int_{-\infty}^{\infty} d\tau \, G(\tau)e^{i\omega\tau} = \int_{0}^{\infty} d\tau \, G(\tau)e^{i\omega\tau}. \qquad (1.39)$$

Thus, the integration over τ extends over only positive values of τ and, for such values, $g(\omega)$ is analytic in the upper half of the complex ω plane. In other words, *causality requires the Fourier transform $g(\omega)$ of the response function $G(\tau)$ to be analytic in the upper half of the complex frequency plane*[3].

Cauchy's theorem states that

$$g(\omega) = \frac{1}{2\pi i} \oint_C \frac{g(\omega')}{\omega' - \omega} d\omega' \qquad (1.40)$$

where the contour C is indicated in figure 1.3. We assume that $g(\omega) \to 0$ faster than $1/|\omega|$ as $|\omega| \to \infty$, so that the contribution from the semicircle vanishes. Then, if ω is on the real axis (figure 1.3),

$$g(\omega) = \frac{1}{\pi i} P \int_{-\infty}^{\infty} \frac{g(\omega')}{\omega' - \omega} d\omega' \qquad (1.41)$$

i.e. the real and imaginary parts of $g(\omega)$ satisfy the relations

$$\text{Re}[g(\omega)] = \frac{1}{\pi} P \int_{-\infty}^{\infty} \frac{\text{Im}[g(\omega')]}{\omega' - \omega} d\omega' \qquad (1.42)$$

$$\text{Im}[g(\omega)] = -\frac{1}{\pi} P \int_{-\infty}^{\infty} \frac{\text{Re}[g(\omega')]}{\omega' - \omega} d\omega' \qquad (1.43)$$

where P denotes the Cauchy principal value[4]. In fact, each of these Hilbert transform relations may be shown to imply the other.

---

[3] If we use $\exp(i\omega t)$ instead of $\exp(-i\omega t)$ for the time dependence of the frequency component ω, causality requires that $g(\omega)$ be analytic in the *lower* half of the complex ω plane.
[4] $P\int_{-\infty}^{\infty} \equiv \lim_{\epsilon\to 0}(\int_{-\infty}^{\omega-\epsilon} + \int_{\omega+\epsilon}^{\infty})$.

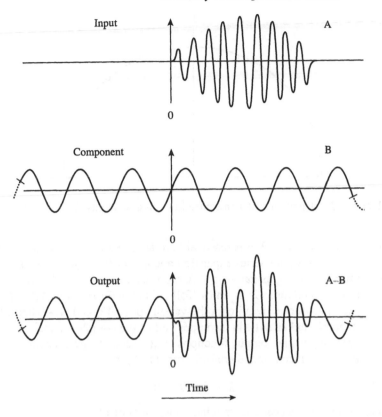

Figure 1.2. From Toll [13], with permission: 'This figure illustrates schematically the basic reason for the logical connection of causality and dispersion. An input A which is zero for times t less than zero is formed as a superposition of many Fourier components such as B, each of which extends from $t = -\infty$ to $t = \infty$. These components produce the zero-input signal by destructive interference for $t < 0$. It is impossible to design a system which absorbs just the component B without affecting other components, for in this case the output would contain the complement of B during times before the onset of the input wave, in contradiction with causality. Thus causality implies that absorption of one frequency must be accompanied by a compensating shift of phase of other frequencies; the required phase shifts are prescribed by the dispersion relation.'

Note that analyticity in the upper half of the complex ω plane is not enough to guarantee that the Hilbert transform relations are satisfied: we also require that $g(\omega)$ falls off faster than $1/|\omega|$ as $|\omega| \to \infty$. Suppose that $G(\tau)$ has a Taylor series expansion about $\tau = 0$. Then [14]

$$g(\omega) = \int_0^\infty d\tau\, [G(0) + G'(0)\tau + \cdots]e^{i\omega\tau} = \frac{iG(0)}{\omega} - \frac{G'(0)}{\omega^2} + \cdots. \quad (1.44)$$

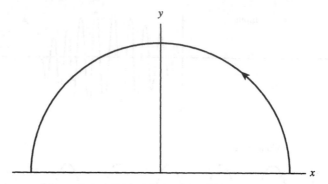

Figure 1.3. Integration contour assumed in writing equations (1.40) and (1.41).

Since $G(\tau) = 0$ for $\tau < 0$, it is reasonable to assume that $G(0) = 0$. It follows that, if $G(\tau)$ has a Taylor series expansion about $\tau = 0^+$, $g(\omega)$ does, in fact, go to zero faster than $1/|\omega|$ as $|\omega| \to \infty$. A rigorous discussion of the basis for the Hilbert transform relations is given by Toll [13] and Nussenzveig [15]. For our purposes, it will suffice to require for the validity of (1.42) and (1.43) that $g(\omega)$ is square-integrable and analytic in the upper half of the complex plane[5].

For real input and output functions $F_{in}(t)$ and $F_{out}(t)$, equation (1.35) requires that $G(\tau)$ is real and, therefore, from (1.39), that

$$g^*(\omega) = g(-\omega) \tag{1.45}$$

or, in terms of the real (g_R) and imaginary (g_I) parts of $g(\omega)$,

$$g_R(-\omega) = g_R(\omega) \tag{1.46}$$

$$g_I(-\omega) = -g_I(\omega). \tag{1.47}$$

These relations allow us to write (1.42) and (1.43) in a different form:

$$g_R(\omega) = \frac{1}{\pi} P \int_0^\infty \frac{g_I(\omega')}{\omega' - \omega} d\omega' + \frac{1}{\pi} P \int_{-\infty}^0 \frac{g_I(\omega')}{\omega' - \omega} d\omega' = \frac{2}{\pi} P \int_0^\infty \frac{\omega' g_I(\omega')}{\omega'^2 - \omega^2} d\omega' \tag{1.48}$$

and, similarly,

$$g_I(\omega) = -\frac{2}{\pi} P \int_0^\infty \frac{\omega g_R(\omega')}{\omega'^2 - \omega^2} d\omega'. \tag{1.49}$$

The (causal) relation $P(\omega) = \epsilon_0 [n^2(\omega) - 1] E(\omega)$ between the polarization and the electric field implies that $g(\omega) = n^2(\omega) - 1$ satisfies these dispersion relations.

[5] The more rigorous treatments are based on the Titchmarsh theorem, which can be stated as follows [13]. If $g(\omega)$ is square-integrable over the real ω-axis, then any one of the following three conditions implies the other two: (1) The Fourier transform $G(\tau)$ of $g(\omega)$ [equation (1.39)] vanishes for $\tau < 0$. (2) $g(\omega_R + i\omega_I)$ is analytic in the upper half of the complex ω plane, approaches $g(\omega_R)$ almost everywhere as $\omega_I \to 0$, and is square-integrable along any line above and parallel to the real axis. (3) $g_R(\omega)$ and $g_I(\omega)$ satisfy the Hilbert transform relations (1.42) and (1.43).

Of more interest to us are dispersion relations involving $n(\omega)$. If the function $\sqrt{n^2(\omega)}$ has no branch points and is analytic in the upper half-plane, it follows that $n(\omega)$ is analytic in the upper half-plane. However, $n(\omega)$ does not satisfy the other condition necessary for it to satisfy the previous dispersion relations: it is not square-integrable. In fact, physical considerations [as well as formulas like (1.30)] suggest that $n(\omega) \to 1$ as $\omega \to \infty$: a material medium cannot respond to infinitely large frequencies, which therefore propagate as if there were no medium. But it is still possible to derive a dispersion relation 'with subtraction' for $n(\omega)$, as follows.

Let Ω be some real frequency and consider $[n(\omega) - n(\Omega)]/(\omega - z)$, where z is a point in the lower half of the complex ω plane. This function satisfies the conditions for equation (1.41) to apply:

$$n(\omega) - n(\Omega) = \frac{\omega - z}{\pi i} P \int_{-\infty}^{\infty} \frac{[n(\omega') - n(\Omega)]}{(\omega' - z)(\omega' - \omega)} \, d\omega'. \qquad (1.50)$$

Subtracting the corresponding expression with $\omega = \Omega$, we have

$$n(\omega) - n(\Omega) = \frac{1}{\pi i} P \int_{-\infty}^{\infty} \frac{[n(\omega') - n(\Omega)]}{\omega' - z} \left[\frac{\omega - z}{\omega' - \omega} - \frac{\Omega - z}{\omega' - \Omega} \right] d\omega'$$

$$= \frac{\omega - \Omega}{\pi i} P \int_{-\infty}^{\infty} \frac{[n(\omega') - n(\Omega)]}{(\omega' - \omega)(\omega' - \Omega)} \, d\omega' \qquad (1.51)$$

so that the function $[n(\omega) - n(\Omega)]/(\omega - \Omega)$ satisfies (1.41). In particular, taking $\Omega \to \infty$,

$$n(\omega) = n(\infty) + \frac{1}{\pi i} P \int_{-\infty}^{\infty} \frac{[n(\omega') - n(\infty)]}{\omega' - \omega} \, d\omega' \qquad (1.52)$$

or, assuming $n(\infty) = 1$ for the reasons mentioned earlier,

$$n(\omega) = 1 + \frac{1}{\pi i} P \int_{-\infty}^{\infty} \frac{[n(\omega') - 1]}{\omega' - \omega} \, d\omega' \qquad (1.53)$$

$$n_R(\omega) = 1 + \frac{1}{\pi} P \int_{-\infty}^{\infty} \frac{n_I(\omega')}{\omega' - \omega} \, d\omega' \qquad (1.54)$$

$$n_I(\omega) = -\frac{1}{\pi} P \int_{-\infty}^{\infty} \frac{[n_R(\omega') - 1]}{\omega' - \omega} \, d\omega'. \qquad (1.55)$$

We also have the symmetry relations

$$n^*(\omega) = n(-\omega)$$
$$n_R(-\omega) = n_R(\omega)$$
$$n_I(-\omega) = -n_I(\omega) \qquad (1.56)$$

analogous to (1.45)–(1.47), which allow us to rewrite (1.54) and (1.55) alternatively as

$$n_R(\omega) = 1 + \frac{2}{\pi} P \int_0^{\infty} \frac{\omega' n_I(\omega')}{\omega'^2 - \omega^2} \, d\omega' \qquad (1.57)$$

$$n_I(\omega) = -\frac{2\omega}{\pi} P \int_0^\infty \frac{[n_R(\omega') - 1]}{\omega'^2 - \omega^2} \, d\omega'. \tag{1.58}$$

We can obtain the dispersion relations (1.54) and (1.55) in another way, without having to exclude branch points. Consider a plane wave propagating in the z direction and write the electric field amplitude at $z = 0$ as

$$E(0, t) = \int_{-\infty}^\infty d\omega \, A(\omega) e^{-i\omega t}. \tag{1.59}$$

According to equation (1.24), the effect of propagation over a distance z is to multiply each Fourier component by $\exp(i\omega n(\omega)z/c)$[6]:

$$E(z, t) = \int_{-\infty}^\infty d\omega \, A(\omega) e^{-i\omega t} e^{i\omega n(\omega)z/c} \tag{1.60}$$

or, using the Fourier inverse of (1.59),

$$E(z, t) = \int_{-\infty}^\infty d\omega \, e^{-i\omega t} e^{i\omega n(\omega)z/c} \frac{1}{2\pi} \int_{-\infty}^\infty dt' \, E(0, t') e^{i\omega t'}$$
$$= \int_{-\infty}^\infty dt' \, G(z, t - t') E(0, t') \tag{1.61}$$

where

$$G(z, \tau) = \frac{1}{2\pi} \int_{-\infty}^\infty d\omega \, e^{-i\omega \tau} e^{i\omega n(\omega)z/c} \tag{1.62}$$

$$e^{i\omega n(\omega)z/c} = \int_{-\infty}^\infty d\tau \, G(z, \tau) e^{i\omega \tau}. \tag{1.63}$$

Equation (1.61) states that the field at $z > 0$ at time t is determined by the field at $z = 0$ at times t'. Since the field does not propagate instantaneously from $z = 0$ to $z > 0$, $G(z, t - t')$ must vanish for $t - t' < T$, where $T \, (> 0)$ is some finite time. Thus, $G(z, \tau) = 0$ for $\tau < T$ and, therefore,

$$e^{i\omega n(\omega)z/c} = \int_T^\infty d\tau \, G(z, \tau) e^{i\omega \tau} \tag{1.64}$$

or

$$e^{i\omega n(\omega)z/c} = e^{i\omega T} \int_0^\infty dt' \, G(z, t' + T) e^{i\omega t'}. \tag{1.65}$$

This is of the form (1.39) and we conclude, since z is arbitrary, that $\omega n(\omega)$ is analytic in the upper half of the complex ω plane. Then the previous subtraction procedure leads to the dispersion relations (1.54) and (1.55).

---

[6] If the field is incident from vacuum onto a medium of refractive index $n(\omega)$, we should include the Fresnel transmission coefficient $2/[n(\omega) + 1]$ in the integrand of (1.60). This modification is of no consequence for the derivation of the dispersion relations.

We close this section with a few remarks about the Kramers–Krönig dispersion relations. Note first that the *approximations* (1.33) and (1.34) do not satisfy the last two symmetry relations (1.56). However, they do satisfy the Hilbert transform relations (1.54) and (1.55). They also illustrate the fact (figure 1.1) that a peak in the function $n_I(\omega)$ is accompanied by a rapid variation and a sign change in $n_R(\omega)$. Physically, this means that the strongest departures of the group velocity from c are to be found in the vicinity of absorption lines (section 1.5).

Rayleigh scattering provides another illustrative example. Consider the classical non-relativistic equation of motion for a bound electron in an applied monochromatic field (see, for instance, Jackson [14] or reference [8]):

$$\ddot{x} + \omega_0^2 x - \tau\, \dddot{x} = \frac{e}{m} E_0 e^{-i\omega t} \tag{1.66}$$

where $\tau = 2e^2/3mc^3 \sim 6 \times 10^{-24}$ s. The polarizability is

$$\alpha(\omega) = \frac{e^2/m}{\omega_0^2 - \omega^2 - i\tau\omega^3} \tag{1.67}$$

and the refractive index in the case of N such atoms per unit volume is given by

$$n^2(\omega) = n_R^2 - n_I^2 + 2in_Rn_I = 1 + \frac{Ne^2/m\epsilon_0}{\omega_0^2 - \omega^2 - i\tau\omega^3}. \tag{1.68}$$

Thus,

$$n_R^2 \cong 1 + \frac{(Ne^2/m\epsilon_0)(\omega_0^2 - \omega^2)}{(\omega_0^2 - \omega^2)^2 + \tau^2\omega^6} \tag{1.69}$$

and

$$n_I \cong \frac{Ne^2\tau\omega^3/2n_Rm\epsilon_0}{(\omega_0^2 - \omega^2)^2 + \tau^2\omega^6} \cong \frac{\omega^3}{c^3}\frac{(n_R^2 - 1)^2}{12\pi n_R N} \tag{1.70}$$

in the approximation $|\omega_0^2 - \omega^2| \gg \tau\omega^3$. The intensity extinction coefficient due to scattering is, therefore,[7]

$$a_s(\omega) = 2\omega n_I(\omega)/c = \frac{\omega^4}{c^4}\frac{(n_R^2 - 1)^2}{6\pi N n_R} \tag{1.71}$$

which is the well-known extinction coefficient due to Rayleigh scattering. It has recently been checked experimentally [16].

What is interesting about this derivation of $a_s(\omega)$ is that the polarizability (1.67) and the refractive index are *not* analytic in the upper half of the complex ω plane. The ω^3 in the denominator of (1.67) or, in other words, the third derivative

---

[7] The 'radiation reaction' damping term in equation (1.66) accounts for the loss of energy due to radiation by the bound electron driven by the applied field. The energy loss is, therefore, associated with *scattering* rather than *absorption* of the applied field.

of x in equation (1.66) leads to a pole in the upper half-plane. And yet it is precisely this ω^3 dependence that gives rise, via (1.71), to the well-known ω^4 dependence of Rayleigh scattering.

The model leading to equation (1.66) is well known to be acausal; but the acausality occurs on such a short time scale that relativistic quantum effects must be taken into account. In other words, the non-relativistic model equation (1.66), even if it is regarded as a quantum-mechanical, Heisenberg operator equation, is fundamentally flawed, though it leads in this example to correct results.

Consider equation (1.57) in the limit $\omega \rightarrow \infty$:

$$n_{\mathrm{R}}(\omega) - 1 = -\frac{2}{\pi \omega^2} \int_0^\infty \omega' n_{\mathrm{I}}(\omega') \, d\omega'. \tag{1.72}$$

In this limit, equation (1.28) gives

$$n_{\mathrm{R}}(\omega) - 1 = -\frac{e^2}{2m\epsilon_0 \omega^2} \sum_i N_i \sum_j f_{ij} = -\frac{e^2 N Z}{2m\epsilon_0 \omega^2} \tag{1.73}$$

where $N = \sum_i N_i$ is the total density of atoms and we have used the Thomas–Reiche–Kuhn sum rule, $\sum_j f_{ij} = Z$, for Z-electron atoms (section 1.2). Thus,

$$\int_0^\infty \omega' n_{\mathrm{I}}(\omega') \, d\omega' = \frac{\pi N Z e^2}{4m\epsilon_0}. \tag{1.74}$$

Now $\exp(i\omega n(\omega)z/c) = \exp(i\omega n_{\mathrm{R}}(\omega)z/c)\exp(-\omega n_{\mathrm{I}}(\omega)z/c)$ means that the intensity decreases as $\exp(-2\omega n_{\mathrm{I}}(\omega)z/c) = \exp(-a(\omega)z)$, where $a(\omega)$ is the absorption coefficient. In terms of the absorption coefficient, the sum rule (1.74) becomes

$$\int_0^\infty a(\omega) \, d\omega = \frac{\pi N Z e^2}{2m\epsilon_0 c} \tag{1.75}$$

which is useful in the analysis of absorption spectra. Various other sum rules and relationships can be obtained from the dispersion relations.

1.4　Signal velocity and Einstein causality

The arguments leading to the Hilbert transform relations in the preceding section were based on causality in the sense that the 'output' of a linear and time-invariant system cannot precede the 'input'. The different requirement that no signal can be transmitted with a velocity exceeding c is often referred to as *Einstein causality*. Suppose an event at (x, t) were to cause an event at $(x + \Delta x, t + \Delta t)$ via some signal with velocity u. In some other reference frame with relative velocity v $(< c)$, the time interval between the two events is

$$\Delta t' = \frac{\Delta t - (v/c^2)\Delta x}{\sqrt{1 - v^2/c^2}} = \frac{\Delta t(1 - uv/c^2)}{\sqrt{1 - v^2/c^2}}. \tag{1.76}$$

Then a superluminal signal velocity ($u > c$) would imply that there are velocities v for which Δt and $\Delta t'$ have opposite signs: the temporal order of 'cause' and 'effect' would be different for different observers. Special relativity and causality, therefore, forbid superluminal signal velocities. But what, precisely, defines a *signal*? We will address this question in the following chapters but for now let us consider a velocity that often characterizes the propagation of an electromagnetic pulse and is often—erroneously—assumed to be a signal velocity.

1.5 Group velocity

The concept of group velocity was evidently first introduced by William Rowan Hamilton in 1839 [17]. John Scott Russell, in his paper reporting the first observation of soliton water waves in 1844 [18], may also have been the first to observe and identify the group velocity of a wave. In an article 'On the Velocity of Light' in 1881, Lord Rayleigh clearly distinguished between the phase velocity (V) and the group velocity (U) [19]:

> If the crest of an ordinary water wave were observed to travel at the rate of a foot per second, we should feel no hesitation in asserting that this was the velocity of the wave; and I suppose that in the ordinary language of undulationists the velocity of light means in the same way the velocity with which an individual wave travels. It is evident, however, that in the case of light, or even of sound, we have no means of identifying a particular wave so as to determine its rate of progress. What we really do in most cases is to impress some peculiarity, it may be of intensity, or of wave-length, or of polarization, upon a part of an otherwise continuous train of waves, and determine the velocity at which this *peculiarity* travels. Thus in the experiments of Fizeau and Cornu, as well as in those of Young and Forbes, the light is rendered intermittent by the action of a toothed wheel; and the result is the velocity of the group of waves, and not necessarily the velocity of an individual wave ... I have investigated the general relation between the group-velocity U and the wave-velocity V. It appears that if k be inversely proportional to the wave-length,

$$U = \frac{d(kV)}{dk} \tag{1.77}$$

> and is identical with V only when V is independent of k, as has hitherto been supposed to be the case for light in vacuum. If, however, as Young and Forbes believe, V varies with k, then U and V are different. The truth is however that these experiments tell us nothing in the first instance about the value of V. They relate to U; and if V is to be deduced from them it must be by the aid of the above given relation.

Consider two plane waves with the same polarization and amplitude but differing slightly in frequency ω and wavenumber k. The sum of the two waves is proportional to

$$
\begin{aligned}
S(z, t) &= \cos[(\omega + \Delta\omega)t - (k + \Delta k)z] + \cos[(\omega - \Delta\omega)t - (k - \Delta k)z] \\
&= 2\cos(\omega t - kz)\cos\Delta\omega\left(t - \frac{\Delta k}{\Delta\omega}z\right).
\end{aligned}
\tag{1.78}
$$

The factor $\cos(\omega t - kz)$ is a carrier wave with phase velocity $v_p = \omega/k$. The second cosine factor gives the wave modulation, or 'envelope'. If $\Delta\omega$ and Δk are small compared with ω and k, the envelope is slowly varying compared with the carrier and propagates with the velocity $\Delta\omega/\Delta k$. If we add a group of waves with a small spread in frequencies and wavenumbers around ω and k, we obtain similarly a carrier wave with phase velocity ω/k and an envelope with *group velocity*

$$
v_g = \frac{d\omega}{dk}.
\tag{1.79}
$$

This is equivalent to Rayleigh's expression (1.77) when we set the phase velocity $V = \omega/k$ in the latter. For the sake of completeness, we now carry out a derivation of this expression for the group velocity.

Consider a plane-wave electric field

$$
E(z, t) = \mathcal{E}(z, t)\exp[-\mathrm{i}(\omega_L t - k_L z)]
\tag{1.80}
$$

where $k_L = k(\omega_L)$. We are assuming a carrier wave with frequency and wavenumber ω_L and k_L, respectively, and that there is a modulation function or envelope that we denote by $\mathcal{E}(z, t)$. We use this expression for $E(z, t)$ in equation (1.61) and differentiate once with respect to z to obtain

$$
\frac{\partial\mathcal{E}}{\partial z} = \frac{\mathrm{i}}{2\pi}\int_{-\infty}^{\infty} dt'\,\mathcal{E}(0, t')\int_{-\infty}^{\infty} d\omega\,[k(\omega_L + \omega) - k_L]\mathrm{e}^{-\mathrm{i}\omega(t-t')}\mathrm{e}^{\mathrm{i}[k(\omega_L+\omega)-k_L]z}.
\tag{1.81}
$$

The Taylor expansion

$$
k(\omega_L + \omega) = k_L + \left(\frac{dk}{d\omega}\right)_{\omega_L}\omega + \frac{1}{2}\left(\frac{d^2k}{d\omega^2}\right)_{\omega_L}\omega^2 + \dots
\tag{1.82}
$$

in equation (1.81) gives

$$
\frac{\partial\mathcal{E}}{\partial z} + \left(\frac{dk}{d\omega}\right)_{\omega_L}\frac{\partial\mathcal{E}}{\partial t} + \frac{\mathrm{i}}{2}\left(\frac{d^2k}{d\omega^2}\right)_{\omega_L}\frac{\partial^2\mathcal{E}}{\partial t^2} + \dots = 0.
\tag{1.83}
$$

It often happens that $(d^2k/d\omega^2)_{\omega_L}\partial^2\mathcal{E}/\partial t^2$ and all the higher-derivative terms in (1.83) are negligible compared with the first-derivative terms. If, furthermore, absorption at frequency ω_L is sufficiently weak that $(dk/d\omega)_{\omega_L}$ may be taken to be real, then

$$\frac{\partial \mathcal{E}}{\partial z} + \frac{1}{v_g}\frac{\partial \mathcal{E}}{\partial t} = 0 \tag{1.84}$$

where the group velocity

$$v_g = \left(\frac{d\omega}{dk}\right)_{\omega_L} = \frac{c}{(n_R + \omega\, dn_R/d\omega)_{\omega_L}} \tag{1.85}$$

n_R again being the real part of the refractive index. In this approximation, the pulse envelope propagates *without change of shape or amplitude* at the group velocity:

$$E(z,t) = \mathcal{E}(0, t - z/v_g)e^{-i(\omega_L t - k_L z)}. \tag{1.86}$$

In a spectral region of 'normal dispersion', $(dn_R/d\omega)_{\omega_L} > 0$, the group velocity is less than the phase velocity c/n_R. Because v_g can exceed c in a region of anomalous dispersion, $(dn_R/d\omega)_{\omega_L} < 0$ (section 1.2), and because v_g was generally thought to be the velocity of energy propagation, $v_g > c$ was, in the past, thought to be in conflict with the special theory of relativity. This conflict was resolved in large part when Sommerfeld and Brillouin proved that the *signal* velocity cannot exceed c even in a region of anomalous dispersion.

The work of Sommerfeld and Brillouin is discussed in the following chapter. The point we wish to emphasize here is that the group velocity of an electromagnetic pulse is not, in general, the velocity of a signal. Unfortunately, the confusion surrounding the meaning of group velocity did not end with the work of Sommerfeld and Brillouin, as one might have expected. Even today, and even in deservedly well-known textbooks and monographs, one finds assertions that the concept of group velocity loses physical significance when the group velocity exceeds c or has one of the other interesting values discussed in the next chapter. The claim is that, contrary to the approximate result (1.86), a pulse will become highly distorted in shape as a consequence of the strong dispersion implied by $v_g > c$. This is a sensible but incorrect conclusion in general, as we shall see in chapter 2. Worse, it is often implied that there would be a conflict with special relativity if the group velocity were larger than c. This is wrong because, to repeat, group velocity is not, in general, a signal velocity: Sommerfeld and Brillouin had no problems with the possibility that the group velocity of a pulse could exceed c, although there was no direct experimental evidence in their time that this actually happened. The notion of group velocity represents an *approximation* to the actual state of affairs in which *signals* cannot propagate faster than c.

Experimental studies of 'superluminal' group velocities are discussed in chapter 2. These studies received considerable 'popular press' while some physicists felt that the experiments added nothing new to our understanding of group velocity, signals, or relativity[8]. In any case, it seems fair to say that this work has at least helped to correct some long-standing misconceptions, among them being that (a) a group velocity greater than c would violate special relativity and (b) group velocity ceases to have physical significance in the case of anomalous dispersion, when it can exceed c. We quote from two of the classic texts to which we have alluded[9]:

> There is no cause for alarm that our ideas of special relativity are violated [when the group velocity exceeds c]; group velocity is just not a useful concept here. A large value of $dn/d\omega$ is equivalent to a rapid variation of ω as a function of k. Consequently the approximations [made in deriving (1.86)] are no longer valid. The behaviour of the pulse is much more involved [14].

> ...in regions of anomalous dispersion the group velocity may exceed the velocity of light or become negative, and in such cases it has no longer any appreciable physical significance [9].

Note that a large value of $dn/d\omega$ does *not* necessarily imply that the approximations leading to (1.86) break down. In particular, it is clear from figure 1.1 that $dn/d\omega$ can be large while $d^2n/d\omega^2$ is small, so that the principal approximation made in deriving (1.86), namely the approximation that second and higher derivatives of $n(\omega)$ can be neglected, does not necessarily break down in a region of anomalous dispersion.

To cite an example where group velocity is implicitly assumed to be a signal velocity, consider the propagation of x-rays in glass, in which case we can take $\omega \gg \omega_0$ and ignore the damping term in equation (1.28):

$$n(\omega) \cong 1 - \frac{e^2}{2m\epsilon_0\omega^2} \sum_i \sum_j N_i f_{ij} = 1 - a/\omega^2 \qquad (1.87)$$

where $a = (e^2/2m\epsilon_0) \sum_i N_i \sum_j f_{ij} = Ne^2/2m\epsilon_0$. Then the phase velocity $c/n(\omega) > c$ while the group velocity

$$v_g = \frac{c}{1 + a/\omega^2} \qquad (1.88)$$

is less than c. Thus, it has been concluded that 'although the phases can travel faster than the speed of light, the modulation signals travel slower, and that is the resolution of the apparent paradox!' [20]. But in a different medium, where a

---

[8] Unfortunately this work was also belittled by some who somehow thought that the experiments involved nothing more subtle than a *phase* velocity greater than c! See *Physics Today* **54** 14 (February 2001).

[9] The pertinent discussion is corrected in the third edition of reference [14] as well as in a later printing of [9].

could be *negative* and the group velocity greater than c, the same argument would lead to a serious 'paradox' indeed!

Even in the more specialized literature, one finds statements that the standard expression for the group velocity loses its physical significance in a region of anomalous dispersion:

> the ...standard expression for the group velocity fails to describe the motion of the peak of the pulse envelope in a region wherein there is either significant absorption or amplification, and hence, no real physical significance may be associated with it in such regions [21].

The relation

$$v_p v_g = c^2 \tag{1.89}$$

between the group velocity v_g and the phase velocity v_p appears frequently in the theory of waveguides [22]. It is not a general result. It requires that

$$\frac{\omega}{k}\frac{d\omega}{dk} = c^2 \tag{1.90}$$

or $\omega^2 - k^2 c^2 = $ constant.

1.6   Maxwell's equations and special relativity: an example

It has been said that Maxwell was lucky: his equations for the electromagnetic field turned out to be Lorentz invariant, gauge invariant, and correct quantum mechanically when interpreted as Heisenberg operator equations of motion. It is primarily the first property of Maxwell's equations that is of interest in this chapter.

Einstein was aware of the Michelson–Morley experiment before he published his epiphanic paper on special relativity in 1905 [23] but the phenomena that most influenced him were the aberration of starlight and the propagation of light in moving media. Fresnel had predicted that there would be no change in the apparent position of a star if a telescope were filled with water and this was confirmed experimentally by Airy (1871). According to Fresnel, a material with refractive index n and velocity v causes light to propagate with the *additional* velocity fv with respect to the stationary ether, where the 'drag coefficient' $f = 1 - 1/n^2$. Fizeau's experimental results (1851) on the velocity of light in flowing water were consistent with a drag coefficient $f = 0.48$, in fair agreement with the theoretical value for $n = 1.33$.

Einstein showed that Fizeau's result, though seeming to support the ether drag hypothesis, could easily be explained by the theory of special relativity (SR), wherein 'the propagation of light always takes place with the same velocity $[c/n]$ *with respect to the liquid*, whether the latter is in motion with reference to other bodies or not. The velocity of light relative to the liquid and the velocity of the latter relative to the tube are thus known, and we require the velocity [u] of light relative to the tube' [24]. Thus, according to the SR velocity addition formula,

$$u = \frac{c/n + v}{1 + v(c/n)/c^2} \cong \frac{c}{n} + \left(1 - \frac{1}{n^2}\right)v = \frac{c}{n} + fv \qquad (v \ll c). \qquad (1.91)$$

It is interesting to describe the Fizeau experiment using the wave equation and Lorentz transformations. Consider the (scalar) wave equation

$$\frac{\partial^2 E}{\partial z'^2} - \frac{n^2}{c^2}\frac{\partial^2 E}{\partial t'^2} = 0 \qquad (1.92)$$

for the electric field E in a non-dispersive medium with refractive index n at rest in a reference frame K'^{10}. Make a Lorentz transformation

$$z = \gamma(z' + vt') \qquad t = \gamma(t' + vz'/c^2) \qquad [\gamma = (1 - v^2/c^2)^{-1/2}] \qquad (1.93)$$

to a frame moving in the z direction with velocity $-v$ with respect to K'. Then

$$\frac{\partial}{\partial z'} = \gamma\left(\frac{\partial}{\partial z} + \frac{v}{c^2}\frac{\partial}{\partial t}\right) \qquad \frac{\partial}{\partial t'} = \gamma\left(\frac{\partial}{\partial t} + v\frac{\partial}{\partial z}\right) \qquad (1.94)$$

and (1.92) becomes, for $v \ll c$,

$$\frac{\partial^2 E}{\partial z^2} - \frac{2(n^2 - 1)v}{c^2}\frac{\partial^2 E}{\partial x \partial t} - \frac{n^2}{c^2}\frac{\partial^2 E}{\partial t^2} = 0 \qquad (1.95)$$

for the field $E(z, t)$ in a medium moving with velocity $v\hat{z}$. This is easily generalized: for a medium moving with velocity v ($v \ll c$),

$$\nabla^2 E - \frac{2}{c^2}(n^2 - 1)v \cdot \nabla\left(\frac{\partial E}{\partial t}\right) - \frac{n^2}{c^2}\frac{\partial^2 E}{\partial t^2} = 0. \qquad (1.96)$$

For a monochromatic field, $E(r, t) = \mathcal{E}(r)e^{-i\omega t}$,

$$\nabla^2 \mathcal{E} + \frac{2i\omega}{c}(n^2 - 1)\frac{v}{c} \cdot \nabla\mathcal{E} + \frac{n^2\omega^2}{c^2}\mathcal{E} = 0. \qquad (1.97)$$

If $\mathcal{E}'$ is a solution of equation (1.97) with $v = 0$, then a solution with $v \neq 0$ is

$$\mathcal{E}(r) = \mathcal{E}'(r)\exp\left[-i\frac{\omega}{c^2}(n^2 - 1)\int^{s(r)} v \cdot ds'\right] \qquad (1.98)$$

where the line integral is over *any* path such that (a) the end point $s(r) = r$ and (b) $\nabla \times v = 0$. The proof that this is so is easily established using the fact that, if $\nabla \times v = 0$, v can be written as the gradient of a scalar function.

10 The effect of dispersion on the drag coefficient is discussed in [27].

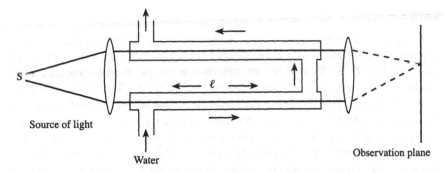

Figure 1.4. Schematic illustration of Fizeau's experiment.

Consider now the simple version of the Fizeau experiment indicated in figure 1.4. The field in the observation plane is the sum of two fields, one having passed through the tube with water flowing in the direction of propagation and the other with the water flowing against the direction of propagation. For either path, $v \neq 0$ and $\nabla \times v = 0$, so that equation (1.98) applies. The intensity at r in the observation plane has an interference part proportional to the real part of

$$\mathcal{E}_1'(r)\mathcal{E}_2'^*(r) \exp\left[i\frac{\omega}{c^2}(n^2-1)\left(\int_{\text{path 1}}^{s(r)} v \cdot ds' - \int_{\text{path 2}}^{s(r)} v \cdot ds'\right)\right]$$

$$= \mathcal{E}_1'(r)\mathcal{E}_2'^*(r) \exp\left[i\frac{\omega}{c^2}(n^2-1)\int \mathbf{\Omega} \cdot n \, dS\right]$$

$$= \mathcal{E}_1'(r)\mathcal{E}_2'^*(r)e^{i\phi} \tag{1.99}$$

where $\mathbf{\Omega} = \nabla \times v$ is the vorticity of the flow and the integral is over the surface bounded by the two paths to the observation plane in figure 1.4.

The vorticity in the idealized Fizeau experiment indicated in figure 1.4 is just $2v\ell$, where ℓ is the length of the tubes through which the light passes. Thus,

$$\phi = \frac{\omega}{c^2}(n^2-1)(2v\ell) \tag{1.100}$$

or

$$\delta = \frac{\phi}{2\pi} = \frac{2\ell}{\lambda c}n^2 fv \tag{1.101}$$

where $f = 1 - n^{-2}$ is the Fresnel drag coefficient. This fringe displacement is identical to that obtained using the formulas $u = c/n \pm fv$ [equation (1.91)] for the light velocities in the two tubes of flowing water and defining the fringe displacment in terms of the difference in time Δt it takes for light to propagate through the tubes with oppositely flowing water:

$$\Delta t = \frac{\ell}{c/n - fv} - \frac{\ell}{c/n + fv} \cong \frac{2\ell}{c^2}n^2 fv \tag{1.102}$$

and $\delta = c\,\Delta t/\lambda$.

The Fizeau experiment as just described has a formal similarity to the Aharonov–Bohm effect in quantum mechanics [25–27]. The velocity v and the vorticity $\mathbf{\Omega} = \nabla \times v$ in our description of the Fizeau experiment are analogous to the vector potential A and the magnetic field B, respectively, in the Aharonov–Bohm effect. The essence of the Aharonov–Bohm effect is that electron interference can be observed even though the electrons pass only through regions with $B = 0$: what matters is $\int B \cdot n\,dS$. In the Fizeau experiment, similarly, the fringe displacement is determined by $\int \mathbf{\Omega} \cdot n\,dS$ and there are interference fringes even though the light only (ideally) passes through regions with $\mathbf{\Omega} = 0$. Of course, the Aharonov–Bohm effect is more subtle for several reasons [27] but the analogy between it and the Fizeau experiment seems interesting nevertheless.

The fact that Maxwell's equations are Lorentz invariant hardly needs mention here and the transformation properties of the fields are treated exhaustively in many books such as *Jackson* [14]. There is probably no end to interesting analyses involving Maxwell's equations and Lorentz transformations but it is beyond our scope here to delve further into this topic. The most important thing for our purposes is the fact that the speed of light c in vacuum is the same in all reference frames and that no *signal* can be transmitted with a velocity greater than this (section 1.4).

1.7   Group velocity can be very small—or zero

It is well known that the group velocity of a pulse can be very *small*. Experiments in self-induced transparency, for instance, have demonstrated group velocities thousands of times smaller than the speed of light in vacuum. This is a nonlinear effect associated with the absorption of the leading part of a pulse in a resonant medium, followed by the amplification, by stimulated emission, of its back part.

The definition (1.85) implies that, if $n + \omega\,dn/d\omega$ is large, the group velocity can be very small even in a medium that responds *linearly* to the field. Experiments in which $\omega\,dn/d\omega$ is large and positive in the vicinity of a narrow resonance have demonstrated that the group velocity of a pulse can not only be reduced to values on the order of metres per second but that it can even be zero; in other words, a pulse can be *stopped*. In fact, the group velocity can be controlled to such an extent that the pulse can be *regenerated* by increasing the group velocity from zero to some finite value. These phenomena, which could conceivably have applications in 'quantum information' storage or in optical communications, are discussed in chapters 5 and 6.

1.8   The refractive index can be negative

Variations of the refractive index that produce unusual group velocities are not the only aspect of dispersion that will be of interest. A recent development is the growing recognition that the refractive index itself can be *negative*: nothing presented in this chapter suggests this possibility (or precludes it).

Veselago [28, 29] was evidently the first to consider seriously the possibility, and some consequences, of having a negative refractive index. In particular, he showed that a negative electric permittivity ϵ *and* a negative magnetic permeability μ implies a negative refractive index. The possibility that the refractive index (or phase velocity) could be negative was briefly touched upon by Mandelstam in connection with Snell's law [30, 31].

Associated with negative refractive indices are highly unusual propagation effects that could conceivably have important applications in the imaging of electromagnetic sources, and one of our goals is to provide a basic introduction (chapters 7 and 8) to negative refraction and its possible applications.

One of the consequences of a negative index of refraction is that the electric field vector E, the magnetic field vector H, and the wavevector k of a plane waveform a *left*-handed triad. For this reason, a medium with negative refractive index is called *left-handed* or *doubly negative* (because ϵ and μ are both negative) or simply a *metamaterial*. I prefer to call the *light* left-handed, because calling the *medium* left-handed might be misconstrued as saying something about its chirality. In any event, there are no known naturally occurring doubly negative materials and it is only recently that such materials have been purposefully fabricated.

1.9   The remainder of this book

The first six chapters of this book will be concerned with dispersive media that are 'unusual' in that the group velocity of light pulses can be greater than c, infinite, negative, much less than c, or zero. All of these situations have been realized experimentally. It will be of interest to understand in some detail why 'superluminal' pulses are consistent with Einstein causality, to see how these unusual propagation effects are described by quantum theory, and to understand some of the basic physics underlying situations where light pulses can be slowed down to velocities far less than c or even brought to a complete stop. In the last two chapters, we will focus on left-handed light, how it can be realized, and what some of its applications might be. The purpose of this chapter has been to set the stage by reviewing some of the most basic historical, conceptual, and theoretical aspects of electromagnetic propagation and dispersion.

Chapter 2

Fast light

In this chapter, we will be concerned primarily with group velocities that exceed c or are even infinite or negative. Such group velocities have been observed in various experiments. Consideration of such abnormal group velocities forces us to consider more carefully what is meant by a 'signal'.

We begin by proving that a front—a sudden, discontinuous turn-on of a field—propagates with the speed of light c in vacuum. Then we review some general theorems about where 'abnormal' group velocities should appear as the frequency of a field is varied across an absorption (or gain) spectrum. The possibility that a Gaussian pulse can have a group velocity that is greater than c or negative is considered, followed by a description of various experiments in which superluminal, negative, and infinite group velocities have been observed. It is argued that these experimental observations are not in conflict with Einstein causality, the statement that no signal can propagate faster than c. We explain why there is nothing surprising in observations of 'superluminal effects' in the propagation of Bessel beams and why what is propagating superluminally is not a signal. Next we consider the velocity with which electromagnetic energy is transported when the group velocity is superluminal. Finally, we list the various propagation velocities introduced in the chapter.

2.1 Front velocity

The fact that the group velocity could exceed c in a spectral region of anomalous dispersion was cause for great concern in the early days of special relativity. Brillouin's frequently cited monograph [32] describes in detail his work with Sommerfeld. The problem and its solution are summarized in the preface:

> [*group velocity*] seems to have been first discovered by Lord Rayleigh, who characterized this velocity in sound waves. It is now known to apply to practically all kinds of waves. Let us use the vocabulary of radio engineers and consider a carrier wave with a superimposed

modulation. The *phase velocity* yields the motion of elementary wavelets in the carrier, while the group velocity gives the propagation of the modulation. Lord Rayleigh considered that the group velocity corresponds to the velocity of energy or signals.

This however raised difficulty with the theory of relativity which states that no velocity can be higher than c, the velocity of light in vacuum. Group velocity, as originally defined, became larger than c or even negative within an absorption band. Such a contradiction had to be resolved and was extensively discussed in many meetings about 1910. Sommerfeld stated the problem correctly and proved that no signal velocity could exceed c. I discussed the problem in great detail and gave a complete answer.

Brillouin went on to mention three different velocities that would be introduced in the book: 'the *group velocity* of Lord Rayleigh', 'the *signal velocity* of Sommerfeld', and 'the *velocity of energy transfer*' and noted that 'These three velocities are identical for nonabsorbing media, but they differ considerably in an absorption band'[1]. Later in the book (p 15) he remarked that 'a galaxy of eminent scientists, from Voigt to Einstein, attached great importance to these fundamental definitions'.

We have already alluded to the fact that group velocity is not, in general a signal velocity. But just what is meant by a *signal*? According to Sommerfeld [32, p 18],

> In order to be able to say something about propagation, we must . . . have a limited wave motion: nothing until a certain moment in time, then, for instance, a series of regular sine waves, which stop after a certain time or which continue indefinitely. Such a wave motion will be called a *signal*.

According to this definition, a signal involves new information or an element of 'surprise' that could not have been predicted from the wave motion at an earlier time. Thus, a wave motion of the form

$$f(t) = \theta(t) \sin \omega t \tag{2.1}$$

where $\theta(t)$ is the Heaviside step function, can be called a signal. It must then be the case that the wave discontinuity or *front* at $t = 0$ propagates with a velocity $\leq c$ in *any* medium; otherwise we would have a signal, or information, propagating with a velocity greater than c, in violation of special relativity.

A proof of this important result can be given using equations (1.61) and (1.62):

$$E(z, t) = \int_{-\infty}^{\infty} dt' \, G(z, t - t') E(0, t') \tag{2.2}$$

---

[1] The italics in these quotations are Brillouin's. As mentioned in section 1.5, group velocity was evidently 'first discovered' by Hamilton, not Rayleigh.

$$G(z, \tau) = \frac{1}{2\pi} \int_{-\infty}^{\infty} d\omega \, e^{-i\omega\tau} e^{i\omega n(\omega)z/c}. \qquad (2.3)$$

As remarked in section 1.3, it is reasonable to assume, as did Sommerfeld and Brillouin, that $n(\omega) \to 1$ as $\omega \to \infty$. With this assumption, it follows that the integrand in (2.3) is $\exp i\omega(z/c - \tau)$ for large ω. Therefore, for $z > c\tau$, the integral can be replaced by a contour integral involving a large semicircle in the upper half of the complex ω plane as in figure 1.3. Now we recall from section 1.3 that causality—in the sense that there can be no polarization before there is an electric field to induce the polarization—requires that the refractive index $n(\omega)$ be analytic in the upper half of the complex ω plane. Therefore, for $z > c\tau$, the integral (2.3) vanishes:

$$G(z, \tau) = 0 \qquad \text{for } \tau < z/c. \qquad (2.4)$$

It follows from (2.2) that if the incident field $E(0, t) = 0$ for all $t < 0$, then the field $E(z, t) = 0$ for all $t < z/c$. In other words, we have the desired result that a wavefront such as (2.1) cannot propagate with a velocity greater than c. In fact, the *front velocity* is exactly c.

This result is so important that we should remind ourselves of some of the assumptions made in deriving it. First, as noted in section 1.3, we have ignored the Fresnel transmission coefficient for the boundary at $z = 0$ in the case when there is a change in refractive index at $z = 0$. This simplification does not affect our conclusion because the transmission coefficient is analytic in the upper half of the complex ω plane. This follows from the analyticity of $n(\omega)$, which is another assumption—or actually a theorem based on the causal relation between the electric field and the induced electric polarization—used in obtaining (2.4).

Second, we have assumed that $n(\omega) \to 1$ as $\omega \to \infty$. As already noted, this is a physically reasonable assumption: the particles that scatter radiation from the incident wave have finite mass and inertia and cannot respond sufficiently rapidly to (infinitely) high-frequency waves to have any effect on them and are, therefore, not 'seen' at all by these waves. Similarly, the particles cannot respond *instantaneously* to a wavefront. In a classical picture, they can scatter radiation only after they are displaced from their initial positions by the wave. Their accelerated motion after this displacement causes them to radiate and the fields from all the particles can then add coherently to (a) 'extinguish' the incident field and (b) replace it by a field having the phase velocity $c/n(\omega)$ (section 1.2). This, incidentally, raises a more subtle assumption in the derivation of (2.4) [32, p 38]:

> Concerning the validity of dispersion theory, we wish to mention one restriction which underlies all calculations of dispersion: that there must be very many particles within a wavelength. Only under this condition can we reckon on a continuous distribution of displacement vectors ... and disregard the molecular discontinuities. This condition is, as is well known, satisfied for wavelengths as short as ultraviolet, but

not for very high (x-ray) frequencies. In so far as we must include these frequencies in our analytical method, we are applying an extrapolation of the formulas of dispersion theory in a realm where their validity is not physically justified.

Brillouin does not discuss this point any further. The assumption that there are very many particles within a wavelength is made in formulating the Ewald–Oseen extinction theorem in the case of continuous media [10]. To deal rigorously and generally with the very high frequencies, one must go beyond the dipole approximation and also treat multiple scattering effects explicitly.

The continuous-medium assumption, as well as the assumption that $n(\infty) = 1$ for any medium, seems innocuous enough, although the latter has not been rigorously established in general in quantum electrodynamics.

To summarize: Sommerfeld and Brillouin introduced the notion of a *signal* as a 'limited wave motion: nothing until a certain moment in time, then, for instance, a series of regular sine waves, which stop after a certain time or which continue indefinitely'. A signal so defined has a sharp front, or step-function behaviour, at some time $t = 0$. The velocity at which this front propagates is exactly c in *any* medium. Thus, the velocity of a signal cannot exceed c no matter how $n(\omega)$ varies with ω and, in particular, no matter what the group velocity is.

2.2 Superluminal group velocity

As noted in section 1.5, the definition

$$v_g = \frac{d\omega}{dk} = \frac{c}{n_R + \omega \, dn_R/d\omega} \tag{2.5}$$

implies that the group velocity v_g can be 'superluminal', or greater than c. It can also be infinite or negative. As discussed in the following sections, such 'abnormal' values for the group velocity have been observed.

Equation (2.5) shows that the group velocity differs most significantly from the phase velocity $v_p = c/n_R$ when $|(\omega/n_R)\, dn_R/d\omega|$ is large. In a region of anomalous dispersion, $dn_R/d\omega < 0$ and $v_g > v_p$. In the case of a homogeneously broadened absorbing medium, where the real part of the refractive index is given approximately by equation (1.33), $v_g > v_p$ near the centre of the absorption line. At the centre of the absorption line ($\omega = \omega_0$), for instance,

$$v_g = \frac{c}{1 - K\omega_0/\gamma^2}. \tag{2.6}$$

Our discussion thus far has implicitly assumed an absorbing medium in which the number density N is approximately the density of atoms in the ground state. More generally, N should be replaced by the difference of lower-state and upper-state populations for the transition of frequency ω_0 (section 1.2). Thus, in the case of an *amplifying* medium, where this difference is negative, K is negative and we have normal dispersion (and $v_g < c$) close to the resonance frequency ω_0. The curve of $n_R(\omega) - 1$ *versus* ω in this case is simply reversed in sign compared to the curve in figure 1.1, so that the dispersion becomes anomalous when the field frequency is tuned *away* from ω_0. Thus, depending on the detuning $\omega - \omega_0$ in the vicinity of a resonance, the group velocity can exceed c in amplifiers as well as absorbers.

More generally, Bolda *et al* [33] have shown that, for any dispersive dielectric medium, there must be a frequency for which the group velocity is 'abnormal', i.e. larger than c, infinite, or negative. Consider the difference

$$\Delta t(\omega) = \frac{z}{v_g} - \frac{z}{c} \tag{2.7}$$

between the time it takes for the peak of a pulse with group velocity v_g to travel a distance z and the time it takes to travel the same distance in vacuum. For abnormal group velocities, this 'group delay' is negative:

$$\Delta t(\omega) = \left(\frac{z}{v_g} - \frac{z}{c}\right) = \frac{d}{d\omega}\left(\omega \frac{z}{c}[n_R(\omega) - 1]\right) \equiv \frac{d}{d\omega}\phi(\omega) < 0. \tag{2.8}$$

Bolda *et al* show, from the Kramers–Krönig relation between $n_R(\omega)$ and $n_I(\omega)$, that $\Delta t(\omega)$ must be negative for at least one frequency ω.

Consider first the high-frequency limit of $\phi(\omega)$. According to equation (1.73), $n_R(\omega) \to 1 - \omega_p^2/2\omega^2$ in this limit, where the plasma frequency ω_p is defined by $\omega_p^2 = NZe^2/m\epsilon_0$. Thus

$$\phi(\omega) = -\frac{z}{c}\frac{\omega_p^2}{2\omega} \qquad (\omega \to \infty) \tag{2.9}$$

and is always negative in this high-frequency limit. In the zero-frequency limit, we have, from (1.57),

$$n_R(0) = 1 + \frac{2}{\pi}P\int_0^\infty \frac{n_I(\omega)}{\omega}\,d\omega \tag{2.10}$$

and

$$\Delta t(0) = \frac{z}{c}[n_R(0) - 1] = \frac{2z}{\pi c}P\int_0^\infty \frac{n_I(\omega)}{\omega}\,d\omega. \tag{2.11}$$

The conclusion that $\Delta t(\omega)$ must be negative for at least one ω, i.e. that there must be at least one frequency for which the group velocity is greater than c, infinite, or negative, now follows from (2.9)–(2.11) and continuity. Consider the case of an absorbing medium, for which $n_I(\omega) > 0$ and, therefore, $\Delta t(0) > 0$ [equation (2.11)]. Since $\phi(0) = 0$ and $\Delta t(\omega) = (d/d\omega)\phi(\omega)$, there must be some small frequency ω at which $\phi(\omega) > 0$, whereas at sufficiently high frequencies $\phi(\omega) < 0$. Now $\phi(\omega)$, like $n_R(\omega)$, must be continuous and differentiable and so, at some intermediate frequency, $\Delta t(\omega) = (d/d\omega)\phi(\omega)$ must be negative, or in other words the group velocity at some frequency in an absorber must be 'abnormal'. In the case of an amplifying medium such that $n_I(\omega) < 0$ and the integral in equation (2.11) is negative, the group velocity is seen to be abnormal at zero frequency in particular [34].

The variation of $n_R(\omega)$ with ω shown in figure 1.1 for the important case of a homogeneously broadened absorber indicates that abnormal group velocities in absorbers are most likely to be observed at exactly those frequencies for which the absorption is strongest. Bolda *et al* [33] show more generally that, in any absorber, the group velocity is, in fact, abnormal at the frequency at which the absorption is greatest, whereas $v_g < c$ at the frequency at which the absorption is weakest. To establish this result, they consider the complex phase difference

$$F(\omega) = \frac{z}{c}\omega[n(\omega) - 1]. \tag{2.12}$$

For real ω, the real part of $F(\omega)$ reduces to $\phi(\omega)$, whereas

$$\mathrm{Im}[F(\omega)] = \frac{z}{c}\omega n_I(\omega) = z\kappa(\omega) \tag{2.13}$$

where $\kappa(\omega)$ is the field attentuation coefficient. $F(\omega)$ is analytic in the upper half of the complex ω plane, and so gives

$$\frac{dF(\omega)}{d\omega} = \frac{1}{2\pi i}\oint \frac{F(\omega')}{(\omega' - \omega)^2}\,d\omega' \tag{2.14}$$

where the path of integration is that of figure 1.3 except that the principal part of the integral is taken at $\omega' = \omega$. $n(\omega) \to 1$ as $\omega \to \infty$ implies that the integral over the large semicircle vanishes, so that

$$\frac{dF(\omega)}{d\omega} = \frac{1}{2\pi i}P\int_{-\infty}^{\infty} d\omega' \frac{F(\omega')}{(\omega' - \omega)^2}. \tag{2.15}$$

This can be rewritten as [33]

$$\begin{aligned}
\frac{dF(\omega)}{d\omega} &= \lim_{\epsilon \to 0} \frac{1}{2\pi i}P\int_{-\infty}^{\infty} d\omega' \frac{F(\omega') - F(\omega) - (\omega' - \omega - i\epsilon)\,dF(\omega)/d\omega}{(\omega' - \omega - i\epsilon)^2} \\
&\quad + F(\omega)\frac{1}{2\pi i}P\int_{-\infty}^{\infty} \frac{d\omega'}{(\omega' - \omega - i\epsilon)^2} \\
&\quad + \frac{dF(\omega)}{d\omega}\frac{1}{2\pi i}P\int_{-\infty}^{\infty} \frac{d\omega'}{(\omega' - \omega - i\epsilon)}
\end{aligned} \tag{2.16}$$

where the second and third terms have been added and subtracted. Thus, performing the simple integrals in these terms, one obtains

$$\frac{dF(\omega)}{d\omega} = \frac{1}{\pi i}P\int_{-\infty}^{\infty} d\omega' \frac{F(\omega') - F(\omega) - (\omega' - \omega)\,dF(\omega)/d\omega}{(\omega' - \omega)^2} \tag{2.17}$$

or, taking the real parts of both sides,

$$\Delta t(\omega) = \frac{d\phi(\omega)}{d\omega} = \frac{z}{\pi}P\int_{-\infty}^{\infty}d\omega'\,\frac{\kappa(\omega') - \kappa(\omega) - (\omega' - \omega)\,d\kappa(\omega)/d\omega}{(\omega' - \omega)^2}. \quad (2.18)$$

Equation (2.18) implies that, at an absorption maximum ($d\kappa(\omega)/d\omega = 0$), $\Delta t(\omega) < 0$, i.e. the group velocity is abnormal at the frequency at which the absorption is strongest. Similarly, $\Delta t(\omega) > 0$ and, therefore, $v_g < c$ at the frequency at which the absorption is weakest. In the case of an amplifying medium, with field amplification coefficient $-\kappa(\omega)$, $v_g < c$ at the frequency at which the gain is greatest. These results generalize the conclusions reached on the basis of a homogeneously broadened absorption (or amplification) line characterized by a Lorentzian lineshape.

2.3 Theoretical considerations of superluminal group velocity

The literature relating to the theoretical possibility of superluminal group velocity is large and diverse and certainly did not end with the work of Sommerfeld and Brillouin. The recent interest in the subject stems from experimental work made possible with lasers and we will accordingly describe some of the relevant theoretical work in laser physics that was done before the most recent experiments.

Basov *et al* [35] concluded from their analysis of pulse propagation in an amplifying medium that 'the velocity of pulse propagation may prove to be much higher than the speed of light in vacuum'. It is now well known that the *group* velocity of a pulse in an amplifier can exceed c due to a pulse reshaping in which the front part of the pulse leaves less gain available for the back part, resulting in an advancement of the peak of the pulse. Whether Basov *et al* actually intended to imply that a *signal* could be propagated superluminally is not clear but, in any case, their conclusion was criticized by Icsevgi and Lamb [36], who argued that Basov *et al* in their analysis assumed 'an unphysical input pulse extending to infinity at both ends'. The measurement of pulse velocities greater than c does not contradict special relativity because 'an experimental apparatus can only trace the bulk of the pulse' which does not constitute a signal. Thus, Icsevgi and Lamb implied, in agreement with Sommerfeld and Brillouin, that a signal must involve a discontinuous wavefront and that the (group) velocity characterizing the 'bulk of the pulse' is not the velocity at which a signal can be propagated.

An early and important contribution for later developments was made by Garrett and McCumber [37] in 1970. They showed that superluminal group velocities can appear when a Gaussian pulse propagates in an absorbing medium, provided that the pulse bandwidth is much smaller than the width of the absorption line and the medium is sufficiently short. Under these circumstances, 'the pulse remains substantially Gaussian and unchanged in width for many exponential absorption depths, and . . . the locus of instants of maximum amplitude follows the classical expression for the group velocity, even if this is greater than the velocity of light, or negative'. This results from a pulse reshaping and advancement process in which the back part of the pulse is more strongly absorbed than the front part.

Consider a Gaussian pulse

$$E(0, t) = E_0 e^{-i\bar{\omega}t} e^{-t^2/2\tau^2} \tag{2.19}$$

incident on a dielectric of refractive index $n(\omega)$ occupying the half-space $z \geq 0$. The spectrum of this field is

$$A(\omega) = \frac{1}{2\pi} \int_{-\infty}^{\infty} dt \, E(0, t) e^{i\omega t} = E_0 \frac{\tau}{\sqrt{2\pi}} e^{-\frac{1}{2}(\omega-\bar{\omega})^2\tau^2} \tag{2.20}$$

and, according to equation (1.60), the field at z in the medium is

$$E(z, t) = E_0 \frac{\tau}{\sqrt{2\pi}} \int_{-\infty}^{\infty} d\omega \, e^{-i\omega t} e^{i\omega n(\omega)z/c} e^{-\frac{1}{2}(\omega-\bar{\omega})^2\tau^2}. \tag{2.21}$$

This is the starting point of the analysis of Garrett and McCumber [37]. They assume a refractive index appropriate to a homogeneously broadened medium with absorption frequency ω_0 and linewidth γ:

$$n(\omega) \cong n_\infty - \frac{\omega_0\omega_p}{\omega(\omega - \omega_0 + i\gamma)} \qquad (|\omega_p/\gamma| \ll n_\infty) \tag{2.22}$$

where n_∞ is the refractive index far from the absorption line.

Assuming that the spectral width of the pulse is much less than the absorption linewidth, i.e. $\gamma\tau \gg 1$, the principal contribution to the integral in (2.21) is from frequencies $\omega \approx \bar{\omega}$. In this case, $\omega n(\omega)$ in the integral can be approximated by the first three terms of its Taylor series:

$$\omega n(\omega) \cong \bar{\omega}n(\bar{\omega}) + (\omega - \bar{\omega})\left[\frac{d(n\omega)}{d\omega}\right]_{\bar{\omega}} + \frac{1}{2}(\omega - \bar{\omega})^2\left[\frac{d^2(n\omega)}{d\omega^2}\right]_{\bar{\omega}}. \tag{2.23}$$

(The fact that higher derivatives of $\omega n(\omega) = k(\omega)c$ can be ignored for sufficiently long pulses can also be seen from equation (1.83).) Then

$$E(z, t) \cong E_0 \frac{\tau}{\sqrt{2\pi}} e^{-i\bar{\omega}(t-n(\bar{\omega})z/c)} \int_{-\infty}^{\infty} du \, e^{-iu(t-\alpha z/c)} e^{-\frac{1}{2}u^2(\tau^2 - i\beta z/c)}$$

$$= \frac{E_0}{\sqrt{1 - i\beta z/c\tau^2}} e^{-i\bar{\omega}(t-n(\bar{\omega})z/c)} \exp\left[\frac{-(t - \alpha z/c)^2}{2\tau^2(1 - i\beta z/c\tau^2)}\right] \tag{2.24}$$

where

$$\alpha \equiv \left[\frac{d(n\omega)}{d\omega}\right]_{\bar{\omega}} \qquad \beta \equiv \left[\frac{d^2(n\omega)}{d\omega^2}\right]_{\bar{\omega}}. \tag{2.25}$$

It has been assumed here that

$$\mathrm{Re}\left(1 - \frac{i\beta z}{c\tau^2}\right) > 0 \tag{2.26}$$

so that the integral in (2.24) converges. Assuming, furthermore, that $(1 - i\beta z/c\tau^2) \sim 1$, we have

$$E(z,t) \cong E_0 e^{-i\overline{\omega}(t - n(\overline{\omega})z/c)} \exp\left[\frac{-(t - \alpha z/c)^2}{2\tau^2(1 - i\beta z/c\tau^2)}\right]. \tag{2.27}$$

After some algebra, we obtain, in this approximation,

$$|E(z,t)|^2 \cong |E_0|^2 e^{-X(z,t)} \tag{2.28}$$

where

$$X(z,t) = 2\overline{\omega}n_\mathrm{I}(\overline{\omega})\frac{z}{c} - \alpha_\mathrm{I}^2\frac{z^2}{c^2} + \frac{\tau^2 + \beta_\mathrm{I}z/c}{(\tau^2 + \beta_\mathrm{I}z/c)^2 + \beta_\mathrm{R}^2 z^2/c^2}$$

$$\times\left(\left[t - \left(\alpha_\mathrm{R}\frac{z}{c} - \frac{\frac{1}{2}\alpha_\mathrm{I}\beta_\mathrm{R}z^2/c^2}{\tau^2 + \beta_\mathrm{I}z/c}\right)\right]^2\right.$$

$$\left.+ \frac{\alpha_\mathrm{I}\beta_\mathrm{R}z^3/c^3}{\tau^2 + \beta_\mathrm{I}z/c}\left[\alpha_\mathrm{R} - \frac{\frac{1}{4}\alpha_\mathrm{I}\beta_\mathrm{R}z/c}{\tau^2 + \beta_\mathrm{I}z/c}\right]\right). \tag{2.29}$$

This expression simplifies to

$$X(z,t) = \frac{2\omega_0\omega_\mathrm{p}}{\gamma}\frac{z}{c} + \frac{(t - z/v_\mathrm{g})^2/\tau^2}{1 - 2\omega_0\omega_\mathrm{p}z/\gamma^3 c\tau^2} \tag{2.30}$$

$$v_\mathrm{g} = \frac{c}{n_\infty - \omega_0\omega_\mathrm{p}/\gamma^2} \tag{2.31}$$

when the central frequency of the pulse coincides with the resonance frequency of the medium ($\overline{\omega} = \omega_0$). The expression (2.31) for the group velocity at line centre is a slight generalization of (2.6).

The first term on the right-hand side of (2.30) gives just the attenuation factor for light of frequency ω_0 that has propagated a distance z into the medium. (For the case of an amplifier, where $\omega_\mathrm{p} < 0$, it gives the amplification factor.) The second term shows that, at a given z, the peak intensity of the pulse occurs at the time $t = z/v_\mathrm{g}$, i.e. the pulse propagates with the group velocity v_g. In fact, if

$$z \ll \gamma^3 c\tau^2/2\omega_0\omega_\mathrm{p} \tag{2.32}$$

the pulse propagates at the group velocity and without change of shape aside from the overall attenuation described by the first term on the right-hand side of (2.30):

$$|E(z,t)|^2 \cong |E_0|^2 e^{-az} e^{-(t - z/v_\mathrm{g})^2/\tau^2} = e^{-az}|E(0, t - z/v_\mathrm{g})|^2 \tag{2.33}$$

where $a = 2\omega_0\omega_p/\gamma c$. The condition (2.32) is consistent with the assumption that $(1 - i\beta z/c\tau^2) \sim 1$ and that (2.26) is satisfied. In terms of the absorption length $\ell_{abs} = 1/a$, we can write (2.32) as

$$z/\ell_{abs} \ll \gamma^2\tau^2 \tag{2.34}$$

which states that the propagation distance measured in units of the absorption length is small compared with $(\Delta\nu_{abs}/\Delta\nu_L)^2$, where $\Delta\nu_{abs}$ and $\Delta\nu_L$ are the spectral widths of the absorption line and the light pulse, respectively.

Note that, for an absorber ($\omega_p > 0$), the group velocity (2.31) can be superluminal or negative. Garrett and McCumber [37] described this as

> quite a paradoxical result ... not only can the pulse appear to travel (in the sense of tracing the locus of instants of maximum amplitude) faster than c: it can even appear to travel backwards .... The nervous reader may perhaps feel reassured if we point out (i) that, in any time snapshot, the amplitude decreases monotonically with z in a lossy medium, and (ii) that the Poynting vector is always directed toward increasing z. Nevertheless, it is still true that the output-pulse peak can sometimes emerge from the far side of a parallel-sided slab of medium *before* the peak of the input pulse enters the near side. This output pulse will be greatly attenuated (or greatly amplified, as the case may be) but still of substantially the same Gaussian shape as the input pulse.

In fact, equation (2.31) shows that the group velocity can be infinite as well as superluminal or negative. An infinite group velocity means that the peak of the pulse emerging at the end of the medium occurs at the same instant as the peak of the pulse at the entrance to the medium. A negative group velocity means that the peak of the emerging pulse occurs at an earlier time than the peak of the pulse at the entrance to the medium (figure 2.1).

Garrett and McCumber went on to study numerically the propagation of an initially Gaussian pulse over distances sufficiently large that the approximations leading to (2.27) break down and the pulse no longer propagates at the group velocity without significant distortion. They noted, however, that these numerical simulations were of limited relevance to experiments, since, for $\gamma\tau \gg 1$, 'significant distortion will generally not occur until the overall gain or attenuation is enormously large'.

Crisp [38] interpreted the possibility of a superluminal group velocity in terms of an 'asymmetric absorption of energy from the light pulse. More energy is absorbed from the trailing half of the pulse than from the front half, causing the centre of gravity of the pulse to move at a velocity greater than the phase velocity of light.' Such an interpretation appears frequently in discussions of superluminal group velocity and can be based as follows on an approximate relation between the polarization P of the medium and the electric field

$$E(z,t) = \mathcal{E}(z,t)e^{-i\bar{\omega}t} = \int_{-\infty}^{\infty} d\Delta \, \tilde{\mathcal{E}}(z,\Delta)e^{-i(\bar{\omega}+\Delta)t}. \tag{2.35}$$

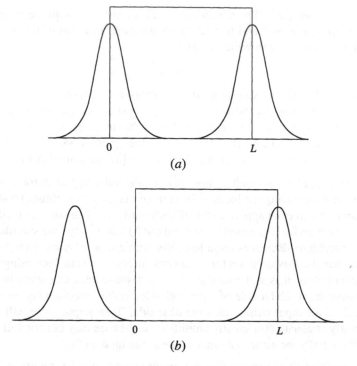

Figure 2.1. Intensity profiles at a fixed time when a light pulse is incident on a medium of length L such that the group velocity is (a) infinite and (b) negative. For schematic purposes, the input and output pulses are shown with the same normalized amplitude.

The polarization is

$$P(z, t) = \int_{-\infty}^{\infty} d\Delta \, \chi(\bar{\omega} + \Delta) \tilde{\mathcal{E}}(z, \Delta) e^{-i(\bar{\omega} + \Delta)t} \qquad (2.36)$$

where $\chi(\omega)$ is the susceptibility. If the spectrum of the field is peaked sufficiently sharply at the central frequency $\bar{\omega}$, i.e. if the envelope $\mathcal{E}(z, t)$ varies sufficiently slowly on a time scale $\sim 1/\bar{\omega}$, we can make the approximation

$$\chi(\bar{\omega} + \Delta) \cong \chi(\bar{\omega}) + \Delta \chi'(\bar{\omega}) \qquad (2.37)$$

in (2.36):

$$P(z, t) \cong e^{-i\bar{\omega}t} \int_{-\infty}^{\infty} d\Delta \, [\chi(\bar{\omega}) + \Delta \chi'(\bar{\omega})] \tilde{\mathcal{E}}(z, \Delta) e^{-i\Delta t}$$

$$= \chi(\bar{\omega}) E(z, t) + i \chi'(\bar{\omega}) e^{-i\bar{\omega}t} \frac{\partial \mathcal{E}(z, t)}{\partial t}. \qquad (2.38)$$

Based on a similar expression derived from the theory of resonant pulse propagation in a medium with absorption linewidth much greater than the spectral width of the pulse, Crisp [38] uses the fact that the sign of $\partial \mathcal{E} / \partial t$ is positive during the front half of the pulse and negative during the trailing half to conclude that more energy is absorbed from the trailing half of the pulse than the front half.

Chiao *et al* [39] have suggested that superluminal group velocities could be observed in optical amplifiers whose relaxation times are long compared with the pulse duration. Recalling equation (1.31) for the refractive index near an atomic resonance frequency ($\omega_{21} = \omega_0$, $f_{12} = f$), we have, for the real part of the index,

$$n_R(\omega) \cong 1 + \frac{e^2 f}{4m\epsilon_0\omega_0} \frac{N_1 - N_2}{\omega_0 - \omega} \tag{2.39}$$

if the field is sufficiently far from resonance that $(\omega_0 - \omega)^2 \gg \gamma^2$. For an inverted (amplifying) medium with $w = (N_2 - N_1)/N > 0$, where N is the number density of atoms,

$$n_R(\omega) \cong 1 - \frac{Nwe^2 f}{4m\epsilon_0\omega_0} \frac{1}{\omega_0 - \omega} \tag{2.40}$$

and

$$k = n_R(\omega)\frac{\omega}{c} = \frac{\omega}{c}\left[1 - \frac{Nwe^2 f}{4m\epsilon_0\omega_0} \frac{1}{\omega_0 - \omega}\right] = \frac{\omega}{c}\left[1 - \frac{\omega_p^2 w/4\omega_0}{\omega_0 - \omega}\right] \tag{2.41}$$

$$k - k_0 = \frac{1}{c}(\omega - \omega_0) - \frac{\omega}{c}\frac{\omega_p^2 w/4\omega_0}{\omega_0 - \omega} \tag{2.42}$$

$$(k - k_0)c \cong (\omega - \omega_0) - \frac{\omega_p^2 w/4}{\omega_0 - \omega}. \tag{2.43}$$

Here the plasma frequency ω_p is defined by

$$\omega_p^2 = \frac{Ne^2 f}{m\epsilon_0} = \frac{2Nd^2\omega_0}{\hbar\epsilon_0} \tag{2.44}$$

where d is the electric dipole transition matrix element. Defining $K = k - k_0$ and $\Omega = \omega - \omega_0$, we write (2.43) as

$$\Omega^2 - Kc\Omega + \tfrac{1}{4}w\omega_p^2 = 0 \tag{2.45}$$

which is the 'tachyonic' dispersion relation derived by different methods by Chiao *et al*.

This dispersion relation is discussed in detail in [39]. Here we simply note that equation (2.43) implies the group velocity

$$v_g = \frac{d\omega}{dk} = c\left(1 - \frac{\omega_p^2 w/4}{(\omega_0 - \omega)^2}\right)^{-1} \tag{2.46}$$

so that, in the case of an amplifier ($w > 0$), an off-resonant pulse can propagate with a group velocity $v_g > c$.

Chiao *et al* argue that a superluminal group velocity might be observed even at the single-photon level. They refer to a single-photon wave packet propagating with superluminal group velocity as an 'optical tachyon'. To avoid spontaneous emission noise, they require that the radiative lifetime of the atoms be large compared with the propagation time through the amplifier, which, in turn, should be large compared with the pulse duration. In order to avoid cooperative spontaneous emission, or superfluorescence (SF), the time delay before the peak of any SF pulse should be larger than all these other times. We discuss the possibility of an optical tachyon in more detail in chapter 4. We now turn our attention to some observations of superluminal and other abnormal group velocities.

2.4   Demonstrations of superluminal group velocity

Experimental demonstrations of superluminal group velocity were reported shortly after the paper by Garrett and McCumber. In the intervening years, various other reports of superluminal group velocity have appeared. In this section, we briefly describe the earliest experiments as well as more recent experiments that have stimulated other work to be discussed later.

2.4.1   Repetition frequency of mode-locked laser pulses

Faxvog *et al* [40] found evidence of a superluminal group velocity in the propagation of mode-locked pulse trains in a resonant absorber. A mode-locked laser producing a train of pulses with a repetition frequency $c/2L$, where L is the length of the laser cavity, will have a pulse repetition frequency of $v_g/2L$ if an absorber in which the group velocity is v_g is put in the cavity. Using a mode-locked He–Ne laser cavity containing an Ne absorption cell, an increase in the pulse repetition frequency when the absorption and dispersion were increased by increasing the current in the absorption cell was observed and found to be consistent with a group velocity exceeding c by about three parts in 10^4 [2].

2.4.2   Pulse propagation in linear absorbers

Chu and Wong [41] confirmed the predictions of Garrett and McCumber by showing that pulses could propagate in an absorber with group velocities that are greater than c, negative, or infinite, and could do so without substantial distortion of their initial shape. In these experiments, pulses from a tunable dye laser were divided by a beam splitter, so that one part of a pulse passed through an absorber while the other propagated essentially in vacuum. Measured cross correlation

---

[2] We quote the value stated by Faxvog *et al*. The cavity length L in the experiments was about 118 cm. For a particular cell current and laser power, an increase in the pulse repetition frequency of about 5 kHz was observed. This implies $(v_g - c)/2L = 5 \times 10^3$ s^{-1} or $v_g/c - 1 \cong 4 \times 10^{-5}$. Faxvog *et al*, however, state that $v_g/c - 1 \cong 3 \times 10^{-4}$.

signals of the recombined pulses gave the delay time of the pulse that passed through the absorber, based on the fact that the pulse velocity for far off-resonance pulses is equal to the phase velocity $c/n_0 = 8.57 \times 10^9$ cm s^{-1}. The laser frequency was varied across the absorption line to determine the delay time and, therefore, the group velocity as a function of frequency. The intensities were far below the saturation level, so that the pulse propagation was very close to linear, as assumed in the analysis of Garrett and McCumber, and the condition (2.34) was also well satisfied.

Figure 2.2 shows the measured pulse delay time and group velocity as the laser frequency was varied across the Lorentzian absorption curve. Chu and Wong remarked: 'Although the delay as a function of laser frequency is a smooth, well behaved function, the pulse velocity goes through some rather counter-intuitive singularities' implied by the formula $v_g(\omega) = c/[n(\omega) + \omega \, dn(\omega)/d\omega]$. In fact, the data show that the delay time of the pulse passing through the absorption layer goes through zero as the laser frequency is varied. In other words, the peak of the pulse emerging from the absorber occurs at the same time as the peak of the pulse incident on the absorber: the group velocity $v_g(\omega) = \infty$ in this case. The data also show, in agreement with the theory (the full curve in figure 2.2), that the group velocity can be negative (figure 2.1). In either case, the group velocity is obviously superluminal in that the time delay in traversing the absorber is less than the time delay for traversing the same distance in vacuum. The cross-correlation data indicated that the pulse emerging from the absorber had nearly the same shape as the incident pulse, albeit with some pulse compression.

2.4.3 Photon tunnelling experiments

More recent interest in abnormal group velocities is due in considerable part to the experimental and theoretical work of Chiao's group [42–44]. This work, *inter alia*, answered some important and long-standing questions about tunnelling times. It is also noteworthy that these were essentially *single-photon* experiments.

MacColl [45] in 1932 considered the transmission and reflection of a wave packet incident on the potential barrier defined by $V(x) = 0$ for $x < 0$ and $x > a$ and $V(x) = V_0 > 0$ for $0 < x < a$. It is important to note that he chose an initial wave packet that did not vanish but was very small for $x > a$: this initial wave packet, as opposed to one that identically vanishes for $x > a$, was chosen so that none of the energy components E making up the incident packet exceed V_0. He found that 'the transmitted packet appears at the point $x = a$ at about the time at which the incident packet reaches the point $x = 0$, so that there is no appreciable delay in the transmission of the packet through the barrier'. Defining the tunnelling time in terms of how long it takes for the peak of the wave packet at $x = a$ to occur relative to the peak of the incident packet at $x = 0$, one would conclude that the tunnelling is superluminal.

It is well known that evanescent waves in optics are analogous to

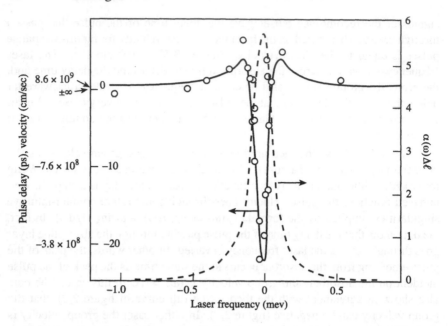

Figure 2.2. Data of Chu and Wong for the pulse delay and group velocity (circles) for 48-ps pulses passing through an absorption layer of length $\Delta\ell = 76$ μm. The broken curve is the measured absorption lineshape and the full curve is the theoretical group velocity obtained from the absorption lineshape, using the Kramers–Krönig relation to numerically determine the real part of the refractive index, $n_R(\omega)$, and from that the group velocity. Note that the length of the absorption layer is less than six times the absorption length $1/\alpha(\omega)$ at the peak of the absorption curve. From [41], with permission.

tunnelling wavefunctions in wave mechanics[3]. The Helmholtz equation $\nabla^2 \mathcal{E} + (n^2\omega^2/c^2)\mathcal{E} = 0$ for the amplitude of a scalar monochromatic wave has the same form as the time-independent Schrödinger equation, $\nabla^2\psi + (2m/\hbar^2)(E - V)\psi$, making the occurrence of an evanescent optical wave (imaginary n) analogous to particle tunnelling ($E < V$).

   Chiao *et al* performed experiments in which the central frequency of a single-photon wave packet was that for minimum transmission through a multilayered dielectric consisting of alternating layers of high and low n. The exponentially decaying, evanescent behaviour of the transmitted wave is analogous to quantum tunnelling—in fact, the situation is analogous to the Krönig–Penney model for the propagation of electrons in a crystal. Near the frequency of minimal transmission in the experiments, the group velocity approximation is quite accurate and, therefore, the tunnelling wave packet should suffer little distortion, although the transmission is, of course, very small.

---

[3] This analogy is discussed in some detail by Zhu *et al* [46].

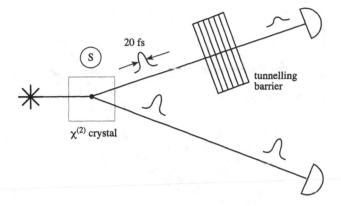

Figure 2.3. Schematic illustration of the experiment by Chiao *et al.* From [43], with permission.

Spontaneous parametric down-conversion was used to simultaneously generate pairs of photons at a wavelength of 702 nm (figure 2.3). Two pinholes used to select the photon pairs determine, by their size and the phase-matching condition, the bandwidth of the down-converted light. [See, for example, [47] for a simplified treatment.] In the experiments, the resulting photon wave packets had a bandwidth of ∼ 6 nm and a temporal width ∼ 20 fs. Using coincidence photon-counting and an application of a Hong–Ou–Mandel interferometer [48] to measure the femtosecond-scale delays between photons that traversed the tunnel barrier and their twins that passed through air, Chiao *et al* were able to determine the photon tunnelling times. Figure 2.4 shows photon coincidence rate data *versus* the path delays with and without the tunnel barrier in place. The negative delay found with the barrier means that a single photon tends to pass through the barrier faster than it would propagate through an equal distance in air. Effective tunnelling velocities of about 1.7*c* were inferred from the measured photon coincidence rates. Thus, these experiments demonstrated that the tunnelling process can indeed be 'superluminal', as predicted by MacColl [45].

2.4.4 Gain-doublet experiments

In the experiments of Chiao *et al*, the tunnelling probability is very small; and, similarly, in the experiments of Chu and Wong [41], there was considerable attenuation of the pulses propagating in the absorbing medium. The observation of distortionless pulses propagating with superluminal group velocity *and* relatively small change in amplitude was reported by Wang *et al* [49]. In these experiments, a gain doublet is employed, such that in the spectral region between two gain peaks there is strong anomalous dispersion but little gain (or absorption) [50], as shown in figure 2.5. In such a spectral region, the group velocity can differ significantly from *c* while the pulse suffers little change in either amplitude or shape.

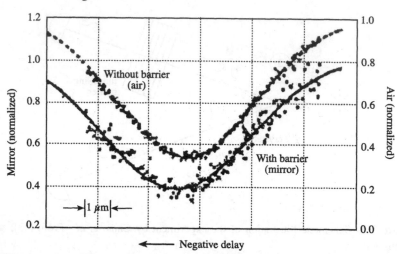

Figure 2.4. Data of Chiao *et al* for the coincidence rate *versus* the path difference of twin photons produced by spontaneous parametric down-conversion, one photon passing through a tunnel barrier and the other through a column of air of the same length as the barrier. From [44], with permission.

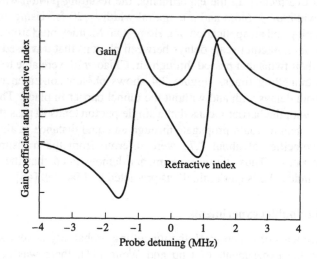

Figure 2.5. Gain and refractive index in the vicinity of a gain doublet. From [49], with permission.

Wang *et al* used a 6-cm cesium cell coated with paraffin, which allows atoms to maintain their spin polarization when they collide with the walls. They prepared the cesium atoms in the three-state system shown in figure 2.6 by optical pumping with polarized light. Two right-hand circularly polarized, continuous-

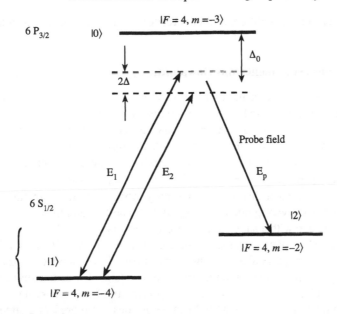

Figure 2.6. Approximate level diagram of cesium in the experiments of Wang *et al.* From [49], with permission.

wave Raman pump beams, shifted in frequency by 2.7 MHz, are incident on the cell; and their electric field amplitudes E_1 and E_2 are indicated in figure 2.6. The state $|2\rangle$ in figure 2.6 is the final state of the Raman transition, while the state $|0\rangle$ serves as the primary intermediate state. A continuous-wave probe field (E_p in figure 2.6) was varied in frequency with an acousto-optical modulator and used to measure the gain and refractive index as a function of frequency: results conforming accurately to the curves in figure 2.5 were obtained. The predicted group velocity was $v_g = -c/330$. Then a weak, nearly Gaussian probe pulse (3.7 μs FWHM) was used to measure transmitted pulse intensity profiles and propagation times. It was verified that the peak of the transmitted pulse appears at the end of the cell before the peak of the incident pulse appears at the entrance to the cell, consistent with the prediction of a negative group velocity. The transmitted peak occurred 62 ns before the incident peak. Thus, the transmitted peak goes about 3×10^8 m s$^{-1} \times 62$ ns = 18 m from the cell before the incident peak even arrives. Since the time for light travelling at c to traverse the 6-cm cell is about 0.2 ns, the 62-ns advance implies a group velocity in the cell of about $-c/310$, consistent with the measured refractive index and the formula $v_g = c/[n_R + \omega \, dn/d\omega]$ for the group velocity.

Note that the 62-ns advance of the pulse peak is a small fraction of the pulsewidth. A similar remark applies to the data of Chu and Wong [41] and Chiao *et al* [44] (figure 2.4). The pulse advance is small compared with the pulsewidth in all 'superluminal' light experiments reported thus far. We return to this point in chapter 4.

2.4.5   Other experiments and viewpoints

A series of elegant experiments by Nimtz and others (see [51] and references therein) has demonstrated evanescent-wave superluminal group velocities in microwave waveguides. To see how this can come about, consider the TE_{01} mode of a waveguide of width a in which the refractive index is n except for a rectangular air gap ($n \approx 1$) [52]. The dispersion relation is then $\omega_c^2 + K_z^2 c^2 = \omega^2/c^2$ in the gap and $\omega_c^2 + k_z^2 c^2 = n^2 \omega^2/c^2$ elsewhere, with $\omega_c = \pi c/a$. Thus, ω can be chosen such that K_z is imaginary while k_z is real. In this case, the propagation in the waveguide is analogous to tunnelling and the evanescent wave can cross the gap with a superluminal group velocity.

Nimtz *et al* have observed superluminal group velocities in microwave waveguides. In particular, in one experiment they encoded Mozart's 40th Symphony on a microwave and reported that this 'signal' was transmitted at $4.7c$. As discussed later, the superluminal group velocity of the transmitted waveform does not violate Einstein causality because it does not represent a superluminal transmission of information. In particular, as noted by Nimtz *et al*, there is no superluminal transmission here of a sharp wave front. Whether an actual 'signal' is transmitted becomes partly a question of semantics but, according to the definition of a signal as a carrier of new information, there is no superluminal signal propagation in these experiments and, therefore, no violation of Einstein causality. As discussed in the following section and, as noted by Chiao and Steinberg [43], '[the] appearance of a waveform faster than c is in itself nothing surprising'.

Suppose there is destructive interference between a wave $\psi(t)$ and a retarded and attenuated portion $\eta \psi(t - \Delta t)$. The superposition of these waves in the first-order approximation is $\psi(t) - \eta \psi(t - \Delta t) \approx (1 - \eta)\psi(t + \chi \Delta t)$, where $\chi = \eta/(1 - \eta) > 0$: the destructive interference, therefore, provides an extrapolation of ψ from t to $t + \chi \Delta t$. This simple observation is relevant to the tunnelling experiments: Chiao and Steinberg [43] discuss the fact that 'the interference at work in tunnelling has the effect of advancing the incident waveform due to the first derivative term of Taylor's theorem', and that this advancement occurs 'without any need for *information* about the later behavior of the incident field'. Chiao and Steinberg [43] observe that

> [T]he time advance being discussed is well under 1 ns in Nimtz's experiments. An acoustic waveform, on the other hand, has a useful bandwidth on the order of 20 kHz, which is to say that no significant deviation from a low-order Taylor expansion occurs in less than about 50 μs. To predict where the wave form would be 50 μs in advance requires little more than a good eye; to predict it 1 ns in advance hardly even requires a steady hand.

Nimtz *et al* [51, 53] have argued that their evanescent fields do not satisfy Einstein causality because, as a practical matter, a 'signal' is limited in its frequency spread or, in other words, a real-world signal cannot have a sharp turn-on or turn-off. Obviously the disagreement over whether a signal can or cannot propagate faster than c hinges on the definition of a signal. We adhere to what seems to be the prevailing view—which is not always made explicit in discussions of Einstein causality—that a signal is something that conveys information and, as such, must involve a *discontinuity* in a waveform or one of its derivatives. A signal defined as such does not violate Einstein causality. This is the viewpoint advocated many years ago by Icsevgi and Lamb [36] in their criticism of the work of Basov *et al* [35]. However, the arguments of Nimtz *et al* raise the valid point that the term 'signal' needs to be better defined, especially when, as discussed in chapter 4, imperfect detectors and quantum effects are considered.

2.5   No violation of Einstein causality

Chu and Wong [41] remarked that the subject of propagation of pulses in dispersive media 'continues to be plagued by widely held misconceptions'. Misconceptions persist even after a century since Sommerfeld and Brillouin [32] proved that the group velocity could exceed c without being in violation of special relativity. This may be due, in part, to the fact that Brillouin's work [32] involves a lot of technical detail and notation that differs considerably from what is conventional today. Moreover, Sommerfeld and Brillouin could cite no relevant *experimental* literature on abnormal group velocities. About the closest they get to what might actually happen in 'the real world' when light propagates in a highly dispersive medium is to mention a remark by W Wien that the group velocity could exceed c. Their motivation centred, according to Sommerfeld, on the fact that 'this apparent contradiction to the theory of relativity had to be resolved'. In addition, it is often the case, depending on the assumed values of parameters like the pulsewidth and the absorption linewidth, that a pulse is greatly distorted in a dispersive medium [21]. For whatever reasons, it has generally been asserted, as discussed in section 1.5, that the concept of group velocity loses meaning when v_g is greater than c or negative. The experiments just described, however, demonstrate that pulses can, in fact, propagate with abnormal group velocities *and* without significant distortion.

Garrett and McCumber [37] observed that the Gaussian pulse (2.19) 'really has no true beginning or end. The $t < 0$ envelope maximum seen by an observer at $z > 0$ is not a direct reflection of the maximum of the input-pulse envelope, but arises from the action of the dispersive medium on the weak early components of that envelope.' Crisp [38], as already mentioned, attributed the possibility of a superluminal group velocity in an absorber to a reshaping of the pulse such that the leading part is less attenuated than the trailing part. Experiment shows that this pulse reshaping can, remarkably, leave the pulse shape and duration essentially unchanged.

It has sometimes been argued that the derivation of the result (2.33) does not require that the pulse 'has no true beginning or end'. However, if this were strictly correct, it would mean, in particular, that a sharply defined wavefront could propagate faster than c, contradicting the Sommerfeld–Brillouin proof that the front velocity is c. The approximation (2.23), from which (2.33) follows, requires that the pulse spectrum not be too large, whereas the spectrum of a (step-function) front includes infinitely large frequencies.

The trivial result (1.78) for the sum of two equal-amplitude monochromatic waves with slightly differing frequency and wavelength already suggests that the pulse envelope, or modulation, can propagate with a velocity that is greater than c or negative. Similarly, the superposition of many monochromatic waves with frequencies and wavelengths lying in narrow bands implies the possibility of such abnormal modulation (group) velocity: to get a group velocity $v_g > c$, all one has to imagine is a dispersive medium in which the phase velocity v_p increases sufficiently rapidly with frequency, since

$$v_g = \frac{c}{n + \omega \, dn/d\omega} = \frac{v_p}{1 - (\omega/v_p) \, dv_p/d\omega}. \tag{2.47}$$

This can be illustrated by a movie in which monochromatic waves in a narrow band are added and the higher-frequency components are given larger phase velocities than the lower-frequency components: the modulation will be seen under appropriate circumstances to propagate faster than the phase velocity of the carrier wave. Such demonstrations, of course, only illustrate what the formula for the group velocity is already telling us. The same is true of interpretations of abnormal group velocities in terms of interference and pulse reshaping: since the problem is linear, what else could be happening other than interference causing the centroid of the pulse to move with a velocity, abnormal or otherwise, that differs from the phase velocity of the carrier? While such interpretations might be helpful conceptually, they do not quite seem to help very much to explain why none of the intriguing experimental results discussed in the preceding section violates the principle of Einstein causality—the principle that no signal can propagate faster than c.

For that purpose, it is useful to cast the group-velocity approximation in a different form [54]. Assume $E(0, t) = A(t) \exp(-i\omega_L t)$, where $A(t)$ varies sufficiently slowly on times scales $\sim \omega_L^{-1}$ that we can replace $k(\omega) = n(\omega)\omega/c$ in equations (2.2) and (2.3) by $k(\omega_L) + (dk/d\omega_L)_{\omega_L}(\omega - \omega_L) = k_L + (\omega - \omega_L)/v_g$. Then, in this approximation,

$$E(z, t) = \frac{1}{2\pi} e^{-i(\omega_L t - k_L z)} \int_{-\infty}^{\infty} dt' \, A(t') \int_{-\infty}^{\infty} d\omega \, e^{-i(\omega - \omega_L)(t - t')} e^{i(\omega - \omega_L)z/v_g}$$

$$= e^{-i(\omega_L t - k_L z)} A(t - z/v_g) \tag{2.48}$$

which, of course, is equivalent to (1.86). But now let us express the integration over ω and t' in (2.48) as

$$\int_{-\infty}^{\infty} dt' \, A(t') \int_{-\infty}^{\infty} d\omega \, \exp^{-i(\omega-\omega_L)(t-t'-z[v_g^{-1}-c^{-1}])} \exp^{(i(\omega-\omega_L)z/c)}$$

$$= \int_{-\infty}^{\infty} dt' \, A(t') \sum_{n=0}^{\infty} \frac{(iz)^n}{n!} (v_g^{-1} - c^{-1})^n$$

$$\times \int_{-\infty}^{\infty} d\omega \, (\omega - \omega_L)^n \exp^{-i(\omega-\omega_L)(t-t'-z/c)}$$

$$= \int_{-\infty}^{\infty} dt' \, A(t') \sum_{n=0}^{\infty} \frac{z^n}{n!} (v_g^{-1} - c^{-1})^n \frac{\partial^n}{\partial t'^n} \int_{-\infty}^{\infty} d\omega \, \exp^{i(\omega-\omega_L)(t'-t+z/c)}$$

$$= 2\pi \int_{-\infty}^{\infty} dt' \, A(t') \sum_{n=0}^{\infty} \frac{z^n}{n!} (v_g^{-1} - c^{-1})^n \frac{\partial^n}{\partial t'^n} \delta(t' - t + z/c)$$

$$= 2\pi \exp^{(c^{-1}-v_g^{-1})z\partial/\partial t} A(t - z/c) \tag{2.49}$$

so that

$$E(z, t) = e^{-i(\omega_L t - k_L z)} e^{[c^{-1}-v_g^{-1}]z\partial/\partial t} A(t - z/c)$$
$$= e^{-i(\omega_L t - k_L z)} A(t - z/v_g) \tag{2.50}$$

which is equivalent to the result derived by Diener [54].

Equation (2.50) states that, in the group-velocity approximation in which second and higher derivatives of the refractive index with respect to frequency are neglected, propagation over a distance z corresponds to an analytic continuation over the time $z/c - z/v_g$ of the vacuum-propagated pulse envelope $A(t - z/c)$. In other words, a superluminal group velocity does not imply a superluminal propagation of new *information*, since there is no information in $A(t - z/v_g)$ that is not already contained in $A(t - z/c)$.

Evidently, new information is propagated only if $A(t - z/c)$ does not have an analytic continuation. In this case, the second equality in (2.50) is invalid, while the first equality holds up to a time at which $A(t - z/c)$ or one of its derivatives has a discontinuity. Thereafter, the pulse evolution becomes much more complicated than a simple undistorted propagation at the superluminal group velocity. The point of singularity behaves like a Sommerfeld–Brillouin front, which propagates at c. That is, *a true signal evidently requires non-analyticity and cannot transmit information at a velocity $> c$.*

The case of an analytic (e.g. Gaussian) waveform propagating without distortion at a superluminal group velocity is, nevertheless, remarkable when one considers that (1) points at time t on the transmitted pulse are causally determined by points at $t < z/c$ on the incident pulse and yet (2) the transmitted pulse advances at the velocity $v_g > c$. This means, as indicated in figure 2.7,

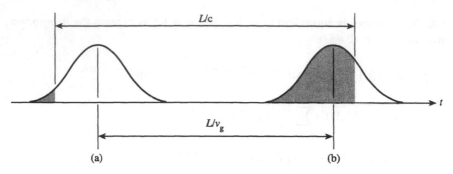

Figure 2.7. Incident (*a*) and transmitted (*b*) pulses for a propagation length L and group velocity $v_g > c$. The shaded portion of (*b*) is completely determined by the shaded portion of (*a*).

that if $(z/c - z/v_g)$ is much larger than the pulse duration, the peak of the transmitted signal is reconstructed entirely from a small tail of the incident pulse: if $(z/c - z/v_g)$ is large enough, nearly the entire transmitted pulse is reconstructed by analytic continuation of a tiny portion of the incident pulse!

Note that the peak of the transmitted pulse is not causally connected to the peak of the incident pulse, so that, in particular, the observation that the pulse peak moves superluminally does not contradict Einstein causality. A simple analogy would be the motion of a spot of light made by shining a rotating flashlight onto a distant wall. The spot can, in principle, move superluminally but there is no violation of causality because the spot at one instant is not the source of the spot at a later instant. In this same sense, the experimental observations of pulse peaks propagating faster than c do not contradict Einstein causality.

The recent experimental observations of superluminal group velocities involve a linear response of the medium to the field and the exchange of energy between the medium and the front and back parts of the pulse, leading to the pulse advancement, can be interpreted using classical spectral arguments [55], as already mentioned.

Linearity also makes it easy to perform a quantum-mechanical calculation using a simple model in which an absorbing dielectric is described as a collection of identical two-level atoms and the source of light is a single excited atom outside the dielectric. It is shown in section 3.3 that, if the source atom of transition frequency ω_0 is suddenly excited at time $t = 0$, then the probability of detecting a photon at a point inside the dielectric and at a distance z from the source atom is zero before the time z/c. This is the analogue of the classical result that a sharp wavefront cannot propagate faster than c. However, if the probability $P(t)$ that the source atom is excited varies smoothly in time, then the photon-counting rate $R(t)$ at an ideal detector located at z is proportional to $\exp(-2\omega_0 n_I(\omega_0)z/c)P(t - z/v_g)$, where n_I is again the imaginary part of the refractive index. Thus, if $v_g > c$, the peak probability of producing a 'click' at the detector can occur earlier than is possible when there is no medium between the detector and the source atom. This is analogous to what has been observed in the optical tunnelling experiments of Chiao *et al.*

In section 2.1, we reviewed the Sommerfeld–Brillouin proof that a sharp wavefront, one that jumps discontinuously from zero to a finite value, cannot propagate faster than c. We noted that the front velocity is determined by the infinite-frequency response of the propagation medium. (The front velocity is $c/n_R(\infty) = c$.) We also noted that Sommerfeld and Brillouin [32] defined a signal as a train of oscillations that starts from zero at some instant and that, acccording to this definition, no signal can propagate faster than c, the front velocity. From the discussion following equation (2.50), one concludes more generally that the propagation of new information—a signal—requires a discontinuity in a waveform or one of its derivatives [43, 54]. The discontinuity involves infinite-frequency components and, therefore, propagates at c. Chiao and Steinberg [43] define an idealized *signal* as 'the complete set of all the points of nonanalyticity $\{t_0, t_1, t_2, \ldots\}$, together with the values of the input function $f_{in}(t)$ in a small but finite interval of time inside the domain of analyticity immediately following these points'.

The signal velocity of an optical pulse is sometimes defined as the velocity of propagation of the half-the-peak-intensity point on the leading part of the pulse. Such a 'signal' velocity can exceed c but, as we have seen, it is not really a signal velocity because it does not necessarily convey information that is not already contained in the leading edge of the pulse. Moreover, as noted by Brillouin [32],

> [this] definition of the signal velocity is somewhat arbitrary ... The signal does not arrive suddenly; there is a quick but still continuous transition from the very weak intensity of the [precursors] to that corresponding to the signal. A detector set to detect an intensity equal to $\frac{1}{4}$ the final intensity will detect the arrival of the signal in agreement with the above arbitrary definition; if the detector is more or less sensitive, then it will detect the arrival of the signal a little earlier or later.

Regarding superluminal tunnelling, it should be noted that MacColl's assumption for the initial wave packet (section 2.4.3) means there is no sharp front and no *signal* in the sense of Sommerfeld and Brillouin. That is, there is no violation of Einstein causality implied by MacColl's zero delay, even though the peak of the wave packet appears to cross the barrier superluminally. The velocity of the peak of the wave packet for the particle is analogous to the group velocity of an electromagnetic pulse.

As discussed in chapters 3 and 4, the concept of a signal must, in general, be extended beyond the completely classical considerations of this chapter to include quantum effects. There are situations where quantum effects appear to 'protect' special relativity against the possibility of superluminal communication of information.

2.6 Bessel beams

The faster-than-c effects described thus far arise from the dispersion of light in a material medium. We now describe some work in which 'superluminal' effects derive solely from the nature of the propagating field. More specifically, we will describe some work on the propagation of so-called Bessel beams [56,57].

Let us first summarize a few salient features of Bessel beams. Consider a scalar wave of frequency ω propagating in the z direction and having the azimuthally symmetric form

$$E(z,t) = \mathcal{E}(\rho)e^{-i(\omega t - k_z z)} \qquad \rho = \sqrt{x^2 + y^2}. \qquad (2.51)$$

Such a wave is 'diffractionless' in that the intensity ($\propto |\mathcal{E}(\rho)|^2$) is independent of the propagation distance z. The wave equation $\nabla^2 E - c^{-2}\partial^2 E/\partial t^2 = 0$ implies

$$\frac{d^2\mathcal{E}}{d\rho^2} + \frac{1}{\rho}\frac{d\mathcal{E}}{d\rho} + (k^2 - k_z^2)\mathcal{E} = 0 \qquad (k = \omega/c) \qquad (2.52)$$

which has the solution

$$\mathcal{E}(\rho) = J_0(k_\rho \rho) \qquad (k^2 = k_z^2 + k_\rho^2) \qquad (2.53)$$

where J_0 is the zeroth-order Bessel function. Writing $k_\rho = k\sin\theta$ and $k_z = k\cos\theta$, and using the integral representation of J_0, we have

$$
\begin{aligned}
E(z,t) &= J_0(k\rho\sin\theta)e^{-i(\omega t - kz\cos\theta)} \\
&= \frac{1}{2\pi}\int_0^{2\pi} d\phi\, e^{-i(\omega t - kz\cos\theta)}e^{i(kx\sin\theta\cos\phi + ky\sin\theta\sin\phi)} \\
&= \frac{1}{2\pi}\int_0^{2\pi} d\phi\, e^{-i(\omega t - \boldsymbol{q}\cdot\boldsymbol{r})}
\end{aligned}
\qquad (2.54)
$$

where the wavevector $\boldsymbol{q} = k(\hat{x}\sin\theta\cos\phi + \hat{y}\sin\theta\sin\phi + \hat{z}\cos\theta)$ intersects the z-axis at angle θ. Thus, a (zeroth order) Bessel beam of frequency ω comprises all possible plane waves with wavevectors of magnitude $|\boldsymbol{q}| = \omega/c$ making an angle θ to the z-axis. It follows that a Bessel beam can be produced by illuminating a narrow annulus lying in the focal plane of a lens [56,57].

The phase velocity of the wave (2.54) is $v_p = c/\cos\theta$. The group velocity along the z direction is $v_g = \partial\omega/\partial k_z = ck_z/k = c\cos\theta$. Some authors [58,59] state that the group velocity is $c/\cos\theta$. The argument for this is evidently that $v_g = d\omega/dk_z = (d\omega/dk)(dk/dk_z) = c/\cos\theta$, i.e. that $k_z = k\cos\theta$ with $\cos\theta$ a constant parameter. The definition of group velocity, however, presumes a wave *packet*, in which case one cannot assume such a fixed relation between k and k_z.

The fact that the plane-wave components of the Bessel beam intersect the z-axis at the angle θ means that the point of contact with the z-axis of all of these waves (or, more precisely, their planes of constant phase) move along the

z-axis with the velocity $c/\cos\theta$. The z-axis, therefore, appears to 'light up' superluminally. As noted by Saari and Reivelt [58], however, '[T]his speed is superluminal in a similar way as one gets a faster-than-light movement of a bright stripe on a screen when a plane wave light pulse is falling at the angle θ onto the screen plane.' In other words, the 'superluminal' propagation along z is merely a geometrical effect, analogous to the old 'scissors paradox' in which the point of contact of the two blades could move faster than c while the ends of the blades move with velocity less than c. The points of contact are not causally connected nor are points of contact along the z-axis causally connected in the case of the Bessel beam. There is certainly no violation of Einstein causality, although it has been claimed [59] that the Bessel beam 'superluminality' calls this fundamental principle into question. This claim has been challenged on both theoretical and experimental grounds [60, 61].

In order to propagate a signal from, say, $(0, 0, 0)$ to $(0, 0, z)$, information would first have to be sent from $(0, 0, 0)$ to points in the $z = 0$ plane at a distance $\rho = z \tan\theta$ away; and this distance gives the location in the $z = 0$ plane of the conical surface on which lie the wavevectors of the plane-wave components that propagate to $(0, 0, z)$ to produce the Bessel beam at $(0, 0, z)$. The information must then propagate over the distance $z/\cos\theta$ from these points to $(0, 0, z)$. The time it takes for the information to be propagated from $(0, 0, 0)$ to $(0, 0, z)$, assuming that the information can be sent at velocity c, is then $t = (z \tan\theta + z/\cos\theta)/c$. Thus, the velocity with which information can be propagated from $(0, 0, 0)$ to $(0, 0, z)$ is $z/t = c/(\tan\theta + \sec\theta) \le c$.

2.7 Propagation of energy

As remarked earlier, the work of Sommerfeld and Brillouin and others was motivated in part by the association of group velocity with the velocity at which electromagnetic energy propagates. The fact that the group velocity can exceed c in a region of anomalous dispersion led them to more careful considerations of the meaning of group and signal velocities.

Let us begin by considering the cycle-averaged electromagnetic energy density u_ω in a dielectric medium for which absorption (or amplification) at frequency ω is negligible [62]. Poynting's theorem states that $\nabla \cdot S + \partial u/\partial t = 0$ in the absence of any currents, where $S = E \times H$ and

$$\frac{\partial u}{\partial t} = E \cdot \frac{\partial D}{\partial t} + H \cdot \frac{\partial B}{\partial t}. \tag{2.55}$$

For a narrow band of frequencies about a frequency ω, within which absorption is negligible, we write

$$E(r, t) = E_\omega(r, t)e^{-i\omega t} = \int_{-\infty}^{\infty} d\Delta\, e_\omega(r, \Delta)e^{-i(\omega+\Delta)t} \tag{2.56}$$

where E_ω is slowly varying in time compared with $\exp(-i\omega t)$ and, as usual, the real part of the complex expression for the field is implicit. Thus,

$$
\begin{aligned}
D(r, t) &= \int_{-\infty}^{\infty} d\Delta\, \epsilon(\omega + \Delta) e_\omega(r, \Delta) e^{-i(\omega+\Delta)t} \\
&\cong \int_{-\infty}^{\infty} d\Delta \left[\epsilon(\omega) + \Delta \frac{d\epsilon}{d\omega} \right] e_\omega(r, \Delta) e^{-i(\omega+\Delta)t} \\
&\cong \epsilon(\omega) E(r, t) + i \frac{d\epsilon}{d\omega} e^{-i\omega t} \frac{\partial E_\omega}{\partial t}
\end{aligned}
\tag{2.57}
$$

$$
\begin{aligned}
\frac{\partial D}{\partial t} &\cong \epsilon \frac{\partial E}{\partial t} + \omega \frac{d\epsilon}{d\omega} \frac{\partial E_\omega}{\partial t} e^{-i\omega t} \\
&= \left[\epsilon \frac{\partial E_\omega}{\partial t} - i\omega\epsilon E_\omega + \omega \frac{d\epsilon}{d\omega} \frac{\partial E_\omega}{\partial t} \right] e^{-i\omega t} \\
&= \left[\frac{d}{d\omega}(\epsilon\omega) \frac{\partial E_\omega}{\partial t} - i\omega\epsilon E_\omega \right] e^{-i\omega t}
\end{aligned}
\tag{2.58}
$$

where we use the assumption that only a narrow band of frequencies is significant. Thus,

$$
E \cdot \frac{\partial D}{\partial t} \cong \frac{1}{4} \frac{d}{d\omega}(\epsilon\omega) \frac{\partial}{\partial t} |E_\omega|^2.
\tag{2.59}
$$

A similar calculation for $H \cdot \partial B/\partial t$ then yields, from (2.55) [62],

$$
u_\omega = \frac{1}{4} \left[\frac{d}{d\omega}(\epsilon\omega)|E_\omega|^2 + \frac{d}{d\omega}(\mu\omega)|H_\omega|^2 \right]
\tag{2.60}
$$

for the field energy density at frequency ω.

Using $|H_\omega|^2 = (\epsilon/\mu)|E_\omega|^2$ for plane waves, we obtain

$$
u_\omega = \frac{1}{2\mu}\epsilon_0\mu_0 \frac{d}{d\omega}(n\omega)|E_\omega|^2 = \frac{n}{2\mu c v_g}|E_\omega|^2 \qquad (n^2 = \epsilon\mu/\epsilon_0\mu_0).
\tag{2.61}
$$

Similarly, the cycle-averaged Poynting vector at frequency ω has magnitude

$$
|S_\omega| = \frac{n}{2\mu c}|E_\omega|^2 = v_g u_\omega.
\tag{2.62}
$$

Defining the electromagnetic energy propagation velocity v_E as the magnitude of the Poynting vector divided by the energy density [32], we have

$$
v_E \equiv |S_\omega|/u_\omega = v_g.
\tag{2.63}
$$

Thus, the energy transport velocity defined in this manner can, like the group velocity, be superluminal or negative, in apparent contradiction with special relativity. Obviously, this is a consequence of the appearance of the group velocity in the energy density (2.61). However, it should be borne in mind that (2.61) assumes that absorption is negligible and that v_g is, in fact, positive and less than c at the frequency at which absorption is weakest (section 2.2). To conclude from (2.61) that the energy density and energy velocity can be greater than c or negative is to assume that the formula is applicable to frequencies outside its domain of validity.

In considering the velocity of energy transport, it is also imperative to recognize the fact that part of the energy density is stored for a finite time in the propagation medium [63, 64]. In this connection, it is instructive to consider a simple model that is essentially that of Loudon [63] for a dielectric medium consisting of Lorentzian electron oscillators coupled to the electric field according to the equation of motion

$$m(\ddot{r} + \Gamma\dot{r} + \omega_0^2 r) = eE. \tag{2.64}$$

The polarization density in this model is $P = Ner$, where N is the number density of oscillators ('atoms') and this implies, for an electric field of frequency ω, the dielectric constant

$$\kappa(\omega) = \epsilon(\omega)/\epsilon_0 = 1 + \frac{\omega_p^2}{\omega_0^2 - \omega^2 - i\Gamma\omega} \tag{2.65}$$

where $\omega_p^2 = Ne^2/m\epsilon_0$. Writing $\kappa^{1/2} = n = n_R + in_I$, we have

$$n_R^2 - n_I^2 = 1 + \frac{\omega_p^2(\omega_0^2 - \omega^2)}{(\omega_0^2 - \omega^2)^2 + \Gamma^2\omega^2} \tag{2.66}$$

$$2n_R n_I = \frac{\omega_p^2 \Gamma\omega}{(\omega_0^2 - \omega^2)^2 + \Gamma^2\omega^2}. \tag{2.67}$$

Let us write Poynting's theorem in the integral form

$$\oint S \cdot \hat{n}\, da = -\int \left[E \cdot \frac{\partial D}{\partial t} + \mu_0 H \cdot \frac{\partial H}{\partial t} \right] dV$$

$$= -\int \left[\frac{1}{2}\frac{\partial}{\partial t}(\epsilon_0 E^2 + \mu_0 H^2) + E \cdot \frac{\partial P}{\partial t} \right] dV \tag{2.68}$$

for a non-magnetic medium ($\mu = \mu_0$), where the integral of the normal component of S on the left-hand side is over a surface enclosing the volume V. From (2.64),

$$E \cdot \frac{\partial P}{\partial t} = \frac{m}{e}(\ddot{r} + \Gamma\dot{r} + \omega_0^2 r) \cdot Ne\dot{r}. \tag{2.69}$$

Then

$$\oint S \cdot \hat{n}\, da + \int Nm\Gamma\dot{r}^2\, dV = -\int \dot{W}\, dV \tag{2.70}$$

where

$$W \equiv \tfrac{1}{2}\epsilon_0 E^2 + \tfrac{1}{2}\mu_0 H^2 + N(\tfrac{1}{2}m\dot{r}^2 + \tfrac{1}{2}m\omega_0^2 r^2). \tag{2.71}$$

Equation (2.70) is, of course, a statement of energy conservation. The first term on the left-hand side is the rate at which energy flows out of the volume V, while the second is the rate of loss of internal energy of the atoms in V due to the damping mechanism characterized by Γ. The integral on the right-hand side gives the rate at which the energy in V increases. Using (2.64), (2.66), and (2.67), we can write the cycle average $\overline{W}$ of W as

$$
\begin{aligned}
\overline{W} &= \frac{1}{4}\epsilon_0 |E|^2 \left[\frac{\omega_p^2(\omega^2 + \omega_0^2)}{(\omega_0^2 - \omega^2)^2 + \Gamma^2\omega^2} + 1 + n_R^2 + n_I^2 \right] \\
&= \frac{1}{4}\epsilon_0 |E|^2 \left[\frac{2n_R n_I}{\Gamma \omega}(\omega^2 + \omega_0^2) + 1 + n_R^2 + n_I^2 \right] \\
&= \frac{1}{2}\epsilon_0 |E|^2 \left[\frac{2\omega n_R n_I}{\Gamma} + n_R^2 \right]
\end{aligned}
\tag{2.72}
$$

while the cycle-averaged magnitude of the Poynting vector has the familiar form

$$
\overline{S} = \tfrac{1}{2} n_R \epsilon_0 |E|^2.
\tag{2.73}
$$

$\overline{W}$ is always positive. The definition $v_E = \overline{S}/\overline{W}$ of the energy velocity gives an expression that is always positive and less than c [63][4]:

$$
v_E = \frac{c}{n_R + 2\omega n_I/\Gamma} = \frac{v_p}{1 + 2\omega n_I/n_R \Gamma}.
\tag{2.74}
$$

In the limit of zero absorption, i.e. for a field frequency ω far from any absorption resonance, we have

$$
\overline{W} = \frac{1}{4}\epsilon_0 |E|^2 \left[\frac{\omega_p^2(\omega^2 + \omega_0^2)}{(\omega_0^2 - \omega^2)^2} + n_R^2 + 1 \right]
\tag{2.75}
$$

$$
n_R^2 = 1 + \frac{\omega_p^2}{\omega_0^2 - \omega^2}
\tag{2.76}
$$

$$
n_I/\Gamma = \frac{\omega_p^2\omega/2n_R}{(\omega_0^2 - \omega^2)^2}
\tag{2.77}
$$

and the energy transport velocity (2.74) reduces to the group velocity:

$$
v_E = c \left[n_R + \frac{\omega^2\omega_p^2/n_R}{(\omega_0^2 - \omega^2)^2} \right]^{-1} = \frac{c}{n_R + \omega \, dn_R/d\omega} = v_g.
\tag{2.78}
$$

The total energy is the energy W, plus the interaction energy between the field and the medium, plus the energy in the 'bath' associated with the damping

[4] Loudon [63] points out that the corresponding result in chapter 5 of *Brillouin* [32] is incorrect because of algebraic errors.

of the internal energy of the 'atoms'. In the absence of damping the latter is zero and expression (2.61) is the total energy density, which propagates at the velocity $v_g \leq c$. When damping is included, the total energy $\overline{W}$ in the field and the atoms propagates at the positive and subluminal velocity (2.74).

Chu and Wong [41] note that equation (2.74) implies a pulse delay rather than an advance in their experiments and, therefore, that v_g, not v_E, is 'the measured quantity in this type of pulse propagation experiment'.

The internal energy of the medium in this model is

$$W_m = N[\tfrac{1}{2}m\dot{r}^2 + \tfrac{1}{2}m\omega_0^2 r^2] \tag{2.79}$$

and, from (2.64), is found to have the cycle-averaged value

$$\overline{W}_m = \frac{1}{4}\epsilon_0 |E|^2 \frac{\omega_p^2(\omega^2 + \omega_0^2)}{(\omega_0^2 - \omega^2)} \tag{2.80}$$

when damping is negligible ($\Gamma = 0$). Thus, in the absence of damping, we can use this expression and (2.75) to define the difference [65]

$$u_\omega^{(F)} \equiv \overline{W} - \overline{W}_m = \tfrac{1}{4}\epsilon_0 |E|^2 (n_R^2 + 1) \tag{2.81}$$

for a non-magnetic and non-dissipative medium. $u_\omega^{(F)}$ is identified by Diener [65] as the 'energy [density] of the electromagnetic field in the proper sense'. He defines the *energy transport velocity* as

$$v_E^{(F)} = |S_\omega|/u_\omega^{(F)} \tag{2.82}$$

which, using equations (2.61), (2.62), and (2.81), is found to be

$$v_E^{(F)} = \left(\frac{2n_R}{n_R^2 + 1}\right)c \tag{2.83}$$

which never exceeds c. It is, thus, possible to define an energy transport velocity that is never superluminal, although, as acknowledged by Diener [65], this velocity is more 'interpretive' than measurable. Note also that (2.83) assumes that damping is negligible ($n_I = 0$), in which case Loudon has shown that the more conventional definition of the energy transport velocity implies non-superluminal propagation of total energy.

Peatross *et al* [66] have taken a different approach based on a measurable quantity, the Poynting vector. They define the pulse arrival time expectation integral

$$\langle t \rangle_r \equiv \frac{\hat{u} \cdot \int_{-\infty}^{\infty} t S(r, t)\, dt}{\hat{u} \cdot \int_{-\infty}^{\infty} S(r, t)\, dt} \tag{2.84}$$

where $\hat{u}$ is the unit vector in the direction in which the energy flux is detected. In the frequency domain, the expected arrival time has the form

$$\langle t \rangle_r = T[E(r, \omega)] \equiv -i \frac{\hat{u} \cdot \int_{-\infty}^{\infty} [\partial E(r, \omega)/\partial \omega] \times H^*(r, \omega) \, d\omega}{\hat{u} \cdot \int_{-\infty}^{\infty} S(r, \omega) \, d\omega}. \tag{2.85}$$

Using this expression, and without any essential approximations, Peatross *et al* [66] show that the time delay $\Delta t \equiv \langle t \rangle_{r+\Delta r} - \langle t \rangle_r$ associated with the propagation of a pulse from r to $r + \Delta r$ can be expressed as the sum of two distinctly interpretable terms[5]:

$$\Delta t = G_{r+\Delta r} + R_r. \tag{2.86}$$

G_r, which Peatross *et al* call the *net group delay*, is given by

$$G_{r+\Delta r} = \frac{\hat{u} \cdot \int_{-\infty}^{\infty} S(r, \omega)[\partial(\text{Re } k)/\partial \omega] \cdot \Delta r \, d\omega}{\hat{u} \cdot \int_{-\infty}^{\infty} S(r, \omega) \, d\omega} \tag{2.87}$$

i.e. the propagation length divided by the average over all frequencies of the inverse of the group velocity. The *reshaping delay* is given by

$$R_r = T[e^{-\text{Im}(k) \cdot \Delta r} E(r, \omega)] - T[Er, \omega)] \tag{2.88}$$

i.e. the difference between the expected pulse arrival times at the initial point r with and without the change in spectral amplitude due to propagation in the medium. In particular, this reshaping delay vanishes if the pulse spectrum does not change upon propagation.

Peatross *et al* [66] present results of numerical computations for the propagation of Gaussian pulses in a medium described by the Lorentz model with absorption. For pulse bandwidths small compared to the absorption linewidth, they obtain results consistent with those of Garrett and McCumber [37], i.e. the delay time is dominated by the net group delay and can correspond to superluminal and negative propagation velocities. For broadband pulses, the reshaping delay becomes important and, in the extreme broadband limit of a delta-function pulse, $\Delta t \to \Delta r/c$ (since $n(\omega) \to 1$ as $\omega \to \infty$). The latter result is consistent with the fact that a sharp wave front propagates with velocity c.

2.8   Precursors

As shown in section 2.1, if a field with a temporal profile $E(0, t)$ that is zero until the time $t = 0$ is incident at the input plane $z = 0$ of a medium, the field at $z > 0$ inside the medium is zero until the time $t = z/c$. What happens after this time depends on the particular form of $n(\omega)$ and $E(0, t)$. Because the medium cannot

---

[5] The subscripts $r + \Delta r$ and r can be interchanged in equation (2.86) without affecting Δt [66].

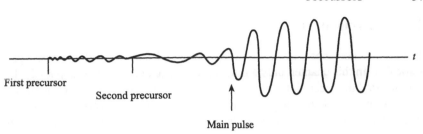

First precursor

Second precursor

Main pulse

Figure 2.8. Schematic illustration of the appearance of first and second precursors followed by the main pulse for the case in which the frequency ω of the signal (2.1) is much less than the resonance frequency ω_0 of the dielectric medium. After figure 20 in Brillouin [32].

respond instantaneously to the applied field, the wave*front* velocity is always equal to c; and, after $t = z/c$, there are *precursors*, or 'forerunners', to the main part of the propagating field, the first precursor arising from the high-frequency parts of the field. A detailed discussion of precursors is given by Oughstun and Sherman [21]. These authors provide a very large and useful list of references on the subject of electromagnetic wave propagation in dispersive media. The reader is also referred to the books by Jackson [14] and Stratton [67] for good introductions to the theory of precursors.

The problem of determining the precursors at times after $t = z/c$ is 'simply' that of solving equations (2.2) and (2.3), i.e. finding a solution of the equation

$$E(z, t) = \int_{-\infty}^{\infty} d\omega \, \mathcal{E}(\omega) e^{-i\omega[t - n(\omega)z/c]} \qquad (2.89)$$

where

$$\mathcal{E}(\omega) = \frac{1}{2\pi} \int_{-\infty}^{\infty} dt \, E(0, t) e^{i\omega t}. \qquad (2.90)$$

Obviously, the solution depends on the form of both the refractive index $n(\omega)$ and the spectrum $\mathcal{E}(\omega)$ of the incident field.

The nature of the precursors obtained by Sommerfeld and Brillouin is shown schematically in figure 2.8 for a signal of the type (2.1). The field at z in the medium, which is characterized by a dielectric constant of the form (2.65), is zero until the time z/c. At $t = z/c$, the *first precursor* (or *Sommerfeld precursor*) begins as a weak-amplitude, high-frequency oscillation. The amplitude and oscillation period increase with time. After a time $\sim n(0)z/c$, a *second precursor* (or *Brillouin precursor*) characterized by a lower oscillation frequency begins. Its amplitude and oscillation period increase with time until the latter approaches that of the signal; then the amplitude increases rapidly and the field starts to take the form of what is the main part of the pulse.

2.9   Six velocities

There have now been quite a few experimental studies of 'abnormal' group velocities. In this chapter, only a few of these have been discussed, namely those that initiated the recent interest in the subject and that demonstrate rather directly the main points of the chapter. The most recent experiments appear to focus more on the nature of signals and, consequently, we defer a discussion of them to chapter 4.

Six velocities have been identified in this chapter:

(1) $c = 299\,792\,458$ m s^{-1}, the speed of light in vacuum. According to special relativity, no signal, or information, can be communicated at a velocity greater than this.

(2)   Phase velocity. As undergraduates are taught, this is associated with monochromatic waves and, therefore, can be greater than c without violating special relativity.

(3)   Group velocity. This is not, in general, the velocity of signal or energy propagation and it can be greater than c, infinite, or negative while still retaining its meaning as the velocity of nearly *undistorted* pulse propagation, as experiments have shown.

(4)   Front velocity, the velocity of propagation of a step-function discontinuity. A signal, according to Sommerfeld and Brillouin, begins with such a front, which propagates at the velocity c in any medium.

(5)   Signal velocity, which has been defined in several ways. In one context, Sommerfeld and Brillouin defined it operationally as the velocity of propagation of the half-the-peak-intensity point on the leading part of the pulse. In reconciling superluminal group velocities with special relativity, however, they defined a signal as 'a limited wave motion: nothing until a certain moment in time, then, for instance, a series of regular sine waves ...'. The second definition has been generalized by Chiao and Steinberg [43] to include all points of non-analyticity, i.e. a signal represents new information that is not already foretold in an earlier portion of a waveform. This is the fundamental sense in which *signal* and *signal velocity* are used here. According to this meaning, special relativity demands that no signal velocity can exceed c. As discussed in chapter 4, an operational definition of a signal velocity requires careful consideration of noise in the field, the propagation medium, and the detector.

(6)   Energy transport velocity. This is approximately equal to the group velocity when the frequency of the field is far from any absorption (or amplification) resonances but near resonances theoretically it can appear to be superluminal if we do not take proper account of the fact that energy is stored for a finite time in the medium.

Chapter 3

Quantum theory and light propagation

Interesting questions can arise when the propagation of light is considered quantum mechanically: How does light propagate from one atom to another? Does the 'spooky action at a distance' in quantum correlations of the Einstein–Podolsky–Rosen type imply some sort of superluminal communication? How is it that quantum mechanics 'protects' Einstein causality against various proposed superluminal communication schemes based on such correlations? These and related matters are taken up in this chapter.

The question of how light propagates from one atom to another was addressed many years ago by Fermi, who made an approximation that rendered inconclusive his attempted proof of 'causal propagation'. We consider the problem in both the Schrödinger and Heisenberg pictures, discuss Fermi's approximation, and prove that Fermi's result was, in fact, correct. In essence, Fermi made what in current jargon is called the rotating-wave approximation, which is basically the approximation that only energy-conserving processes contribute to transition amplitudes. In the rotating-wave approximation, the field from a source is not properly retarded; and we show how this approximation can lead to acausal consequences in the theory of photodetection, whereas the properly formulated theory is, in fact, causal. Using a fully quantum-mechanical approach to the refractive index of a dielectric medium, we show that an excited source atom outside the medium can produce photon counts earlier than if the emitted photon propagated through the same distance in vacuum: no violation of Einstein causality is implied by this possibility. We review the no-cloning theorem, which has its origin in a proposed superluminal communication scheme based on Einstein–Podolsky–Rosen correlations, recall briefly what is meant by teleportation, and show why a superluminal communication scheme based on phase conjugation must fail. An intriguing conceptual question arising when a cavity that inhibits spontaneous emission by an appropriately placed atom is suddenly modified so that one of the mirrors is replaced by a detector: can a photon be counted instantaneously after this replacement or only after some retardation time? In the appendix to the chapter, we take the opportunity to express an opinion about whether Einstein would have been surprised by experiments that support quantum theory while ruling out a certain class of hidden variable theories.

3.1 Fermi's problem

Suppose we have two identical atoms A and B in vacuum, such that atom A is in the first excited state at time $t = 0$ while atom B is in the ground state. The atoms are stationary and separated by a distance r. Atom A can spontaneously emit a photon and drop to the ground state and there is a chance that the emitted photon can be absorbed by atom B, causing it to jump from the ground state to the first excited state. What is the probability at $t \geq 0$ that atom B is in the first excited state? We will solve a simplified version of this problem in which A and B are assumed to have only two states, a ground state and an excited state (figure 3.1). Actually this is not a serious restriction as long as we are not concerned with radiative level shifts (Lamb shifts). Aside from these shifts, the two-state model yields essentially the same results as a full 'multilevel atom' calculation.

Figure 3.1. Two-state atoms, A and B, at rest and separated by a distance r in vacuum. A is in the excited state at time $t = 0$ while B is in the ground state. What is the probability that B is in its excited state at a time $t > 0$?

This model for light propagation in quantum electrodynamics (QED) has a long history. The first publication to consider it appears to be that of Kikuchi [68], whose work was carried out at the suggestion of Heisenberg. In his well-known review in 1932, Fermi [69] discussed the problem as a model of how light propagation is described by QED. He deduced that atom B has zero probability of being excited at times $t \leq r/c$, consistent with the idea that B can be excited only after radiation from A has had time to propagate from A to B. But, in doing the calculation, Fermi made an approximation without which the desired result would not have been obtained[1]. Following a suggestion by Ferretti and Peierls [71, 72] that QED might not provide a causal solution, the problem was revisited by Hamilton [73], Heitler and Ma [74] and Fierz [75], who again showed that atom B can be excited only after a time $t = r/c$. A quarter-century later, the problem was taken up by Milonni and Knight [76, 77], who improved somewhat on Fermi's approximation and results but did not solve the problem of proving *exactly* that atom B cannot be excited before $t = r/c$. The approximation made by Fermi was also noted by Shirokov [78]. After another 20 years, a paper by Hegerfeldt [79] was highlighted in a piece in *Nature* [80] stating that 'a sixty-year-old calculation

---

[1] Fermi's problem is treated in the same approximation in *Louisell* [70].

by Enrico Fermi is discovered to be in error, and inter-atomic signalling between atoms to be potentially faster than light'. This, in turn, prompted a rebuttal [81]. The problem has continued to attract interest (see [82–86] and references therein).

We will now discuss in more detail the approximation made by Fermi. Then we will show that 'inter-atomic signalling between atoms' is *not* 'faster than light'.

The Hamiltonian for the problem has the usual form[2]

$$\hat{H} = \hat{H}_{\text{Atoms}} + \hat{H}_{\text{Field}} + \hat{H}_{\text{Int}}. \tag{3.1}$$

$\hat{H}_{\text{Atoms}}$ is the Hamiltonian operator for the internal energy of the two atoms. Denoting the difference in energy between the unperturbed excited and ground states of each atom by $\hbar\omega_0$, we can write $\hat{H}_{\text{Atoms}}$ in the form

$$\hat{H}_{\text{Atoms}} = \hbar\omega_0 \hat{\sigma}_A^\dagger \hat{\sigma}_A + \hbar\omega_0 \hat{\sigma}_B^\dagger \hat{\sigma}_B \tag{3.2}$$

where $\hat{\sigma}$ and $\hat{\sigma}^\dagger$ are, respectively, the two-state (Pauli) lowering and raising operators. $\hat{H}_{\text{Field}}$ is the operator corresponding to the energy of the electromagnetic field in vacuum:

$$\hat{H}_{\text{Field}} = \frac{1}{2} \int (\epsilon_0 \hat{E}^2 + \mu_0 \hat{H}^2)\, dV. \tag{3.3}$$

The electric field operator $\hat{E}$ can be expanded in plane-wave modes as

$$\hat{E}(r) = i \sum_{k\lambda} \left(\frac{\hbar\omega_k}{2\epsilon_0 V}\right)^{1/2} [\hat{a}_{k\lambda} e^{ik\cdot r} - \hat{a}_{k\lambda}^\dagger e^{-ik\cdot r}] e_{k\lambda} \tag{3.4}$$

and the corresponding expression for $\hat{H}$ follows from the (operator) Maxwell equation $\nabla \times \hat{E} = -\mu_0 \partial\hat{H}/\partial t$. Here $\hat{a}_{k\lambda}$ and $\hat{a}_{k\lambda}^\dagger$ are the usual annihilation and creation operators, respectively, for the plane-wave mode with wavevector k and polarization index λ (=1,2) and V is the volume of the quantization box. We use a linear polarization basis in which the polarization unit vectors $e_{k\lambda}$ ($k \cdot e_{k\lambda} = 0$, $\lambda = 1, 2$) are real. In terms of annihilation and creation operators, we have

$$\hat{H}_{\text{Field}} = \sum_{k\lambda} \hbar\omega_k (\hat{a}_{k\lambda}^\dagger \hat{a}_{k\lambda} + \tfrac{1}{2}). \tag{3.5}$$

We will drop the zero-point field energy $\sum_{k\lambda} \frac{1}{2}\hbar\omega_k$, which is just an additive constant in the Hamiltonian and plays no role for our purposes[3],

We assume the standard electric dipole form of the atom–field interaction:

$$\hat{H}_{\text{Int}} = -\hat{d}_A \cdot \hat{E}(r_A) - \hat{d}_B \cdot \hat{E}(r_B) \tag{3.6}$$

---

[2] We follow the convention of using a caret (∧) to denote an operator in Hilbert space.
[3] Implications of zero-point field energy are discussed, for instance, in [8].

where $\hat{d}_{A,B}$ are the electric dipole moment operators for the atoms and the positions of the atoms are specified by the coordinate vectors r_A and r_B. We can express the dipole moment operators in terms of the two-state raising and lowering operators as follows. In the two-state Hilbert space of either atom, $|1\rangle\langle 1| + |2\rangle\langle 2| = \hat{I}$, the unit operator. Thus,

$$\hat{d} = \hat{I}\hat{d}\hat{I} = d(|1\rangle\langle 2| + |2\rangle\langle 1|) \qquad (3.7)$$

if we assume that $\langle 1|\hat{d}|1\rangle = \langle 2|\hat{d}|2\rangle = 0$, i.e. that the atom has no permanent dipole moment, and if the dipole matrix element d is taken to be real: $d = \langle 1|\hat{d}|2\rangle = \langle 2|\hat{d}|1\rangle$. Then

$$\hat{\sigma} = |1\rangle\langle 2| \qquad \hat{\sigma}^{\dagger} = |2\rangle\langle 1| \qquad (3.8)$$
$$\hat{d} = d(\hat{\sigma} + \hat{\sigma}^{\dagger}) \qquad (3.9)$$

and

$$\hat{H}_{\text{Int}} = -d_A(\hat{\sigma}_A + \hat{\sigma}_A^{\dagger}) \cdot \hat{E}(r_A) - d_B(\hat{\sigma}_B + \hat{\sigma}_B^{\dagger}) \cdot \hat{E}(r_B). \qquad (3.10)$$

The complete Hamiltonian is, therefore,

$$\hat{H} = \hbar\omega_0\hat{\sigma}_A^{\dagger}\hat{\sigma}_A + \hbar\omega_0\hat{\sigma}_B^{\dagger}\hat{\sigma}_B + \sum_{k\lambda}\hbar\omega_k\hat{a}_{k\lambda}^{\dagger}\hat{a}_{k\lambda}$$

$$- i\hbar \sum_{j=A,B}\sum_{k\lambda}[C_{jk\lambda}(\hat{\sigma}_j + \hat{\sigma}_j^{\dagger})\hat{a}_{k\lambda} - C_{jk\lambda}^{*}(\hat{\sigma}_j + \hat{\sigma}_j^{\dagger})\hat{a}_{k\lambda}^{\dagger}] \qquad (3.11)$$

$$C_{jk\lambda} \equiv \left(\frac{\omega_k}{2\epsilon_0\hbar V}\right)^{1/2} d_j \cdot e_{k\lambda}e^{ik\cdot r_j}. \qquad (3.12)$$

To solve the Schrödinger equation

$$i\hbar\frac{\partial}{\partial t}|\psi(t)\rangle = \hat{H}|\psi(t)\rangle \qquad (3.13)$$

we first write the state vector $|\psi(t)\rangle$ as an expansion in the complete set of eigenstates $|\phi_m\rangle$ of the uncoupled atom–field system:

$$|\psi(t)\rangle = \sum_m b_m(t)|\phi_m\rangle \qquad (\hat{H}_{\text{Atoms}} + \hat{H}_{\text{Field}})|\phi_m\rangle = E_m|\phi_m\rangle. \qquad (3.14)$$

Then, from the orthonormality of the $|\phi_m\rangle$,

$$i\hbar\dot{b}_n(t) = E_n b_n(t) + \sum_m b_m(t)\langle\phi_n|\hat{H}_{\text{Int}}|\phi_m\rangle. \qquad (3.15)$$

The initial state is

$$|\phi_1\rangle = |+\rangle_A|-\rangle_B|0\rangle \qquad E_1 \equiv 0 \qquad (3.16)$$

i.e. atom A is in the excited state ($|+\rangle_A$), atom B is in the ground state ($|-\rangle_B$), and the field is in the vacuum state of no photons ($|0\rangle$). This state is coupled by $\hat{H}_{\text{Int}}$ to the state

$$|\phi_{k\lambda}\rangle = |-\rangle_A|-\rangle_B|1_{k\lambda}\rangle \qquad E_{k\lambda} = \hbar(\omega_k - \omega_0) \qquad (3.17)$$

in which both atoms are in their ground states and there is one photon in the mode (k, λ) of the electromagnetic field. Physically, this state results from the spontaneous emission of a photon from the initial state $|\phi_1\rangle$. $|\phi_{k\lambda}\rangle$, in turn, is coupled to $|\phi_1\rangle$ and the state

$$|\phi_2\rangle = |-\rangle_A|+\rangle_B|0\rangle \qquad E_2 \equiv 0 \qquad (3.18)$$

in which B is excited, A is unexcited, and there are no photons. Obviously the processes coupling the states (3.16)–(3.18) conserve excitation number and, for $\omega_k = \omega_0$, energy. They are described by terms $\hat{\sigma}_j \hat{a}_{k\lambda}^\dagger$ and $\hat{\sigma}_j^\dagger \hat{a}_{k\lambda}$ in $\hat{H}_{\text{Int}}$, i.e. terms associated with the creation of a photon and the 'lowering' of an atom, and the annihilation of a photon and the 'raising' of an atom.

$\hat{H}_{\text{Int}}$ also has energy-non-conserving parts involving $\hat{\sigma}_j^\dagger \hat{a}_{k\lambda}^\dagger$ and $\hat{\sigma}_j \hat{a}_{k\lambda}$, corresponding to the creation of a photon and the raising of an atom, and the annihilation of a photon and the lowering of an atom. Anticipating that energy-non-conserving terms should have a tiny effect on the probability amplitudes $b_1(t)$ and $b_2(t)$ of interest, let us retain only the energy-conserving terms. That is, let us include only the states $|\phi_1\rangle$, $|\phi_2\rangle$, and $|\phi_{k\lambda}\rangle$ in the coupled amplitude equations (3.15):

$$\dot{b}_1(t) = -\sum_{k\lambda} C_{Ak\lambda} b_{k\lambda}(t) \qquad (3.19)$$

$$\dot{b}_2(t) = -\sum_{k\lambda} C_{Bk\lambda} b_{k\lambda}(t) \qquad (3.20)$$

$$\dot{b}_{k\lambda}(t) = -\mathrm{i}(\omega_k - \omega_0) b_{k\lambda}(t) + C_{Ak\lambda}^* b_1(t) + C_{Bk\lambda}^* b_2(t). \qquad (3.21)$$

Using the formal solution of the last equation in the first two, we obtain coupled integro-differential equations for $b_1(t)$ and $b_2(t)$:

$$\dot{b}_1(t) = -\int_0^t \mathrm{d}t'\, b_1(t') \sum_{k\lambda} |C_{k\lambda}|^2 \mathrm{e}^{\mathrm{i}(\omega_k - \omega_0)(t'-t)}$$

$$\qquad\qquad - \int_0^t \mathrm{d}t'\, b_2(t') \sum_{k\lambda} C_{Ak\lambda} C_{Bk\lambda}^* \mathrm{e}^{\mathrm{i}(\omega_k - \omega_0)(t'-t)} \qquad (3.22)$$

$$\dot{b}_2(t) = -\int_0^t \mathrm{d}t'\, b_2(t') \sum_{k\lambda} |C_{k\lambda}|^2 \mathrm{e}^{\mathrm{i}(\omega_k - \omega_0)(t'-t)}$$

$$\qquad\qquad - \int_0^t \mathrm{d}t'\, b_1(t') \sum_{k\lambda} C_{Bk\lambda} C_{Ak\lambda}^* \mathrm{e}^{\mathrm{i}(\omega_k - \omega_0)(t'-t)} \qquad (3.23)$$

where, for the identical atoms assumed here,

$$|C_{k\lambda}|^2 = |C_{Ak\lambda}|^2 = |C_{Bk\lambda}|^2. \tag{3.24}$$

We will first do the mode summations in (3.22) and (3.23), assuming, for simplicity, that $d_A = d_B \equiv d = d^*$ and that d is orthogonal to the vector $r \equiv r_B - r_A$ pointing from A to B. In the limit $V \to \infty$, we replace in the usual fashion[4] the summation $\sum_k$ by the integration $(V/8\pi^3) \int d^3k$. Then[5]

$$\sum_{k\lambda} |C_{k\lambda}|^2 e^{i(\omega_k - \omega)(t'-t)}$$

$$\to \frac{V}{8\pi^3} \int d^3k \, \frac{\omega_k}{2\epsilon_0 \hbar V} e^{i(\omega_k - \omega)(t'-t)} \sum_{\lambda=1,2} (d \cdot e_{k\lambda})^2$$

$$= \frac{d^2}{16\pi^3 \epsilon_0 \hbar} \int d^3k \, \omega_k e^{i(\omega_k - \omega)(t'-t)} [1 - (\bar{d} \cdot \bar{k})^2]$$

$$= \frac{d^2}{16\pi^3 \epsilon_0 \hbar} \int dk \, k^2 \omega_k e^{i(\omega_k - \omega)(t'-t)} 2\pi \int_0^\pi d\theta \, \sin\theta (1 - \cos^2\theta)$$

$$= \frac{d^2}{6\pi^2 \epsilon_0 \hbar c^3} \int_0^\infty d\omega \, \omega^3 e^{i(\omega - \omega_0)(t'-t)}. \tag{3.25}$$

Similarly,

$$\sum_{k\lambda} C_{Bk\lambda} C_{Ak\lambda}^* e^{i(\omega_k - \omega)(t'-t)} = \frac{d^2}{16\pi^3 \epsilon_0 \hbar} \int dk \, k^2 \omega_k e^{i(\omega_k - \omega)(t'-t)}$$

$$\times \int d\Omega_{\bar{k}} [1 - (\bar{d} \cdot \bar{k})^2] e^{ik \cdot r} \tag{3.26}$$

where $\int d\Omega_{\bar{k}}$ is an integration over solid angles about $\bar{k}$. Let the z direction be the direction of r. Then, by assumption, $\bar{d} = a\bar{x} + b\bar{y}$, where $a^2 + b^2 = 1$ and $\bar{x}, \bar{y}$ are the unit vectors in the x, y directions, $\bar{d} \cdot \bar{k} = a(\bar{k} \cdot \bar{x}) + b(\bar{k} \cdot \bar{y})$,

$$\int d\Omega_{\bar{k}} [1 - (\bar{d} \cdot \bar{k})^2] e^{ik \cdot r}$$

$$= \int_0^{2\pi} d\phi \int_0^\pi d\theta \, ([1 - \sin^2\theta(a^2 \cos^2\phi + b^2 \sin^2\phi)]$$

$$- 2ab \sin^2\theta \sin\phi \cos\phi) e^{ikr \cos\theta}$$

---

[4] Based on the box normalization used in writing (3.4), we impose periodic boundary conditions, assuming that space is divided into cubes of volume $V = L^3$. Then $(k_x, k_y, k_z) = (2\pi/L)(n_x, n_y, n_z)$, where the n's are integers, and the factor $L^3/(2\pi)^3$ appears when we convert the summation over the n's to an integration over the k's.

[5] Here $\bar{d}$ and $\bar{k}$ are the unit vectors in the directions of d and k, respectively. Note that, since e_{k1}, e_{k2}, and $\bar{k}$ are mutually orthogonal, $\sum_{\lambda=1,2}(\bar{d} \cdot e_{k\lambda})^2 = 1 - (\bar{d} \cdot \bar{k})^2$.

$$= \pi \int_0^\pi d\theta \, \sin\theta [1 + \cos^2\theta] e^{ikr\cos\theta}$$

$$= 4\pi \left(\frac{\sin kr}{kr} + \frac{\cos kr}{(kr)^2} - \frac{\sin kr}{(kr)^3} \right) \equiv 4\pi \, G\left(\frac{\omega r}{c}\right) \qquad (3.27)$$

and

$$\sum_{k\lambda} C_{Bk\lambda} C_{Ak\lambda}^* e^{i(\omega_k - \omega)(t' - t)} = \sum_{k\lambda} C_{Bk\lambda}^* C_{Ak\lambda} e^{i(\omega_k - \omega)(t' - t)}$$

$$= \frac{d^2}{4\pi^2 \epsilon_0 \hbar c^3} \int d\omega \, \omega^3 G\left(\frac{\omega r}{c}\right) e^{i(\omega - \omega_0)(t' - t)}. \qquad (3.28)$$

With these results, we can write equations (3.22) and (3.23) as

$$\dot{b}_1(t) = -\frac{d^2}{6\pi^2 \epsilon_0 \hbar c^3} \int_0^t dt' \, b_1(t') \int_0^\infty d\omega \, \omega^3 e^{i(\omega - \omega_0)(t' - t)}$$

$$- \frac{d^2}{4\pi^2 \epsilon_0 \hbar c^3} \int_0^t dt' \, b_2(t') \int_0^\infty d\omega \, \omega^3 G\left(\frac{\omega r}{c}\right) e^{i(\omega - \omega_0)(t' - t)} \qquad (3.29)$$

$$\dot{b}_2(t) = -\frac{d^2}{6\pi^2 \epsilon_0 \hbar c^3} \int_0^t dt' \, b_2(t') \int_0^\infty d\omega \, \omega^3 e^{i(\omega - \omega_0)(t' - t)}$$

$$- \frac{d^2}{4\pi^2 \epsilon_0 \hbar c^3} \int_0^t dt' \, b_1(t') \int_0^\infty d\omega \, \omega^3 G\left(\frac{\omega r}{c}\right) e^{i(\omega - \omega_0)(t' - t)}. \qquad (3.30)$$

The first term on the right-hand side of either equation is associated with single-atom spontaneous emission. We will make an approximation characteristic of virtually every approach to the theory of spontaneous emission, namely the 'Weisskopf–Wigner' or 'Markovian' approximation. In the present formulation, this approximation is tantamount to the replacement of $b_1(t')$ by $b_1(t)$ in the first term on the right-hand side of (3.29) and of $b_2(t')$ by $b_2(t)$ in the first term on the right-hand side of (3.30). This approximation assumes that the time evolution of the excited-state probability in free-space spontaneous emission is 'memory-less' or Markovian and it leads to the exponential decay of the excited-state probability that agrees extremely well with experiment. The approximation of exponential decay results from the additional replacement

$$\int_0^t dt' \, e^{i(\omega - \omega_0)(t' - t)} \rightarrow \pi \delta(\omega - \omega_0) - iP \frac{1}{\omega - \omega_0} \qquad (3.31)$$

where P indicates that the Cauchy principal part is to be taken in integrals over ω. This replacement comes from

$$\int_0^t dt' \, e^{i(\omega - \omega_0)(t' - t)} = \frac{\sin(\omega - \omega_0)t}{\omega - \omega_0} - i\left[\frac{1 - \cos(\omega - \omega_0)t}{\omega - \omega_0} \right]. \qquad (3.32)$$

The first term oscillates rapidly and effectively vanishes unless $\omega = \omega_0$, in which case it is equal to t. The second term vanishes if $\omega = \omega_0$ but it is otherwise effectively $1/(\omega - \omega_0)$ owing again to rapid oscillations. Thus, for sufficiently long times, we make the replacement (3.31).

With these approximations, the first term on the right-hand side of (3.29) is

$$-\frac{d^2}{6\pi^2\epsilon_0\hbar c^3}\left[\pi\int_0^\infty d\omega\,\omega^3\delta(\omega - \omega_0) - iP\int_0^\infty \frac{d\omega\,\omega^3}{\omega - \omega_0}\right]b_1(t)$$

$$= -\left[\frac{\gamma}{2} - i\Delta\right]b_1(t) \tag{3.33}$$

where

$$\gamma = \frac{d^2\omega_0^3}{3\pi\epsilon_0\hbar c^3} \tag{3.34}$$

and

$$\Delta = \frac{d^2}{6\pi^2\epsilon_0\hbar c^3}P\int_0^\infty \frac{d\omega\,\omega^3}{\omega - \omega_0} \tag{3.35}$$

γ is just the Einstein A coefficient for the rate of decay of the excited-state probability in spontaneous emission. Δ corresponds to a radiative level shift and, as mentioned earlier, is not correctly accounted for by the two-state model for an atom[6]. Partly for this reason, but mainly because it is simply irrelevant for our purposes, we ignore the single-atom radiative level shift: it can be assumed to have been included in the definition of the transition frequency ω_0. We make the same approximations in the first term on the right-hand side of (3.30) and are left with

$$\dot{b}_1(t) = \frac{\gamma}{2}b_1(t) - \frac{3\gamma}{4\pi\omega_0^3}\int_0^t dt'\,b_2(t')\int_0^\infty d\omega\,\omega^3 G\left(\frac{\omega r}{c}\right)e^{i(\omega-\omega_0)(t'-t)} \tag{3.36}$$

$$\dot{b}_2(t) = -\frac{\gamma}{2}b_2(t) - \frac{3\gamma}{4\pi\omega_0^3}\int_0^t dt'\,b_1(t')\int_0^\infty d\omega\,\omega^3 G\left(\frac{\omega r}{c}\right)e^{i(\omega-\omega_0)(t'-t)}. \tag{3.37}$$

If we now make the Markovian approximation of replacing $b_2(t')$ and $b_1(t')$ by $b_2(t)$ and $b_1(t)$ in the terms coupling b_1 and b_2, we obtain the well-known results for the resonant interaction of two identical atoms [76, 87] when we use (3.31). However, this approximation treats the interaction as effectively instantaneous in that $b_1(t)$ is coupled directly to $b_2(t)$. Let us instead evaluate the integral over ω appearing in (3.36) and (3.37). The near-field $(1/r^3)$ contribution to this integral is

$$-\int_0^\infty d\omega\,\omega^3\frac{\sin kr}{(kr)^3}e^{i(\omega-\omega_0)(t'-t)}$$

---

[6] Nor is it properly accounted for in the electric dipole form of the Hamiltonian used here. See, for instance, [8].

$$= -\frac{c^3}{2ir^3}e^{i\omega_0(t'-t)}\int_0^\infty d\omega\,[e^{i\omega(t'-t+r/c)} - e^{i\omega(t'-t-r/c)}]. \quad (3.38)$$

This is not properly retarded—there are contributions from the advanced time $t' = t + r/c$ as well as the retarded time $t' = t - r/c$. But since the dominant contribution is expected to come from field frequencies $\omega \cong \omega_0$, while contributions from frequencies far removed from ω_0 are expected to be unimportant, let us make the approximation of extending the integration in (3.38) to $-\infty$:

$$-\int_0^\infty d\omega\,\omega^3\frac{\sin kr}{(kr)^3}e^{i(\omega-\omega_0)(t'-t)}$$

$$\rightarrow -\frac{c^3}{2ir^3}e^{-i\omega_0(t'-t)}\int_{-\infty}^\infty d\omega\,[e^{i\omega(t'-t+r/c)} - e^{i\omega(t'-t-r/c)}]$$

$$\rightarrow -\frac{\pi c^3}{ir^3}e^{-i\omega_0(t'-t)}\delta(t'-t+r/c) \quad (3.39)$$

since $t' < t$ in (3.36) and (3.37). *This is the approximation made by Fermi.* To express it in a form more specifically related to Fermi's calculation, note that if we make the Markovian approximation in (3.36) and (3.37) we are left with the t' integration

$$\int_0^t dt'\,e^{i(\omega-\omega_0)(t'-t)} = \pi\delta(\omega-\omega_0) - iP\frac{1}{\omega-\omega_0} = \frac{1}{\omega-\omega_0-i\epsilon} \quad (\epsilon \rightarrow 0^+)$$
$$(3.40)$$

and

$$\int_0^\infty d\omega\,\omega^3\frac{\sin kr}{(kr)^3}e^{i(\omega-\omega_0)(t'-t)} = \frac{c^3}{r^3}\int_0^\infty d\omega\,\frac{\sin(\omega r/c)}{\omega-\omega_0-i\epsilon}. \quad (3.41)$$

This form shows more directly that the approximation of extending the integration over ω to $-\infty$ is a sensible one. But it is, nevertheless, an *approximation* and, by invoking it, we have not *proven* that $b_2(t)$ is zero unless $t > r/c$.

The terms going as $1/r$ and $1/r^2$ in the integral over ω in (3.36) and (3.37) are evaluated similarly in this approximation:

$$\int_0^\infty d\omega\,\omega^3\frac{\sin kr}{kr}e^{i(\omega-\omega_0)(t'-t)}$$

$$\rightarrow \frac{c}{2ir}e^{-i\omega_0(t'-t)}\int_{-\infty}^\infty d\omega\,\omega^2[e^{i\omega(t'-t+r/c)} - e^{i\omega(t'-t-r/c)}]$$

$$\rightarrow -\frac{\pi c}{ir}e^{-i\omega_0(t'-t)}\frac{\partial^2}{\partial t'^2}\delta(t'-t+r/c) \quad (3.42)$$

$$\int_0^\infty d\omega\,\omega^3\frac{\cos kr}{(kr)^2}e^{i(\omega-\omega_0)(t'-t)}$$

$$\rightarrow \frac{c^2}{2r^2} e^{-i\omega_0(t'-t)} \int_{-\infty}^{\infty} d\omega\, \omega [e^{i\omega(t'-t+r/c)} + e^{i\omega(t'-t-r/c)}]$$

$$\rightarrow \frac{\pi c^2}{ir^2} e^{-i\omega_0(t'-t)} \frac{\partial}{\partial t'} \delta(t' - t + r/c). \tag{3.43}$$

Then (3.36) and (3.37) become

$$\dot{b}_1(t) = \frac{\gamma}{2} b_1(t) - \frac{3i}{4}\gamma \left[\frac{-i}{k_0 r} + \frac{1}{k_0^2 r^2} + \frac{i}{k_0^3 r^3} \right] e^{ik_0 r} b_2 \left(t - \frac{r}{c} \right) \theta \left(t - \frac{r}{c} \right)$$

$$\tag{3.44}$$

$$\dot{b}_2(t) = \frac{\gamma}{2} b_2(t) - \frac{3i}{4}\gamma \left[\frac{-i}{k_0 r} + \frac{1}{k_0^2 r^2} + \frac{i}{k_0^3 r^3} \right] e^{ik_0 r} b_1 \left(t - \frac{r}{c} \right) \theta \left(t - \frac{r}{c} \right)$$

$$\tag{3.45}$$

where $k_0 = \omega_0/c$, θ is the unit step function and we have used the fact that $b_1(t)$ and $b_2(t)$ vary slowly compared with $\exp(-i\omega_0 t)$ so that, for instance, $(\partial/\partial t)[b_1(t)\exp(-i\omega_0 t)] \cong -i\omega_0 b_1(t)\exp(-i\omega_0 t)$.

Equations (3.44) and (3.45) exhibit the correct dependence on the retardation time $t - r/c$. The solution for the initial state of interest, namely for $b_1(0) = 1$ and $b_2(0) = 0$, is

$$b_1(t) = \sum_{\substack{n=0 \\ n \text{ even}}}^{\infty} \frac{\alpha^n}{n!} \left(t - \frac{nr}{c} \right)^n e^{-\frac{1}{2}\gamma(t-nr/c)} \theta \left(t - \frac{nr}{c} \right) \tag{3.46}$$

$$b_2(t) = \sum_{\substack{n=0 \\ n \text{ odd}}}^{\infty} \frac{\alpha^n}{n!} \left(t - \frac{nr}{c} \right)^n e^{-\frac{1}{2}\gamma(t-nr/c)} \theta \left(t - \frac{nr}{c} \right) \tag{3.47}$$

where

$$\alpha = \frac{3\gamma}{4} \left(\frac{i}{k_0 r} - \frac{1}{k_0^2 r^2} - \frac{i}{k_0^3 r^3} \right). \tag{3.48}$$

These solutions show the expected behaviour of the probability amplitudes. Atom B has zero probability of being excited until after the time $t = r/c$ that it takes light to propagate from A to B after emission by A. Atom A is only affected by atom B after the time it takes light to propagate from A to B and then back to A. This continues indefinitely: the probability amplitude for the initial state $|+\rangle_A|-\rangle_B|0\rangle$ depends only on the even retardation times 0, $2r/c$, $4r/c$, etc. Similarly the probability amplitude for the state $|-\rangle_A|+\rangle_B|0\rangle$ depends only on the odd retardation times r/c, $3r/c$, etc. Exactly the same behaviour is found in the radiative coupling of two *non*-identical atoms [77] as well as two classical dipole oscillators [88]. If we replace $t - nr/c$ by t in (3.46) and (3.47), we can sum the series to obtain well-known results for the resonant interaction of two identical atoms [76, 87], the same interaction one obtains when the Markovian approximation is made in (3.36) and (3.37).

These results are based on the approximation of including only energy-conserving states, together with the other approximation made by Fermi, namely the approximation of extending frequency integrals to $-\infty$. The first approximation without the second does not lead to correctly retarded expressions like (3.44)–(3.47).

As discussed earlier, the Hamiltonian includes energy-non-conserving processes such as the emission of a photon with the simultaneous raising of an atomic state. Thus, the term $\hat{\sigma}_j^\dagger \hat{a}_{k\lambda}^\dagger$ in $\hat{H}_{\text{Int}}$ couples the initial state $|\phi_1\rangle$ to the state $|+\rangle_A|+\rangle_B|1_{k\lambda}\rangle$, no probability amplitude for which has been included in our analysis. This state is, in turn, coupled by $\hat{H}_{\text{Int}}$ to states with more than one photon in the field: going beyond our approximation of accounting for only energy-conserving processes obviously introduces an infinite number of states that we have not accounted for. In a rather complicated analysis, Berman and Dubetsky [86] have shown how properly retarded probability amplitudes can be obtained when energy-non-conserving states are included. In other words, including energy-non-conserving states leads *ipso facto* to frequency integrals extending to $-\infty$.

Another remark relating to the extension of frequency integrals to $-\infty$: if we let $r \to \infty$ equation (3.29) reduces to

$$\dot{b}_1(t) = -\frac{\gamma}{2\pi\omega_0^3} \int_0^t dt'\, b_1(t') \int_0^\infty d\omega\, \omega^3 e^{i(\omega-\omega_0)(t'-t)} \tag{3.49}$$

which describes the time evolution due to spontaneous emission of the excited-state amplitude of atom A in the absence of atom B. The Markovian approximation gives the familiar exponential decay:

$$|b_1(t)|^2 = e^{-\gamma t}. \tag{3.50}$$

But suppose that, instead of the Markovian approximation, we extend the integration over ω in (3.49) to $-\infty$:

$$\dot{b}_1(t) \to -\frac{i\gamma}{2\pi\omega_0^3} \int_0^t dt'\, b_1(t') e^{-i\omega_0(t'-t)} \frac{\partial^3}{\partial t'^3} \int_{-\infty}^\infty d\omega\, e^{i\omega(t'-t)}$$

$$\to \frac{i\gamma}{2\omega_0^3} e^{i\omega_0 t} \frac{\partial^3}{\partial t^3} [b(t) e^{-i\omega_0 t}] \tag{3.51}$$

where, in the second line, we have dropped (divergent) terms associated with radiative level shifts—which as already noted (see footnote 6) are not described correctly in our model anyway. The appearance of a third derivative with respect to time when we extend the ω integral to $-\infty$ is reminiscent of the classical theory of radiation reaction [recall, for instance, equation (1.66)]. Of course the approximation that $b_1(t)$ varies negligibly on a time scale $\sim \omega_0^{-1}$ in equation (3.51) reproduces the exponential decay law (3.50).

3.1.1 Heisenberg picture

The complication of energy-non-conserving states in the Schrödinger picture can be circumvented by taking the Heisenberg-picture approach in which operators rather than states evolve in time. The operator $\hat{\sigma}_B$, for instance, evolves in time according to the Heisenberg equation of motion

$$i\hbar\frac{d\hat{\sigma}_B}{dt} = [\hat{\sigma}_B, \hat{H}]. \tag{3.52}$$

$\hat{\sigma}_B$ commutes with every operator in the Hamiltonian except $\hat{\sigma}_B^\dagger$:

$$[\hat{\sigma}_B, \hat{\sigma}_B^\dagger] = -(\hat{\sigma}_B^\dagger\hat{\sigma}_B - \hat{\sigma}_B\hat{\sigma}_B^\dagger) \equiv -\hat{\sigma}_{zB} \tag{3.53}$$

where we now introduce the Pauli $\hat{\sigma}_z$ operator with algebraic properties familiar from the theory of a spin-$\frac{1}{2}$ system (or any other two-state system):

$$[\hat{\sigma}_{zB}, \hat{\sigma}_B] = -2\hat{\sigma}_B \qquad [\hat{\sigma}_{zB}, \hat{\sigma}_B^\dagger] = 2\hat{\sigma}_B^\dagger. \tag{3.54}$$

From (3.1), (3.2), (3.6), and (3.53), it follows that

$$\begin{aligned}\frac{d\hat{\sigma}_B}{dt} &= -i\omega_0[\hat{\sigma}_B(t), \hat{\sigma}_B^\dagger(t)\hat{\sigma}_B(t)] + \frac{i}{\hbar}\boldsymbol{d}_B \cdot \hat{\boldsymbol{E}}(\boldsymbol{r}_B, t)[\hat{\sigma}_B(t), \hat{\sigma}_B^\dagger(t)] \\ &= -i\omega_0\hat{\sigma}_B(t) - \frac{i}{\hbar}\boldsymbol{d}_B \cdot \hat{\boldsymbol{E}}(\boldsymbol{r}_B, t)\hat{\sigma}_{zB}(t). \end{aligned} \tag{3.55}$$

The expectation value $\langle\hat{\sigma}_B^\dagger(t)\hat{\sigma}_B(t)\rangle$ in the initial state of the atom–field system is the probability at time t that atom B is in the excited state.

The time evolution of the electric field operator $\hat{\boldsymbol{E}}(\boldsymbol{r}_B, t)$ is determined by the time evolution of the mode annihilation and creation operators $\hat{a}_{k\lambda}$ and $\hat{a}_{k\lambda}^\dagger$:

$$\frac{d\hat{a}_{k\lambda}}{dt} = -i\omega_k\hat{a}_{k\lambda} + \sum_{j=A,B} C_{jk\lambda}^*(\hat{\sigma}_j + \hat{\sigma}_j^\dagger) \tag{3.56}$$

where we have used (3.11) and the commutation relations

$$[\hat{a}_{k\lambda}, \hat{a}_{k'\lambda'}] = 0 \qquad [\hat{a}_{k\lambda}, \hat{a}_{k'\lambda'}^\dagger] = \delta_{\lambda\lambda'}\delta_{k,k'}^3. \tag{3.57}$$

Thus,

$$\hat{a}_{k\lambda}(t) = \hat{a}_{k\lambda}(0)e^{-i\omega_k t} + \sum_{j=A,B} C_{jk\lambda}^*\int_0^t dt'\,\hat{\sigma}_{xj}(t')e^{i\omega_k(t'-t)} \tag{3.58}$$

where we have introduced the Pauli $\hat{\sigma}_x$ operator, $\hat{\sigma}_{xj} \equiv \hat{\sigma}_j + \hat{\sigma}_j^\dagger$. The electric field operator at $\boldsymbol{r}_B, t$ is, therefore,

$$\hat{\boldsymbol{E}}(\boldsymbol{r}_B, t) = \hat{\boldsymbol{E}}_0(\boldsymbol{r}_B, t) + \hat{\boldsymbol{E}}_B(\boldsymbol{r}_B, t) + \hat{\boldsymbol{E}}_A(\boldsymbol{r}_B, t) \tag{3.59}$$

$$\hat{E}_0(\boldsymbol{r}_B, t) = i \sum_{k\lambda} \left(\frac{\hbar \omega_k}{2\epsilon_0 V} \right)^{1/2} \hat{a}_{k\lambda}(0) e^{-i(\omega_k t - \boldsymbol{k} \cdot \boldsymbol{r}_B)} + \text{h.c.} \tag{3.60}$$

$$\hat{E}_B(\boldsymbol{r}_B, t) = i \sum_{k\lambda} \frac{\omega_k}{2\epsilon_0 V} (\boldsymbol{d} \cdot \boldsymbol{e}_{k\lambda}) \boldsymbol{e}_{k\lambda} \int_0^t \mathrm{d}t' \, \hat{\sigma}_{xB}(t') e^{i\omega_k(t'-t)} + \text{h.c.} \tag{3.61}$$

$$\hat{E}_A(\boldsymbol{r}_B, t) = i \sum_{k\lambda} \frac{\omega_k}{2\epsilon_0 V} (\boldsymbol{d} \cdot \boldsymbol{e}_{k\lambda}) \boldsymbol{e}_{k\lambda} e^{i\boldsymbol{k} \cdot \boldsymbol{r}} \int_0^t \mathrm{d}t' \, \hat{\sigma}_{xA}(t') e^{i\omega_k(t'-t)} + \text{h.c.} \tag{3.62}$$

where h.c. $\equiv$ hermitian conjugate and again we define $\boldsymbol{r} = \boldsymbol{r}_B - \boldsymbol{r}_A$ and make the simplifying assumption that $\boldsymbol{d}_B = \boldsymbol{d}_A = \boldsymbol{d} = \boldsymbol{d}^*$. $\hat{E}_0(\boldsymbol{r}_B, t)$ is the free field at $\boldsymbol{r}_B$, i.e. the quantum field that exists at $\boldsymbol{r}_B$ even in the absence of any sources. $\hat{E}_B(\boldsymbol{r}_B, t)$ is the field of atom B evaluated at the position of atom B, i.e. it is the 'radiation reaction' field acting on atom B. $\hat{E}_A(\boldsymbol{r}_B, t)$ is the field of atom A evaluated at the position of atom B.

The most interesting part of $\hat{E}(\boldsymbol{r}_B, t)$ for our purposes is $\hat{E}_A(\boldsymbol{r}_B, t)$, for it is through this field that A and B interact. This field is easily evaluated using the mode summation formulas employed in our Schrödinger-picture approach:

$$
\begin{aligned}
\boldsymbol{d} \cdot \hat{E}_A(\boldsymbol{r}_B, t) &= \frac{i d^2}{2\epsilon_0 V} \frac{V}{8\pi^3} \int \mathrm{d}k \, k^2 \omega_k \int \mathrm{d}\Omega_{\overline{k}} [1 - (\overline{\boldsymbol{k}} \cdot \overline{\boldsymbol{d}})^2] e^{i\boldsymbol{k} \cdot \boldsymbol{r}} \\
&\quad \times \int_0^t \mathrm{d}t' \, \hat{\sigma}_{xA}(t') e^{i\omega_k(t'-t)} + \text{h.c.} \\
&= \frac{i d^2}{4\pi^2 \epsilon_0 c^3} \int_0^t \mathrm{d}t' \, \hat{\sigma}_{xA}(t') \int_0^\infty \mathrm{d}\omega \, \omega^3 G\left(\frac{\omega r}{c} \right) e^{i\omega(t'-t)} + \text{h.c.} \\
&= -\frac{d^2}{4\pi\epsilon_0} \int_0^t \mathrm{d}t' \, \hat{\sigma}_{xA}(t') \left[\frac{1}{r^3} \delta\left(t' - t + \frac{r}{c} \right) \right. \\
&\quad \left. - \frac{1}{cr^2} \frac{\partial}{\partial t'} \delta\left(t' - t + \frac{r}{c} \right) + \frac{1}{c^2 r} \frac{\partial^2}{\partial t'^2} \delta\left(t' - t + \frac{r}{c} \right) \right] \\
&= -\frac{d^2}{4\pi\epsilon_0} \left[\frac{1}{r^3} \hat{\sigma}_{xA}\left(t - \frac{r}{c} \right) + \frac{1}{cr^2} \dot{\hat{\sigma}}_{xA}(t - r/c) \right. \\
&\quad \left. + \frac{1}{c^2 r} \ddot{\hat{\sigma}}_{xA}\left(t - \frac{r}{c} \right) \theta\left(t - \frac{r}{c} \right) \right].
\end{aligned} \tag{3.63}
$$

This has exactly the same form as the field from a classical dipole. The fact that it vanishes for times $t < r/c$ ensures that the interaction between the atoms is *exactly* causal and, in particular, that atom B in the Fermi model cannot be excited before times $t \leq r/c$ [85].

Hegerfeldt [79] has stated a theorem to the effect that the initially unexcited atom B 'starts to move out of the ground state immediately and is thus influenced by atom A instantaneously'. It has been noted, however, that the theorem applies *regardless of whether atom A is present* and, therefore, it should not be used as an argument against causality in the two–atom interaction [85]. Such 'immediate influences' are associated with the fact that the assumed initial state is not an eigenstate of the interacting atom–field system: a true eigenstate of the system involves an *admixture* of 'bare' states such as $|\phi_1\rangle$, $|\phi_2\rangle$, and $|\phi_{k\lambda}\rangle$. Such admixtures involving excited, unperturbed atomic (and field) states occur even in the case of a *single* atom coupled to the field, and are associated with phenomena, such as the Lamb shift, involving virtual (energy-non-conserving) transitions. In the two-atom case, however, there are no *interatomic (r-dependent)* 'immediate influences' before the time r/c after the system is presumed to be prepared in an eigenstate of the unperturbed atom–field system. This is implied by the operator equation (3.63), which makes no reference to any specific states, bare or dressed.

To illustrate the nature of the 'immediate influences' when a system is supposed to be in an unperturbed state at $t = 0$, consider the problem of a *single* two-level atom interacting with a single field mode of frequency ω. The single-mode assumption here is made only for simplicity and is certainly not essential for the present discussion. The Hamiltonian for this system is

$$H = \tfrac{1}{2}\hbar\omega_0\hat{\sigma}_z + \hbar\omega\hat{a}^\dagger\hat{a} - \mathrm{i}C(\hat{a} - \hat{a}^\dagger)(\hat{\sigma} + \hat{\sigma}^\dagger) \tag{3.64}$$

where we now express $\hat{H}_{\text{Atom}}$ in terms of the $\hat{\sigma}_z$ operator, as is often done[7]. The Heisenberg equations of motion for $\hat{\sigma}$, $\hat{\sigma}_z$, and $\hat{a}$ are

$$\dot{\hat{\sigma}} = -\mathrm{i}\omega_0\hat{\sigma} + \frac{C}{\hbar}(\hat{a} - \hat{a}^\dagger)\hat{\sigma}_z \tag{3.65}$$

$$\dot{\hat{\sigma}}_z = \frac{2C}{\hbar}(\hat{\sigma}\hat{a} - \hat{a}^\dagger\hat{\sigma}) + \text{h.c.} \tag{3.66}$$

and

$$\dot{\hat{a}} = -\mathrm{i}\omega\hat{a} + \frac{C}{\hbar}(\hat{\sigma} + \hat{\sigma}^\dagger). \tag{3.67}$$

It is convenient to define the slowly-varying operator

$$\hat{S}(t) = \hat{\sigma}(t)e^{\mathrm{i}\omega_0 t} \tag{3.68}$$

in terms of which we obtain the following formal equation for the expectation value of the population difference operator $\hat{\sigma}_z$:

$$\langle\dot{\hat{\sigma}}_z(t)\rangle = \frac{4C^2}{\hbar^2} \,\text{Re} \int_0^t \mathrm{d}t' \,[\langle\hat{S}(t)\hat{S}(t')\rangle e^{\mathrm{i}(\omega-\omega_0)t'} e^{-\mathrm{i}(\omega+\omega_0)t}$$
$$+ \langle\hat{S}(t)\hat{S}^\dagger(t')\rangle e^{\mathrm{i}(\omega+\omega_0)(t'-t)} - \langle\hat{S}^\dagger(t')\hat{S}(t)\rangle e^{-\mathrm{i}(\omega-\omega_0)(t'-t)}$$
$$- \langle\hat{S}(t')\hat{S}(t)\rangle e^{-\mathrm{i}(\omega+\omega_0)t'} e^{\mathrm{i}(\omega-\omega_0)t}]. \tag{3.69}$$

[7] $\hat{\sigma}^\dagger\hat{\sigma} = \tfrac{1}{2}[\hat{\sigma}^\dagger, \hat{\sigma}] + \tfrac{1}{2}(\hat{\sigma}^\dagger\hat{\sigma} + \hat{\sigma}\hat{\sigma}^\dagger) = \tfrac{1}{2}\hat{\sigma}_z + \tfrac{1}{2}\hat{I}$, since $\hat{\sigma}^\dagger\hat{\sigma} + \hat{\sigma}\hat{\sigma}^\dagger$ is the unit operator ($\hat{I}$) in the two-state Hilbert space. Therefore, $\hbar\omega_0\hat{\sigma}^\dagger\hat{\sigma} = \tfrac{1}{2}\hbar\omega_0\hat{\sigma}_z$ plus a constant term that commutes with every operator and, as such, can be dropped from the Hamiltonian.

We have assumed an initial state such that the field is in its vacuum state but no approximations have been made.

The terms involving $\omega + \omega_0$ do not arise when one ignores energy-non-conserving processes. But even with such processes there are no 'real' transitions over times long compared with a few periods of oscillation. For short times, however,

$$\langle \dot{\hat{\sigma}}_z(t) \rangle \cong \frac{2C^2}{\hbar^2} \langle \hat{S}(0) \hat{S}^\dagger(0) \rangle \int_0^t dt' \, e^{i(\omega+\omega_0)(t'-t)} + \text{c.c.} = \frac{4C^2}{\hbar^2} \frac{\sin(\omega+\omega_0)t}{(\omega+\omega_0)} \tag{3.70}$$

where we have used the operator identity $\hat{S}^2(0) = 0$ and also the expectation values $\langle \hat{S}(0) \hat{S}^\dagger(0) \rangle = 1$, $\langle \hat{S}^\dagger(0) \hat{S}(0) \rangle = 0$ appropriate to the case of the atom initially in its lower state. Thus,

$$\dot{P}(t) = \frac{1}{2} \langle \dot{\hat{\sigma}}_z(t) \rangle = \frac{2C^2}{\hbar^2} \frac{\sin(\omega+\omega_0)t}{\omega+\omega_0} \tag{3.71}$$

and the probability of the atom being excited over short times is

$$P(t) = \frac{2C^2}{\hbar^2} \left(\frac{1}{\omega+\omega_0} \right)^2 [1 - \cos(\omega+\omega_0)t]. \tag{3.72}$$

For $\omega = \omega_0$ and $\omega_0 t \ll 1$, for instance,

$$P(t) \cong \frac{C^2}{\hbar^2 \omega_0^2} \omega_0^2 t^2 = \frac{C^2}{\hbar^2} t^2 \tag{3.73}$$

i.e. there is a non-vanishing probability for $t \ll \omega_0^{-1}$ that the atom, initially in its lower state with no photons in the field, is excited. This is consistent with the energy–time uncertainty relation: for short enough times 'energy-non-conserving' transitions are possible. Over times long in the sense of the energy–time uncertainty relation, of course, only energy-conserving processes contribute to real transition rates and the energy-non-conserving terms manifest themselves only through virtual transitions contributing to energy-level shifts.

These result are entirely consistent with Hegerfeldt's theorem based on continuity requirements. The point we wish to make is that 'immediate influences' of the type suggested by Hegerfeldt are present even in the absence of a second atom and that they pose no real difficulty for the proof of causality in the Fermi problem.

3.2 Causality in photodetection theory

The quantum theory of photodetection [89–94] leads to normally ordered field correlation functions such as

$$G_{ij}^{(1)}(r_1, t_1; r_2, t_2) = \langle \hat{E}_i^{(-)}(r_1, t_1) \hat{E}_j^{(+)}(r_2, t_2) \rangle \tag{3.74}$$

$$G_{ijk\ell}^{(2)}(r_1, t_1; r_2, t_2; r_3, t_3; r_4, t_4) = \langle \hat{E}_i^{(-)}(r_1, t_1) \hat{E}_j^{(-)}(r_2, t_2)$$

$$\times \hat{E}_k^{(+)}(r_3, t_3) \hat{E}_\ell^{(+)}(r_4, t_4) \rangle \quad (3.75)$$

where $\hat{E}^{(+)}(r, t)$ and $\hat{E}^{(-)}(r, t)$ are, respectively, the positive- and negative-frequency parts of the field. $\hat{E}^{(+)}(r, t)$ may be defined formally as

$$\hat{E}^{(+)}(r, t) = \lim_{\epsilon \to 0^+} \frac{1}{2\pi i} \int_{-\infty}^{\infty} dt' \, \frac{\hat{E}(r, t - t')}{t' - i\epsilon}. \quad (3.76)$$

Thus, if $\hat{E}(r, t)$ has a Fourier expansion involving $e^{\pm i\omega t}$, $\omega > 0$, the integration of (3.76) over a contour along the real axis and a semicircle in the upper half-plane picks out the 'positive-frequency' $e^{-i\omega t}$ components, whereas closure of the contour in the lower half-plane ensures that there are no 'negative-frequency' components $e^{i\omega t}$.

In the absence of sources, the quantized electric field has the positive-frequency part

$$\hat{E}_0^{(+)}(r, t) = i \sum_{k\lambda} \left(\frac{\hbar \omega_k}{2\epsilon_0 V} \right)^{1/2} \hat{a}_{k\lambda}(0) e^{-i\omega_k t} e^{ik \cdot r} e_{k\lambda}. \quad (3.77)$$

With sources, however, $\hat{a}_{k\lambda}(t) \neq \hat{a}_{k\lambda}(0) e^{-i\omega_k t}$ and

$$i \sum_{k\lambda} \left(\frac{\hbar \omega_k}{2\epsilon_0 V} \right)^{1/2} \hat{a}_{k\lambda}(t) e^{ik \cdot r} e_{k\lambda} \quad (3.78)$$

is not exactly equal to the positive-frequency part of the field defined by (3.76). That is, equations (3.76) and (3.78), in general, define two different fields. *Neither definition gives a 'causal' (retarded) field,* although of course the complete electric field operator $\hat{E} = \hat{E}^{(+)} + \hat{E}^{(-)}$ is retarded and is the same regardless of whether (3.76) or (3.78) is used to define $\hat{E}^{(\pm)}$. The non-retarded character of $E^{(\pm)}(r, t)$ has raised concern about the general validity of the standard theory of photodetection based on normally ordered field correlation functions [95, 96]. We now address the question of causality in the theory of photodetection.

Consider first the simplest case, the measurement of the intensity of an optical field. Our model for the detector will at first be a two-state atom at a point r, so that the Hamiltonian for the system consisting of the field and the detector is

$$\hat{H} = \tfrac{1}{2} \hbar \omega_0 \hat{\sigma}_z + \hat{H}_{\text{Field}} - d_j \hat{E}_j(r) \hat{\sigma}_x \quad (3.79)$$

where we use the summation convention for repeated Cartesian indices. The Heisenberg equation of motion for $\hat{\sigma}_z(t)$ is easily found from the commutation relations for the Pauli two-state operators:

$$\dot{\hat{\sigma}}_z(t) = \frac{1}{i\hbar} [\hat{\sigma}_z, \hat{H}] = -\frac{2i}{\hbar} d_j \hat{E}_j(r) [\hat{\sigma}(t) - \hat{\sigma}^\dagger(t)]$$

$$= \frac{2i}{\hbar} d_j [\hat{\sigma}^\dagger(t) \hat{E}_j(r, t) - \hat{E}_j(r, t) \hat{\sigma}(t)] \quad (3.80)$$

where, in the second line, we have made use of the commutativity of equal-time atom and field operators to put the operator products in an order that happens to be convenient. Similarly,

$$\dot{\hat{\sigma}}(t) = -i\omega_0\hat{\sigma}(t) - \frac{i}{\hbar}d_j\hat{E}_j(r, t)\hat{\sigma}_z(t). \tag{3.81}$$

We use the formal solution of this equation in (3.80):

$$\langle\dot{\hat{\sigma}}_z(t)\rangle = -\frac{4}{\hbar^2}d_jd_k\operatorname{Re}\int_0^t dt'\,\langle\hat{E}_j(r, t)\hat{E}_k(r, t')\hat{\sigma}_z(t')\rangle e^{i\omega_0(t'-t)}. \tag{3.82}$$

Here the expectation value is over an initial state $|\psi\rangle$ with the detector atom in the ground state, so that $\hat{\sigma}(0)|\psi\rangle = \langle\psi|\hat{\sigma}^\dagger(0) = 0$, and this has been assumed in writing (3.82). We are interested, of course, in the more practical situation in which the detector is not a two-state system but, in fact, has a continuum of possible final states, such that stimulated emission from an excited state is negligible compared with absorption from the ground state and the detector is consequently unsaturable. We also assume that absorption is weak enough that the occupation probability of the initial state of the detector remains close to unity. In the context of our idealized two-state atom, this means we can take $\hat{\sigma}_z(t') \cong \hat{\sigma}_z(0)$ in (3.82) and use the assumption $\hat{\sigma}_z(0)|\psi\rangle = -|\psi\rangle$ that the detector atom is initially in its ground state:

$$\langle\dot{\hat{\sigma}}_z(t)\rangle \cong \frac{4}{\hbar^2}d_jd_k\operatorname{Re}\int_0^t dt'\,\langle\hat{E}_j(r, t)\hat{E}_k(r, t')\rangle e^{i\omega_0(t'-t)}. \tag{3.83}$$

Here the field may be written as

$$\hat{E}_j(r, t) = \hat{E}_{0,j}(r, t) + \hat{E}_{RR,j}(r, t) + \hat{E}_{S,j}(r, t). \tag{3.84}$$

$\hat{E}_{0,j}(r, t)$ is again the source-free 'vacuum' field, while $\hat{E}_{RR,j}(r, t)$ is the radiation reaction field of the detector on itself. $\hat{E}_{S,j}(r, t)$ is the 'external' source field due, for instance, to a thermal source or a laser. Since the 'full' source field consisting of both positive- and negative-frequency parts is, of course, retarded [cf (3.63)], we can write $\hat{E}_{S,j}(r, t) = \hat{F}_j(r, t)\theta(t - r/c)$ and, therefore,

$$\hat{E}_j(r, t) = \hat{E}_{0,j}(r, t) + \hat{E}_{RR,j}(r, t) + \hat{F}_j(r, t)\theta\left(t - \frac{r}{c}\right) \tag{3.85}$$

where r is the distance from the external source to the detector. We are simply indicating explicitly here the retarded nature of the field from the source at a distance r from the detector.

We proceed now by writing

$$\hat{E}_{0,j}(r, t) = \hat{E}_{0,j}^{(+)}(r, t) + \hat{E}_{0,j}^{(-)}(r, t) \tag{3.86}$$

$$\hat{E}_{RR,j}(r, t) = \hat{E}_{RR,j}^{(+)}(r, t) + \hat{E}_{RR,j}^{(-)}(r, t) \tag{3.87}$$

$$\hat{F}_j(r, t) = \hat{F}_j^{(+)}(r, t) + \hat{F}_j^{(-)}(r, t) \tag{3.88}$$

where the positive-frequency parts of the fields $\hat{E}_0$, $\hat{E}_{RR}$, and $\hat{F}$ are defined formally by (3.76) and the negative-frequency parts by hermitian conjugation. The positive- and negative-frequency parts, therefore, have Fourier components $e^{-i\omega t}$ and $e^{i\omega t}$, respectively, where all frequencies ω are, of course, positive. From (3.31), it then follows that only the *normally ordered* combination $\langle \hat{E}_j^{(-)}(r, t)\hat{E}_k^{(+)}(r, t')\rangle$ will contribute to (3.83) for times $t \gg \omega_0^{-1}$, i.e. only this combination of positive- and negative-frequency parts of the field produces energy-conserving transitions:

$$\langle \dot{\hat{\sigma}}_z(t)\rangle \cong \frac{4}{\hbar^2}d_j d_k \operatorname{Re}\int_0^t dt'\ \langle \hat{E}_j^{(-)}(r, t)\hat{E}_k^{(+)}(r, t')\rangle e^{i\omega_0(t'-t)}. \tag{3.89}$$

From (3.86)–(3.88), therefore,

$$\begin{aligned}\langle \dot{\hat{\sigma}}_z(t)\rangle \cong \frac{4}{\hbar^2}d_j d_k \operatorname{Re}\int_0^t dt'\ [&\langle \hat{E}_{RR,j}^{(-)}(r, t)\hat{E}_{RR,j}^{(+)}(r, t')\rangle \\ &+ \theta\left(t' - \frac{r}{c}\right)\langle \hat{E}_{RR,j}^{(-)}(r, t)\hat{F}_j^{(+)}(r, t')\rangle \\ &+ \theta\left(t - \frac{r}{c}\right)\langle \hat{F}_j^{(-)}(r, t)\hat{E}_{RR,k}^{(+)}(r, t')\rangle \\ &+ \theta\left(t - \frac{r}{c}\right)\theta\left(t' - \frac{r}{c}\right)\langle \hat{F}_j^{(-)}(r, t)\hat{F}_j^{(+)}(r, t')\rangle] e^{i\omega_0(t'-t)}\end{aligned} \tag{3.90}$$

where we have used $\hat{E}_{0,j}^{(+)}(r, t)|\psi\rangle = \langle\psi|\hat{E}_{0,j}^{(-)}(r, t) = 0$.

$d \cdot \hat{E}_{RR}(r, t)$ is easily seen to be equal to the $r \to 0$ limit of the second line of equation (3.63). In this limit, $G(\omega r/c) \to 2/3$ and

$$\begin{aligned}d \cdot \hat{E}_{RR}(r, t) &= \frac{d^2}{3\pi^2\epsilon_0 c^3}\int_0^t dt'\ \hat{\sigma}_x(t')\operatorname{Re}\int_0^\infty d\omega\, i\omega^3 e^{i\omega(t'-t)} \\ &= -\frac{d^2}{3\pi\epsilon_0 c^3}\int_0^t dt'\ \hat{\sigma}_x(t')\frac{\partial^3}{\partial t'^3}\delta(t'-t) \\ &\to -\frac{d^2}{6\pi\epsilon_0 c^3}\dddot{\hat{\sigma}}_x(t)\end{aligned} \tag{3.91}$$

where, in the last line, we have once again dropped terms associated with radiative level shifts. $d \cdot E_{RR}$ has the same form as the radiation reaction field of a classical point dipole: the divergent terms we are ignoring here correspond classically to the divergent electromagnetic mass. Now since $\hat{\sigma}_x(t) = \hat{\sigma}(t) + \hat{\sigma}^\dagger(t)$ and $\hat{\sigma}(t) \cong \hat{\sigma}(0)\exp(-i\omega_0 t)$, $\hat{\sigma}^\dagger(t) \cong \hat{\sigma}^\dagger(0)\exp(i\omega_0 t)$, we have

$$\dddot{\hat{\sigma}}_x(t) \cong i\omega_0^3[\hat{\sigma}(t) - \hat{\sigma}^\dagger(t)] \tag{3.92}$$

which means that the positive- and negative-frequency parts of $d_j \hat{E}_{RR,j}(r, t)$ are

$$d_j \hat{E}_{RR,j}^{(+)}(r, t) \cong \frac{i\hbar\gamma}{2}\hat{\sigma}(t) \tag{3.93}$$

$$d_j \hat{E}_{RR,j}^{(-)}(r, t) \cong -\frac{i\hbar\gamma}{2}\hat{\sigma}^\dagger(t). \tag{3.94}$$

Consequently,

$$d_j d_k \langle \hat{E}_{RR,j}^{(-)}(\mathbf{r},t) \hat{E}_{RR,k}^{(+)}(\mathbf{r},t') \rangle = \left(\frac{\hbar\gamma}{2}\right)^2 \langle \hat{\sigma}^\dagger(t)\hat{\sigma}(t') \rangle$$

$$\cong \left(\frac{\hbar\gamma}{2}\right)^2 \langle \hat{\sigma}^\dagger(0)\hat{\sigma}(0) \rangle e^{-i\omega_0(t'-t)} = 0 \quad (3.95)$$

and, likewise, $\langle \hat{E}_{RR,j}^{(-)}(\mathbf{r},t)\hat{F}_j^{(+)}(\mathbf{r},t') \rangle \cong 0$, $\langle \hat{F}_j^{(-)}(\mathbf{r},t)\hat{E}_{RR,k}^{(+)}(\mathbf{r},t') \rangle \cong 0$ under the assumption that the detector atom is only weakly perturbed by the external field. Thence,

$$\langle \dot{\hat{\sigma}}_z(t) \rangle \cong \frac{4}{\hbar^2} d_j d_k \theta\left(t - \frac{r}{c}\right) \mathrm{Re} \int_{r/c}^t dt' \, \langle \hat{F}_j^{(-)}(\mathbf{r},t)\hat{F}_j^{(+)}(\mathbf{r},t') \rangle e^{i\omega_0(t'-t)}. \quad (3.96)$$

There are three points worth stressing about the simple result (3.96). The first is that the appearance of the step function $\theta(t - r/c)$ is *exact*, i.e. the influence of the external field on the atom is properly causal independently of the approximations made in going from the exact expression (3.82) to the approximation (3.96). Second, the appearance of a normally ordered field correlation function *is* an approximation—the consequence of considering 'energy-conserving' transitions at times $t \gg \omega_0^{-1}$ long enough for the detector transition frequency to be resolvable in the sense of the energy–time uncertainty relation. Finally, we note that it is important, for the purpose of exhibiting causality, to include the step function $\theta(t - r/c)$ explicitly in equation (3.85) *before* making the approximation leading to the normally ordered field correlation function: without the step function, the result (3.96) for the excitation rate in second-order perturbation theory is *not* manifestly causal, for the positive- and negative-frequency parts of the field are themselves not retarded, as already noted. These points carry over to the case of a more realistic model for a photodetector which we now consider.

For a system with a ground-state energy E_g and a manifold of excited states $\{E_a\}$, (3.96) generalizes to the following expression for the rate $\dot{P}(t) [= \frac{1}{2}\langle \dot{\hat{\sigma}}_z(t) \rangle]$ at which electrons make transitions out of the ground state:

$$\dot{P}(t) \cong \frac{2}{\hbar^2} \sum_a d_{ag,j} d_{ga,k} R(a)\theta\left(t - \frac{r}{c}\right)$$

$$\times \mathrm{Re} \int_{r/c}^t dt' \, \langle \hat{F}_j^{(-)}(\mathbf{r},t)\hat{F}_k^{(+)}(\mathbf{r},t') \rangle e^{i\omega_{ag}(t'-t)} \quad (3.97)$$

where $\omega_{ag} = (E_a - E_g)/\hbar$ and $R(a)$ gives the probability, which will depend on the physical characteristics of the detector, of actually counting a photoelectron of energy E_a. Following Glauber [90], we define a sensitivity function

$$s_{jk}(\omega) \equiv 2\pi \frac{1}{\hbar^2} \sum_a R(a) d_{ag,j} d_{ga,k} \delta(\omega - \omega_{ag}) \quad (3.98)$$

in terms of which

$$\dot{P}(t) \cong \theta\left(t - \frac{r}{c}\right) \frac{2}{2\pi} \operatorname{Re} \int_{r/c}^{t} dt' \, \langle \hat{F}_{j}^{(-)}(\mathbf{r}, t) \hat{F}_{k}^{(+)}(\mathbf{r}, t') \rangle \int_{-\infty}^{\infty} d\omega \, s_{jk}(\omega) e^{i\omega(t'-t)}.$$

(3.99)

　　Two approximations have been made in the derivation of (3.99): (1) the replacement of $\langle \hat{\sigma}_z(t') \rangle$ by its initial value $\langle \hat{\sigma}_z(0) \rangle$, amounting to standard second-order perturbation theory for the calculation of absorption rates; and (2) the restriction to energy-conserving transitions implicit in the assumption $t \gg \omega_0^{-1}$ made in the two-state formulation. Except for the appearance of the step function $\theta(t - r/c)$ and the time r/c appearing as the lower limit of integration in (3.99), our result is essentially identical to that of Glauber. He defines an ideal broadband detector such that $s_{jk}(\omega) \cong$ constant $\equiv s_{jk}$ for all frequencies ω (or actually for all frequencies within the bandwidth of the external field), so that

$$\int_{-\infty}^{\infty} d\omega \, s_{jk}(\omega) e^{i\omega(t'-t)} = 2\pi s_{jk} \delta(t' - t)$$

(3.100)

and

$$\dot{P}(t) \cong \theta\left(t - \frac{r}{c}\right) s_{jk} \langle \hat{F}_{j}^{(-)}(\mathbf{r}, t) \hat{F}_{k}^{(+)}(\mathbf{r}, t) \rangle.$$

(3.101)

3.2.1　Causality

As already noted, the appearance of the step function $\theta(t - r/c)$ in (3.96)–(3.99) is exact. Moreover, this causal feature is implicit in the original formulation by Glauber [89, 90], since he begins with the *full* electric field operator and proceeds to normally ordered field correlation functions involving $\hat{\mathbf{E}}^{(\pm)}$ under the condition of energy-conserving transitions. In other words, properly retarded effects of the external field on the detector responding to it are implicit in the Glauber theory, although, as a practical matter, one simply writes expressions such as

$$\dot{P}(t) \cong s_{jk} \langle \hat{F}_{j}^{(-)}(\mathbf{r}, t) \hat{F}_{k}^{(+)}(\mathbf{r}, t) \rangle$$

(3.102)

instead of (3.101) and similarly for higher-order correlation functions.

　　This contradicts the claim by Bykov and Tatarskii [95, 96] that the Glauber theory violates causality. Their claim is based on the presumption that the photon-counting rate, for instance, fundamentally involves (3.102) rather than (3.101) and, therefore, that causality is violated owing to the non-retarded character of $\mathbf{F}^{(\pm)}(\mathbf{r}, t)$. As we have shown, causality in the sense meant in this context is trivially implicit in the Glauber theory, beginning as it does with the full, properly retarded electric field. Normally ordered field correlation functions appear at a later stage in the theory and follow from a long-time 'energy conservation' approximation much the same as in the derivation of Fermi's golden rule.

　　The putative violation of causality, according to Bykov and Tatarskii [95, 96], 'requires changing the determination of the correlation functions and, in particular, the velocity of photocounting'. They suggest the replacement of (3.102), for instance, by

$$\dot{P}_{BT}(t) = s_{jk}\langle: \hat{E}_{S,j}(\boldsymbol{r}, t)\hat{E}_{S,k}(\boldsymbol{r}, t) :\rangle \tag{3.103}$$

where the colons denote normal ordering. The full external field operator $E_S(\boldsymbol{r}, t)$ is employed in order to guarantee causality which, as we have shown, is, in fact, already implicit in the conventional theory. The normal ordering of the field product in (3.103) is employed 'to avoid an infinite contribution of vacuum fluctuations'.

The replacement of (3.82) by (3.89) involves a *rotating-wave approximation* (RWA), i.e. the approximation of dropping energy-non-conserving terms, which, of course, is not an essential part of the standard theory. Retention of non-RWA terms leads to expressions such as

$$\dot{P}(t) = s_{jk}\langle\hat{E}_{S,j}(\boldsymbol{r}, t)\hat{E}_{S,k}(\boldsymbol{r}, t)\rangle \tag{3.104}$$

which differs from (3.103) in that it includes an antinormally ordered term $\langle\hat{E}_{S,j}^{(+)}(\boldsymbol{r}, t)\hat{E}_{S,k}^{(-)}(\boldsymbol{r}, t)\rangle$. This term does not arise in a formulation of photodetection theory that accounts for dissipation as well as fluctuations in the response of the detector [85]. In other words, the modification of standard photon-counting theory suggested by Bykov and Tatarskii, stemming from the presumption that the standard theory violates causality, amounts only to dropping the RWA, and does not lead to anything essentially new.

The *ab initio* use of the RWA in the Hamiltonian is, thus, seen to be the basis in effect not only of Hegerfeldt's criticism of previous work on the Fermi model but also the criticism of the Glauber theory of photodetection by Bykov and Tatarskii. To underscore this point, recall that the restriction to energy-conserving processes in the Fermi model means that the terms $\hat{a}_{k\lambda}\hat{\sigma}$ and $\hat{a}_{k\lambda}^{\dagger}\hat{\sigma}^{\dagger}$ in the Hamiltonian are ignored. To wit, the interaction term in the Hamiltonian (3.11) is effectively replaced by

$$\hat{H}_{\text{Int}}^{(\text{RWA})} = -i\hbar \sum_{j=A,B} \sum_{k\lambda} [C_{jk\lambda}\hat{\sigma}_j^{\dagger}\hat{a}_{k\lambda} - C_{jk\lambda}^*\hat{a}_{k\lambda}^{\dagger}\hat{\sigma}_j]. \tag{3.105}$$

If we take (3.78) to be the positive-frequency part of the field, this can be written as

$$\hat{H}_{\text{Int}}^{(\text{RWA})} = -\sum_{j=A,B} [\hat{\sigma}_j^{\dagger}(t)\boldsymbol{d}_j \cdot \hat{\boldsymbol{E}}^{(+)}(\boldsymbol{r}_j, t) + \boldsymbol{d}_j \cdot \hat{\boldsymbol{E}}^{(-)}(\boldsymbol{r}_j, t)\hat{\sigma}_j(t)]. \tag{3.106}$$

With such an interaction, we obviously do not obtain (3.82) but rather an expression involving the *non-retarded* fields $\hat{\boldsymbol{E}}^{(\pm)}(\boldsymbol{r}, t)$, each of which involves $\int_0^{\infty} d\omega(\ldots)$ rather than $\int_{-\infty}^{\infty} d\omega(\ldots)$. In neither the Fermi model nor our model for photodetection is there any violation of causality when one works from the start with the complete (non-RWA) Hamiltonian including the possibility of energy-non-conserving (virtual) transitions.

Our derivation of $\dot{P}(t)$ has led to normal ordering as a consequence of the RWA. If we proceed directly from the two-state result (3.82) to its multilevel generalization without any RWA, we obtain instead of (3.97)

$$
\begin{aligned}
\dot{P}(t) &\cong \frac{2}{\hbar^2} \sum_a d_{ag,j} d_{ga,k} R(a) \, \text{Re} \int_{r/c}^{t} dt' \, \langle \hat{E}_j(r,t) \hat{E}_k(r,t') \rangle e^{i\omega_{ag}(t'-t)} \\
&= \frac{2}{2\pi} \text{Re} \int_{r/c}^{t} dt' \, \langle \hat{E}_j(r,t) \hat{E}_k(r,t') \rangle \int_{-\infty}^{\infty} d\omega \, s_{jk}(\omega) e^{i\omega(t'-t)} \\
&\rightarrow s_{jk} \langle \hat{E}_j(r,t) \hat{E}_k(r,t) \rangle
\end{aligned}
\tag{3.107}
$$

where, in the last step, we have gone to the limit of ideal broadband detection. Bykov and Tatarskii introduce normal ordering as in (3.103) to eliminate from (3.107) the infinite quantity $\langle \hat{E}_j^{(+)}(r,t) \hat{E}_k^{(-)}(r,t) \rangle$, the infinity arising from the vacuum field $\hat{E}_{0,j}(r,t)$.

Care must be exercised in applying the non–RWA expression (3.107). For one thing, the vacuum field has an infinite bandwidth and, therefore, the limit of idealized broadband detection is inapplicable. Just as important is the fact that, in dealing with vacuum field contributions, we cannot, in general, ignore radiation reaction, i.e. we cannot completely separate the vacuum fluctuations driving the detector from its own internal dissipation [8]. Indeed equation (3.107) as it stands does not distinguish among the vacuum, radiation reaction, or external fields due to the sources causing the photoabsorption. By proceeding more carefully to an expression for the absorption rate without any RWA, and accounting for radiation reaction, it has been shown that the infinite term that Bykov and Tatarskii propose to eliminate by normal ordering is present even with normal ordering and, moreover, is present regardless of what ordering is employed [85]. More importantly, the term in question has been shown to be without physical significance for photodetection [85].

It is straightforward to generalize the preceding results for photon-counting theory to higher-order field correlation functions. For the rate at which photons are counted jointly at two identical broadband detectors at (r_1, t_1) and (r_2, t_2), for instance, we obtain

$$
\begin{aligned}
R(r_1, t_1; r_2, t_2) = s_{jm} s_{k\ell} \theta \left(t_1 - \frac{r_1}{c} \right) \theta \left(t_2 - \frac{r_2}{c} \right) \langle \hat{E}_j^{(-)}(r_1, t_1) \hat{E}_k^{(-)}(r_2, t_2) \\
\times \hat{E}_\ell^{(+)}(r_2, t_2) \hat{E}_m^{(+)}(r_1, t_1) \rangle
\end{aligned}
\tag{3.108}
$$

where r_1 and r_2 are the distances from the (point) source to r_1 and r_2. Once again causality is manifested explicitly in the appearance of the step functions $\theta(t_j - r_j/c)$. The RWA has been employed in writing this result and again it is the RWA that leads naturally to a normally ordered field correlation function.

To summarize: the RWA can lead to apparent violations of 'causality', i.e. retardation, in theoretical analyses involving the propagation of light. Such violations are artificial and are eliminated by working from the start with a Hamiltonian that accounts for non-RWA terms associated primarily with virtual transitions and level shifts. Causality is implicit in the standard photodetection theory as originally formulated by Glauber. The non-RWA contributions obtained by Bykov and Tatarskii are, in principle, already contained in the standard theory.

Our results concerning the RWA are not inconsistent with those of De Haan [82] or Compagno *et al* [83]. The latter authors, for instance, show that if non-RWA, 'energy-non-conserving' terms are dropped at the outset from the Hamiltonian, then the field operators calculated from the approximate Hamiltonian are not retarded. Our use of the term 'RWA' here is somewhat different in that it refers to the omission of counter-rotating terms only after calculating the field based on the full Hamiltonian including energy-non-conserving terms. In other words, the neglect of counter-rotating terms is simply made at a later point in the calculation. Regardless of the point in the calculations where counter-rotating terms are dropped, such terms are fundamentally necessary in our approach, as in the work of De Haan and Compagno *et al*, for the formal demonstration of causality. We have followed an approach that arguably shows most clearly how a trivial modification of standard photodetection theory exhibits its correctly causal character.

3.3 Microscopic approach to refractive index and group velocity

In the Fermi model, it is assumed that the two atoms are in free space, and, in the model for photodetection just discussed, the light is assumed to propagate in vacuum to the detector. We now consider a model in which the light from an initially excited atom propagates in a dielectric medium to a detector [97]. Under certain approximations, the primary one being that the atoms constituting the dielectric stay with high probability in their ground states, this model can be solved essentially exactly. Such a model is conceptually attractive for our purposes because: (1) it is completely quantum-mechanical; (2) the total electric field operator due to the source atom and all the atoms of the medium is obtained *self-consistently*; (3) it is shown that the radiation from the source atom can register a 'click' at the detector sooner, on average, than it could if there were no medium between the source atom and the detector, i.e. there is an observable 'superluminal' effect; and (4) the classical theorem that the front velocity cannot exceed c has a simple quantum counterpart.

The source atom has transition frequency and electric dipole moment ω_0 and d, respectively, and is located at a point ($r = 0$) outside the dielectric. There are N_T identical atoms making up the dielectric. They have transition frequency ω_d and transition dipole moment μ, respectively, and are located at points r_j. All the atoms are coupled to the quantized electromagnetic field in the electric dipole approximation. The Hamiltonian is

$$
H = \frac{1}{2}\hbar\omega_0\hat{\sigma}_z + \sum_{j=1}^{N_T}\frac{1}{2}\hbar\omega_d\hat{\sigma}_{zj} + \sum_{k\lambda}\hbar\omega_k\hat{a}_{k\lambda}^\dagger\hat{a}_{k\lambda} - i\hbar\frac{d}{\mu}\sum_{k\lambda}D_{k\lambda}[\hat{a}_{k\lambda} - \hat{a}_{k\lambda}^\dagger]\hat{\sigma}_x
$$

$$
- i\hbar\sum_{j=1}^{N_T}\sum_{k\lambda}D_{k\lambda}[\hat{a}_{k\lambda}e^{ik\cdot r_j} - \hat{a}_{k\lambda}^\dagger e^{-ik\cdot r_j}]\hat{\sigma}_{xj}. \tag{3.109}
$$

The coupling constant $D_{k\lambda} = \mu(\omega_k/2\epsilon_0\hbar V)^{1/2}$, where V is again the quantization volume.

It will be convenient for our purposes to simplify the model by restricting the field modes to plane waves propagating along a single (z) direction and with a single polarization, so that $k, \lambda \to k$. (In reality, of course, the field from the source atom will have a dipole radiation pattern. This can be dealt with easily enough [10] but it only complicates the results without affecting our conclusions.) Then, using the formal solution of the Heisenberg equation of motion for the field operators $\hat{a}_k(t)$, we obtain

$$
\hat{E}(z, t) = i\sum_k\left(\frac{\hbar\omega_k}{2\epsilon_0 V}\right)^{1/2}\hat{a}_k(t)e^{i\omega_k z/c} + \text{h.c.}
$$

$$
= \hat{E}_0^{(+)}(z, t) + i\sum_k\left(\frac{d\omega_k}{2\epsilon_0 V}\right)e^{ikz}\int_0^t dt'\,\hat{\sigma}_x(t')e^{i\omega_k(t'-t)}
$$

$$
+ i\sum_k\sum_{j=1}^{N_T}\left(\frac{\mu\omega_k}{2\epsilon_0 V}\right)e^{ik(z-z_j)}\int_0^t dt'\,\hat{\sigma}_{xj}(t')e^{i\omega_k(t'-t)} + \text{h.c.} \tag{3.110}
$$

$$
\hat{E}_0^{(+)}(z, t) = i\sum_k\left(\frac{\hbar\omega_k}{2\epsilon_0 V}\right)^{1/2}\hat{a}_k(0)e^{-i(\omega_k t - kz)} \tag{3.111}
$$

for the electric field operator at any point z. Now the sum over field modes

$$
2\text{Re}\left[i\sum_k\left(\frac{\omega_k}{2\epsilon_0 V}\right)e^{ikz}e^{i\omega_k(t'-t)}\right] \to \text{Re}\,\frac{i}{AL}\frac{L}{4\pi\epsilon_0 c}\int_{-\infty}^\infty d\omega\,(2\pi\omega)e^{i\omega(t'-t+z/c)}
$$

$$
= \frac{1}{2\epsilon_0 Ac}\frac{\partial}{\partial t'}\delta(t' - t + z/c) \tag{3.112}
$$

where A is the cross-sectional area of our quantization volume $V = AL$, which we have allowed to become infinite by taking $L \to \infty$. Thus, for $z > 0$,

$$
\hat{E}(z, t) = \hat{E}_0(z, t) - \frac{d}{2\epsilon_0 Ac}\dot{\hat{\sigma}}_x(t - z/c)\theta(t - z/c)
$$

$$
- \frac{\mu}{2\epsilon_0 Ac}\sum_j\dot{\hat{\sigma}}_{xj}\left(t - \frac{|z - z_j|}{c}\right)\theta\left(t - \frac{|z - z_j|}{c}\right). \tag{3.113}
$$

It is convenient to work with Fourier-transformed operators defined by writing

$$\hat{E}(z, t) = \int_{-\infty}^{\infty} d\omega\, \tilde{E}(z, \omega) e^{-i\omega t} \tag{3.114}$$

$$\hat{\sigma}_x(t) = \int_{-\infty}^{\infty} d\omega\, \tilde{\hat{\sigma}}_x(\omega) e^{-i\omega t} \tag{3.115}$$

$$\hat{\sigma}_{xj}(t) = \int_{-\infty}^{\infty} d\omega\, \tilde{\hat{\sigma}}_{xj}(\omega) e^{-i\omega t}. \tag{3.116}$$

Then equation (3.113) implies

$$
\begin{aligned}
\tilde{E}(z, \omega) &= \tilde{E}_0(z, \omega) + \frac{id}{2\epsilon_0 A c}\omega e^{i\omega z/c}\tilde{\hat{\sigma}}_x(\omega) + \frac{i\mu}{2\epsilon_0 A c}\omega \sum_j e^{i\omega|z - z_j|/c}\tilde{\hat{\sigma}}_{xj}(\omega) \\
&\equiv \hat{F}(z, \omega) + \frac{i\mu}{2\epsilon_0 A c}\omega \sum_j \tilde{\hat{\sigma}}_{xj}(\omega)e^{i\omega|z - z_j|/c} \\
&\to \hat{F}(z, \omega) + i\mu \frac{\omega}{2\epsilon_0 c} N \int_{z_0}^{\infty} dz'\, \tilde{\hat{\sigma}}_x(z', \omega)e^{i\omega|z - z'|/c}
\end{aligned} \tag{3.117}
$$

where the atoms of the dielectric are assumed to be uniformly distributed with density N and to occupy the half-space $z > z_0$.

From the Hamiltonian (3.109), it follows from the Heisenberg equations of motion and the commutation relations for the two-state operators that the $\hat{\sigma}_{xj}$'s satisfy

$$\ddot{\hat{\sigma}}_{xj} + 2\gamma\dot{\hat{\sigma}}_{xj} + \omega_\mu^2 \hat{\sigma}_{xj} = -\frac{\mu\omega_d}{2\epsilon_0 \hbar}\hat{E}(z_j, t)\hat{\sigma}_{zj} \cong \frac{\mu\omega_d}{2\epsilon_0 \hbar}\hat{E}(z_j, t) \tag{3.118}$$

where, in the last step, we have replaced the operator $\hat{\sigma}_{zj}$ by -1 under the assumption that the atoms making up the dielectric remain with high probability in their ground states. 2γ is the rate of spontaneous emission of the dielectric atoms, which, in the present model, undergo no other relaxation. Equation (3.118) implies

$$\mu\tilde{\hat{\sigma}}_x(z', \omega) = \alpha(\omega)\tilde{E}(z', \omega) \tag{3.119}$$

for the induced dipole moment at frequency ω, where

$$\alpha(\omega) = \frac{\mu^2 \omega_d/2\epsilon_0 \hbar}{\omega_d^2 - \omega^2 - 2i\gamma\omega} \tag{3.120}$$

is the polarizability. Then

$$
\begin{aligned}
\tilde{E}(z, \omega) &= \hat{f}(\omega)e^{i\omega z/c} + i\frac{\omega}{2\epsilon_0 c}N\alpha(\omega)\int_{z_0}^{z} dz'\, \tilde{E}(z', \omega)e^{i\omega(z - z')/c} \\
&\quad + i\frac{\omega}{2\epsilon_0 c}N\alpha(\omega)\int_{z}^{\infty} dz'\, \tilde{E}(z', \omega)e^{i\omega(z' - z)/c}
\end{aligned} \tag{3.121}
$$

where we have written $\hat{F}(z, \omega)$ as $\hat{f}(\omega)e^{i\omega z/c}$. This integral equation is basically just a statement of the superposition principle: the total electric field at any point in the medium is the sum of the vacuum field, the field from the source atom, and the field from all the dipoles in the medium, these dipoles being induced by the same total electric field. To solve this equation, we write

$$\tilde{\hat{E}}(z, \omega) = \hat{g}(\omega)e^{in(\omega)\omega z/c} \tag{3.122}$$

and determine $\hat{g}(\omega)$ and $n(\omega)$ by substitution. The result of the algebra is

$$\tilde{\hat{E}}(z, \omega) = \frac{2}{n(\omega) + 1} \hat{f}(\omega)e^{in(\omega)(z-z_0)\omega/c}e^{i\omega z_0/c} \tag{3.123}$$

for $z > z_0$, where $n(\omega)$ is the complex refractive index: $n^2(\omega) = 1 + (N/\epsilon_0)\alpha(\omega)$.

Combining these results, we obtain the following expression for the electric field operator at (z, t) inside the medium:

$$\hat{E}(z, t) = \hat{E}_0(z, t) + \frac{id}{2\epsilon_0 Ac} \int_{-\infty}^{\infty} dt' \, \hat{\sigma}_x(t') \int_{-\infty}^{\infty} \frac{d\omega \, \omega}{n(\omega) + 1} e^{i\omega[t'-t+n(\omega)z/c]} \tag{3.124}$$

where $\hat{E}_0(z, t)$ is the source-free (vacuum) field inside the medium and, for simplicity, we take $z_0 = 0$ for the position of the initially excited source atom.

It may be worthwhile to note that, aside from the approximation that the host atoms remain with probability one in their ground states, the expression (3.124) is exact for the case of a dilute dielectric medium. We began with the Hamiltonian (3.109) for the field and the atoms *in vacuum* and obtained (a) the relation between the refractive index and the polarizability and (b) the electric field operator (3.124), by solving *exactly* the self-consistent integral equation (3.121).

Now the probability that the radiation from the source atom at $z = 0$ will register a count at an ideal broadband detector at $z > 0$ at time t involves an expectation value over the initial state $|\psi\rangle$ in which all the atoms of the dielectric are in their ground states, the field is in the vacuum state, and the state of the source atom is arbitrary. The photon-counting rate is proportional to

$$R(z, t) = \langle \hat{E}^{(-)}(z, t)\hat{E}^{(+)}(z, t)\rangle$$
$$= \int_{-\infty}^{\infty} dt' \int_{-\infty}^{\infty} dt'' \, \langle \hat{\sigma}_x(t')\hat{\sigma}_x(t'')\rangle \int_{-\infty}^{\infty} \frac{d\omega \, \omega}{n^*(\omega) + 1} e^{-i\omega[t'-t+n^*(\omega)z/c]}$$
$$\times \int_{-\infty}^{\infty} \frac{d\omega' \, \omega'}{n(\omega') + 1} e^{i\omega'[t''-t+n(\omega')z/c]}. \tag{3.125}$$

In terms of the two-state lowering and raising operators $\hat{\sigma}$ and $\hat{\sigma}^\dagger$, respectively,

$$\langle \hat{\sigma}_x(t')\hat{\sigma}_x(t'')\rangle = \langle \hat{\sigma}(t')\hat{\sigma}(t'')\rangle + \langle \hat{\sigma}^\dagger(t')\hat{\sigma}(t'')\rangle$$
$$+ \langle \hat{\sigma}(t')\hat{\sigma}^\dagger(t'')\rangle + \langle \hat{\sigma}^\dagger(t')\hat{\sigma}^\dagger(t'')\rangle. \tag{3.126}$$

$\hat{\sigma}(t)$ and $\hat{\sigma}^\dagger(t)$ vary in time predominantly as $e^{-i\omega_0 t}$ and $e^{i\omega_0 t}$, respectively. Therefore, only the second term on the right-hand side of (3.126) will contribute to (3.125) over the time scales of interest (i.e. 'energy-conserving' times, long compared with $1/\omega_0$). The rate at which radiation emitted by the source atom will register counts at the detector at z is, therefore, proportional in this approximation to

$$R(z, t) = \int_{-\infty}^{\infty} dt' \int_{-\infty}^{\infty} dt'' \, \langle \hat{\sigma}^\dagger(t')\hat{\sigma}(t'') \rangle \int_{-\infty}^{\infty} \frac{d\omega \, \omega}{n^*(\omega) + 1} e^{-i\omega[t'-t+n^*(\omega)z/c]}$$
$$\times \int_{-\infty}^{\infty} \frac{d\omega' \, \omega'}{n(\omega') + 1} e^{i\omega'[t''-t+n(\omega')z/c]}. \tag{3.127}$$

Suppose the source atom is excited by a time-dependent mechanism, e.g. by a resonant laser pulse $\hat{E}(t) = \text{Re}[\hat{F}_0(t)\exp(-i\omega_0 t)]$. If the Rabi frequency is small compared with the radiative decay rate and the pulse duration is long compared with the radiative lifetime, then

$$\langle \hat{\sigma}^\dagger(t')\hat{\sigma}(t'') \rangle \propto \langle \hat{F}_0(t')\hat{F}_0(t'') \rangle e^{i\omega_0(t'-t'')} = f_0(t')f_0^*(t'')e^{i\omega_0(t'-t'')} \tag{3.128}$$

and

$$R(z, t) \propto \left| \int_{-\infty}^{\infty} dt' \, e^{-i\omega_0 t'} f_0(t') \int_{-\infty}^{\infty} \frac{d\omega \, \omega}{n(\omega) + 1} e^{i\omega[t' \; t \; | \; n(\omega)z/c]} \right|^2$$
$$\propto \left| \int_{-\infty}^{\infty} dt' \, a(t') \int_{-\infty}^{\infty} \frac{d\omega \, \omega}{n(\omega) + 1} e^{i\omega[t'-t+n(\omega)z/c]} \right|^2 \tag{3.129}$$

where $a(t)$ is the probability amplitude that the source atom is in the upper state at time t and we have assumed that the source atom is initially in the lower state $[\hat{\sigma}(-\infty)|\psi\rangle = 0]$. If $\hat{F}_0(t) = \hat{C}\exp(-t^2/2\tau^2)$, for instance,

$$R(z, t) \propto \left| \int_{-\infty}^{\infty} \frac{d\omega \, \omega}{n(\omega) + 1} e^{-i\omega t} e^{i\omega n(\omega)z/c} e^{-\frac{1}{2}(\omega-\omega_0)^2\tau^2} \right|^2. \tag{3.130}$$

Except for the factor $\omega/[n(\omega) + 1]$, the integral here is the same as that appearing in the work of Garrett and McCumber [37] discussed in section 2.3.

The main contribution to the integral over t' in equation (3.129) comes from frequencies ω near ω_0. Write $n(\omega) = n_R(\omega) + in_I(\omega)$, where $n_R(\omega)$ and $n_I(\omega)$ are real. Then

$$\int_{-\infty}^{\infty} dt' \, f_0(t') \int_{-\infty}^{\infty} \frac{d\omega \, \omega}{n(\omega) + 1} e^{i(\omega-\omega_0)t'} e^{-i\omega[t-n(\omega)z/c]}$$
$$\approx \frac{\omega_0}{n(\omega_0) + 1} e^{-\omega_0 n_I(\omega_0)z/c} \int_{-\infty}^{\infty} dt' \, f_0(t') e^{-i\omega_0 t'} \int_{-\infty}^{\infty} d\omega \, e^{i\omega[t'-t+n_R(\omega)z/c]}$$
$$\tag{3.131}$$

and the double integral is approximately

$$\int_{-\infty}^{\infty} dt' \, f_0(t') e^{-i\omega_0 t'} \int_{-\infty}^{\infty} d\Delta \, e^{i(\omega_0+\Delta)(t'-t)} e^{i(\omega_0+\Delta)[n_R(\omega_0)+\Delta n_R'(\omega_0)]z/c}$$

$$\approx e^{-i\omega_0[t-n_R(\omega_0)z/c]} \int_{-\infty}^{\infty} dt' \, f_0(t') \int_{-\infty}^{\infty} d\Delta \, e^{i\Delta(t'-t)} e^{i\Delta[n_R(\omega_0)+\omega_0 n_R'(\omega_0)]z/c}$$

$$= 2\pi e^{-i\omega_0[t-n_R(\omega_0)z/c]} f_0(t-z/v_g) \tag{3.132}$$

where $v_g = c/[d(\omega n_R)/d\omega]_{\omega=\omega_0}$ is the group velocity and it is assumed that group velocity dispersion is negligible. Then, from (3.129),

$$R(z,t) \propto e^{-2\omega_0 n_I(\omega_0)z/c} P(t-z/v_g) \tag{3.133}$$

where $P(t) = |a(t)|^2$ is the probability at time t that the source atom is excited.

If the group velocity exceeds c near the resonance frequency of the absorbing medium, the result (3.133) implies that the peak probability that a single photon is counted at z can occur sooner than it could if there were no medium between the source atom and the detector, albeit the probability that the photon is counted is reduced by the factor $\exp(-\alpha z)$, where $\alpha = 2\omega_0 n_I(\omega_0)/c$ is the absorption coefficient. Note also that the argument $t - z/v_g$ does not have to be positive, nor does $t - z/c$ have to be positive in the limiting case where the medium of propagation is the vacuum: because of our assumption that the pumping mechanism for exciting the atom is on at all times, there is a finite probability at all times that a photon count will be recorded at the detector. However, if a peak excitation probability for the atom occurs at time T, say, then the photon-counting rate corresponding to this peak will itself peak at time $T + z/v_g$ when the photon propagates in the medium, compared with the *later* time $T + z/c$ at which it would peak were it propagating in vacuum.

Consider instead the case where the source atom is suddenly put in its excited state at $t = 0$, having been in its lower state prior to that time. Then, since the lowering operator $\hat{\sigma}(t'')$ acting on the lower state gives 0 for all times t'' earlier than $t = 0$, we can replace (3.127) by

$$R(z,t) = \int_{-\infty}^{\infty} dt' \int_0^{\infty} dt'' \, \langle \hat{\sigma}^\dagger(t') \hat{\sigma}(t'') \rangle \int_{-\infty}^{\infty} \frac{d\omega \, \omega}{n^*(\omega)+1} e^{-i\omega[t'-t+n^*(\omega)z/c]}$$

$$\times \int_{-\infty}^{\infty} \frac{d\omega' \, \omega'}{n(\omega')+1} e^{i\omega'[t''-t+n(\omega')z/c]}. \tag{3.134}$$

For our purposes, we need only observe a general property of the last integral. Namely, since $n(\omega') + 1$ only has poles in the lower half of the complex ω' plane, and $n(\omega') \to 1$ as $\omega' \to \infty$, the integral over ω' has to vanish unless $t > t'' + z/c$. Consequently, since the integration over t'' starts at $t'' = 0$, $P(z,t)$ vanishes unless $t > z/c$. In other words, a suddenly excited atom cannot cause a photon to be counted at z before the time it takes for light to propagate in vacuum from the atom to the detector.

This is the analog of the classical result that a sharp wavefront cannot propagate faster than c. In our quantum-mechanical model, however, we cannot create a sharp 'front' of photon probability because the radiative lifetime of the excited state is finite and the emitted radiation must have a finite spectral width. And, of course, the emitted 'pulse' in our model represents a probability distribution for finding a single photon. Nevertheless, the limiting case of an atom excited by a delta-function pulse, and with a very short radiative lifetime, provides a quantum analog of the idealized sharp classical wavefront. Our result is then analogous to the classical Sommerfeld–Brillouin proof of Einstein causality, i.e. the proof that the front velocity cannot exceed c.

Regarding the model employed here, we note that the assumption that the atoms of the dielectric remain with high probability in the ground state renders our treatment of propagation very similar to a purely classical one. Our treatment of the source atom, of course, is completely quantum mechanical. It might also be noted that the spontaneous emission rate of the source atom is modified by the presence of the dielectric and can, in general, be either greater than or smaller than the free-space spontaneous emission rate[8].

3.4 EPR correlations and causality

A quantum field $\hat{\phi}(r, t)$ can create or annihilate particles when it acts on a state $|\psi\rangle$. If $c^2(t - t')^2 - (r - r')^2 < 0$, the creation and annihilation events at (r, t) and (r', t') should not affect one another. Thus, $\hat{\phi}(r, t)\hat{\phi}(r', t')|\psi\rangle = \hat{\phi}(r', t')\hat{\phi}(r, t)|\psi\rangle$, i.e. the commutator $[\hat{\phi}(r, t), \hat{\phi}(r', t')] = 0$ for spacelike separations[9].

Any physical process that would violate this condition would be inconsistent with causality and must, therefore, be forbidden. Obviously, this is a rather formal condition and, as in the classical case, it is instructive to consider explicit examples where the requirement of causality puts constraints on what physical processes are possible.

Part of the mystique surrounding Einstein–Podolsky–Rosen (EPR) correlations stems from what Einstein [100] famously called the 'spooky action at a distance' they seem to imply[10]. Consider the example of two photons in the

---

[8] For recent work on the modification of the spontaneous emission rate of an atom in a dielectric, see [98] and the papers cited therein.

[9] Related commutation relations for the free electromagnetic field were used by Bohr and Rosenfeld in several (difficult) papers in which they considered among other things the role of the finite velocity of light in field measurements involving two test bodies; see [99] and the papers cited therein. They showed that the quantum-mechanical uncertainty relations for the test bodies imply uncertainty relations for the field that are just those derivable from the field commutation relations.

[10] In this connection it is frequently said, especially in the popular media, that Einstein would not have expected the experimental results supporting Bell's theorem. A different view is expressed in the appendix to this chapter.

'entangled' polarization state

$$|\Psi\rangle = \frac{1}{\sqrt{2}}(|H_A\rangle|V_B\rangle - |V_A\rangle|H_B\rangle)) \qquad (3.135)$$

where H and V denote 'horizontal' and 'vertical' polarization. Alice and Bob each observe one member of the photon pair: if Alice measures H (V) polarization, then the state of Bob's photon is immediately reduced to V (H). This sort of 'action at a distance' cannot be used for instantaneous *communication of information*, or signalling, simply because Alice cannot *choose* whether to 'send' a V or an H to Bob; she has a 50/50 chance of getting an H or a V herself.

She does, of course, have a choice as to polarization basis. She can, for example, use the circular polarization states $|R\rangle = (1/\sqrt{2})(|H\rangle - i|V\rangle)$ and $|L\rangle = -(1/\sqrt{2})(|H\rangle + i|V\rangle)$ instead of $|H\rangle$ and $|V\rangle$. In this basis, we can write the state (3.135) equivalently as

$$|\Psi\rangle = \frac{1}{\sqrt{2}}(|R_A\rangle|R_B\rangle - |L_A\rangle|L_B\rangle)). \qquad (3.136)$$

Thus, if Alice chooses to work in the H, V basis, her measurement reduces Bob's photon state to V or H, whereas if she chooses to work in the R, L basis, her measurement reduces Bob's photon state to R or L. However, her choice of basis cannot serve to send information to Bob because, given a single photon, Bob cannot distinguish between linear or circular polarization. That is, there is no device that can measure the polarization parameters of a single photon [101]. If such a device were possible, EPR correlations could be used for superluminal (instantaneous) communication.

Consider, alternatively, the density matrix ρ_B describing Bob's photon. Tracing over Alice's states, we obtain

$$\rho_B = \text{Tr}_A[|\Psi\rangle\langle\Psi|] = (1/2(|H_B\rangle\langle H_B| + |V_B\rangle\langle V_B|) \qquad (3.137)$$

in the H, V basis and

$$\rho_B = (1/2)(|R_B\rangle\langle R_B| + |L_B\rangle\langle L_B|) \qquad (3.138)$$

in the R, L basis. These are just different ways of writing the same (reduced) density matrix, and so *Alice's choice of whether to measure linear or circular polarization cannot affect Bob's measurements and, therefore, cannot be used to transmit information.*

3.5　No cloning

But if Bob has *many* particles in the same state he can perform measurements to determine that state. For example, he can let N ($\gg 1$) photons pass through polarization-dependent beam splitters and thereby determine with a high degree of accuracy whether a measurement of linear or circular polarization was made on Alice's EPR-correlated photon. In other words, if the single photon at Bob's end can be sent through an amplifier to produce a large number of photons in the same (arbitrary) state, it would be possible for Alice to communicate superluminally to Bob whether she measured linear or circular polarization [102].

This superluminal communication scheme fails because photons cannot be perfectly copied: the amplifier will produce photons with polarization different from the incident photon by spontaneous emission—we cannot have stimulated emission without spontaneous emission and, therefore, we cannot amplify an arbitrary polarization [103]. Much more generally, quantum theory does not permit the cloning of a single quantum [104, 105].

It is worth noting that this conclusion does not require that an optical amplifier be polarization-dependent. It is possible for an amplifier to produce a final state that is independent of the polarization of the incident photon but such an amplifier does not violate the no-cloning theorem: 'the essential element that prevents cloning is here seen to be spontaneous emission, rather than any dependence of amplifier gain on polarization' [106]. Note also that suppression of spontaneous emission by cavity QED or other methods will not help, since stimulated emission will be likewise suppressed.

A proof of the no-cloning theorem goes as follows [104]. Suppose there is an amplifier that transforms a polarization state $|s\rangle$ of a photon into a state $|ss\rangle$ in which there are two photons in the same state, i.e. the amplifier clones a photon. Thus, $A_0\rangle|s\rangle \rightarrow |A_s\rangle|ss\rangle$, where $|A_0\rangle$ and $|A_s\rangle$ are, respectively, the initial and final states of the amplifier. In particular, for vertically and horizontally polarized photons,

$$|A_0\rangle|V\rangle \rightarrow |A_V\rangle|VV\rangle \qquad (3.139)$$

$$|A_0\rangle|H\rangle \rightarrow |A_H\rangle|HH\rangle \qquad (3.140)$$

and, according to quantum theory, such transformations are linear, so that

$$|A_0\rangle[\alpha|V\rangle + \beta|H\rangle] \rightarrow \alpha|A_V\rangle|VV\rangle + \beta|A_H\rangle|HH\rangle \qquad (3.141)$$

for an arbitrary superposition state $\alpha|V\rangle + \beta|H\rangle$ of the original photon. But, with our hypothetical cloning device, the transformed polarization state should be that in which both photons are described by the polarization state $\alpha|V\rangle + \beta|H\rangle$ and this state cannot be described by (3.141), regardless of the possible final states $|A_H\rangle$ and $|A_V\rangle$ of the amplifier. Simply put, if two states can be cloned, a superposition of them cannot. The linearity of quantum theory, therefore, precludes the replication of an arbitrary polarization state and, obviously, this conclusion follows regardless of the specific quantum system: *a single quantum cannot be cloned* [104].

The word 'arbitrary' here is crucial. If, for instance, we know beforehand that the photon we want to clone is right-hand circularly polarized (RHC), we could prepare the atoms of a laser amplifier by optical pumping so that they radiate only RHC photons by stimulated and spontaneous emission. But, for an *arbitrary* initial state, this cannot be done because we have no *a priori* information to tell us how to prepare the amplifier and spontaneous emission is as likely as not to produce an additional photon in a polarization state that is orthogonal to that we wish to clone.

Glauber [107] has discussed the impossibility of superluminal communication using laser amplifiers and has shown that amplification does not allow Bob to ascertain whether Alice measures linear or circular polarization. In connection with the no-cloning theorem, he remarks that the proof just reviewed 'takes the definition of cloning quite literally, requiring all photons to be identical, and places the further restriction that the initially pure one-photon state has to always remain pure. It is not related, therefore, to the action of any real amplifier.'

Recall that the probability of stimulated emission into a given mode is q times the probability of spontaneous emission into that mode, where q is the expectation value of the photon number. For cloning one member of an EPR pair, $q = 1$ and stimulated emission is as likely as spontaneous emission. It is, therefore, more likely than not that the incoming photon is replicated: stimulated emission will produce a photon with the same polarization as the incoming photon and spontaneous emission is as likely to produce a photon with that polarization as it is to produce an orthogonally polarized photon. There is, therefore, a probability of 2/3 that the additional photon has the same polarization as the incoming photon. The cloning *fidelity* $\mathcal{F}$ is defined as the probability that an outgoing photon is in the same state as the incoming photon:

$$\mathcal{F} = \tfrac{2}{3} \times 1 + \tfrac{1}{3} \times \tfrac{1}{2} = \tfrac{5}{6}. \tag{3.142}$$

Imperfect quantum cloning i.e. cloning with fidelity < 1, has been of interest in connection with quantum information studies. In the present context, it is worth noting that Gisin [108] has obtained a bound on the fidelity of quantum cloning by imposing the requirement of no superluminal communication and finds that the maximum allowed fidelity is equal to the optimal quantum cloning fidelity (3.142) obtained by Bužek and Hillery [109][11]. The fidelity (3.142) is *exactly* the maximum fidelity that is possible without violating Einstein causality. Gisin notes: '[O]nce again, quantum mechanics is right at the border line of contradicting relativity, but does not cross it'.

It is also noteworthy that the optimal fidelity (3.142) has been closely approximated in parametric down-conversion experiments (section 2.4.3) [110]. In these experiments, loosely speaking, incident photons from a laser are split in a nonlinear crystal into two photons that are polarization-correlated [111]. The process occurs spontaneously if there are no incident photons in either of the modes of the down-converted photons; otherwise the down-conversion can

---

[11] Gisin's proof is straightforward but rather technical. The basic idea is that, for superluminal signalling to be impossible, indistinguishable density matrices [such as (3.137) and (3.138)] should remain indistinguishable after (imperfect) cloning. This sets an upper limit on the cloning fidelity that is possible without violating the 'no signalling' criterion.

be stimulated. The spontaneous and stimulated down-conversion processes are much like the spontaneous and stimulated emission in the case of excited atoms and, again, it is the spontaneous process that limits the cloning fidelity.

3.5.1 Teleportation

While quantum theory forbids (perfect) cloning, it does allow Alice to 'teleport' an exact quantum state to Bob. Such teleportation is not inconsistent with the no-cloning theorem because the original state is destroyed in the process, which does not produce twin states. Here we will briefly review teleportation [112] and explain why it cannot be done superluminally.

Using the H, V basis, we write the (arbitrary) polarization state of Alice's photon (labelled as '1') as

$$|\psi_1\rangle = a|H\rangle_1 + b|V\rangle_1. \tag{3.143}$$

Teleportation of this state involves the use two EPR-correlated photons, photon '2' directed to Alice and photon '3' to Bob. These 'shared' photons are described by the state

$$|\psi_{23}\rangle = \frac{1}{\sqrt{2}}(|H\rangle_2|V\rangle_3 - |V\rangle_2|H\rangle_3). \tag{3.144}$$

The 'Bell basis' for the two photons 1 and 2 consists of the four states

$$|\psi_{12}^{(A)}\rangle = 1\sqrt{2}(|H\rangle_1|H\rangle_2 + |V\rangle_1|V\rangle_2)$$
$$|\psi_{12}^{(B)}\rangle = 1\sqrt{2}(|H\rangle_1|H\rangle_2 - |V\rangle_1|V\rangle_2)$$
$$|\psi_{12}^{(C)}\rangle = 1\sqrt{2}(|H\rangle_1|V\rangle_2 + |V\rangle_1|H\rangle_2)$$
$$|\psi_{12}^{(D)}\rangle = 1\sqrt{2}(|H\rangle_1|V\rangle_2 - |V\rangle_1|H\rangle_2). \tag{3.145}$$

Using this basis, the three-photon state $|\psi_{123}\rangle = |\psi_1\rangle|\psi_{23}\rangle$ can be written as

$$|\psi_{123}\rangle = \tfrac{1}{2}|\psi_{12}^{(A)}\rangle * (-b|H\rangle_3 + a|V\rangle_3) + \tfrac{1}{2}|\psi_{12}^{(B)}\rangle * (b|H\rangle_3 + a|V\rangle_3)$$
$$+ \tfrac{1}{2}|\psi_{12}^{(C)}\rangle * (-a|H\rangle_3 + b|V\rangle_3) + \tfrac{1}{2}|\psi_{12}^{(D)}\rangle * (-a|H\rangle_3 - b|V\rangle_3). \tag{3.146}$$

The teleportation process begins with Alice making a 'Bell measurement' on the two photons (1 and 2) available to her. This collapses the state $|\psi_{123}\rangle$ to one of the Bell components. Then Alice signals Bob to tell him which component. What Bob then does is perform a unitary transformation, according to the teleportation table 3.1, on the photon 3 he has received. Inspection of table 3.1 shows that, after Bob's unitary transformation, photon 3 is in the state $|\psi_3\rangle = a|H\rangle_1 + b|V\rangle_1$, the same as the original state of photon 1. In other words, the state of Alice's photon has been teleported to Bob. Alice and Bob can be separated by a large distance but there is no superluminal transmission of information between them because Alice

Table 3.1. Teleportation table. The σ's are Pauli matrices in the standard notation, and I is the unit 2×2 matrix.

Alice's Bell-state projection	Bob's unitary transformation
$|\psi_{12}^{(A)}\rangle$	$i\sigma_y$
$|\psi_{12}^{(B)}\rangle$	σ_x
$|\psi_{12}^{(C)}\rangle$	$-\sigma_z$
$|\psi_{12}^{(D)}\rangle$	$-I$

has to signal Bob by some means to tell him the result of her Bell measurement. In the first experimental demonstration of high-fidelity quantum teleportation, Alice and Bob were a metre apart [113].

3.6 A superluminal quantum Morse telegraph?

Another clever scheme for superluminal communication makes use of EPR polarization-correlated photon pairs and a Michelson interferometer modified such that one of the mirrors is replaced by a phase-conjugating mirror (PCM) [114].

In contrast to an ordinary mirror, where circular (but not linear) polarization is reversed by reflection, polarization does not change upon 'reflection' from a PCM. It would seem, therefore, that, when one of the ordinary mirrors of a Michelson interferometer is replaced by a PCM (figure 3.2), there will be interference of the two propagation paths when the incident light is linearly polarized but not when it is circularly polarized. (In the latter case, the fields from the PCM and the ordinary mirror are orthogonally polarized and so do not interfere.) With a steady stream of polarization-correlated EPR pairs, therefore, an observer A would presumably be able to superluminally communicate to an observer B, equipped with the PCM-modified Michelson interferometer, whether she is measuring linear or circular polarization. This would allow what has been called a 'superluminal quantum Morse telegraph', as discussed in more detail by Garuccio [114].

The interference properties assumed for the PCM-modified Michelson interferometer have, in fact, been verified experimentally for coherent laser fields [115]. A simple calculation explains this observation. The positive-frequency part of the electric field operator at the detector D (figure 3.2) can be written as the sum of three terms. One term relates to the field that is incident on the beam splitter from the left in figure 3.2 and that reflects off M before propagating to D.

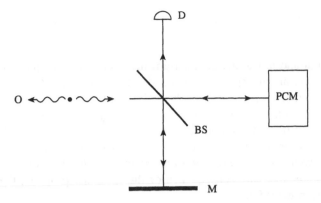

Figure 3.2. Experimental arrangement in which one photon of a polarization-correlated EPR pair is incident upon a Michelson interferometer in which one of the mirrors (M) has been replaced by a phase-conjugating mirror (PCM).

Denoting the polarization component i ($i = 1, 2$) of this field by E_{1i}, we write

$$\hat{E}_{1i}^{(+)} = \sum_{j=1}^{2} \alpha_{ij} \hat{A}_j^{(+)} \qquad (3.147)$$

where $\hat{A}_j^{(+)}$ is the positive-frequency part of the electric field operator associated with the electric field of polarization j and the (complex) number α_{ij} allows for a possible polarization change of this field due to reflection, together with phase changes due to propagation. Another contribution to the field at D arises from reflection off the PCM. We denote this term by $\hat{E}_{2i}^{(+)}$ and write

$$\hat{E}_{2i}^{(+)} = \sum_{j=1}^{2} \beta_{ij} \hat{A}_j^{(-)} \qquad (3.148)$$

β_{ij}, like α_{ij}, accounts for polarization and phase changes, while the hermitian conjugation of the incident field $\hat{A}_j^{(+)}$ ($\hat{A}^{(-)} = A^{(+)\dagger}$) accounts for the effect of the PCM, i.e. phase conjugation corresponds quantum mechanically to hermitian conjugation. The third contribution ($\hat{E}_{3i}^{(+)}$) to the field at D is made up of all other modes, including fields incident on the PCM from the right (figure 3.2) and propagating to D after reflection off the beam splitter.

The total positive-frequency part of the ith polarization component of the field incident on D is $\hat{E}_i^{(+)} = \hat{E}_{1i}^{(+)} + \hat{E}_{2i}^{(+)} + \hat{E}_{3i}^{(+)}$ and the normally ordered field expectation value that determines the photon-counting rate of an ideal detector at D is

$$\langle \hat{E}_i^{(-)} \hat{E}_i^{(+)} \rangle = \sum_{j=1}^{2} \sum_{k=1}^{2} [\alpha_{ij} \alpha_{jk}^* \langle \hat{A}_k^{(-)} \hat{A}_j^{(+)} \rangle + \alpha_{ik}^* \beta_{ij} \langle \hat{A}_k^{(-)} \hat{A}_j^{(-)} \rangle$$

$$+ \alpha_{ij}\beta_{ik}^{*}\langle \hat{A}_{k}^{(+)}\hat{A}_{j}^{(+)}\rangle + \beta_{ik}^{*}\beta_{ij}\langle \hat{A}_{k}^{(+)}\hat{A}_{j}^{(-)}\rangle]. \qquad (3.149)$$

The first and last terms on the right-hand side of this equation obviously arise from non-interfering paths to the detector. The last term accounts for 'spontaneous radiation' noise from the PCM [117, 118]. Only the second and third terms account for any interference—but they vanish for any state of the field with, at most, one photon and, in particular, when the field incident on the interferometer shown in figure 3.2 is one member of an EPR pair. Thus, the alternative paths to D *never interfere, no matter what the polarization of the incident photon* and, as a consequence of this *quantum* property, the proposed superluminal scheme fails.

Note that there is no contradiction with the experiments of Boyd *et al* [115] in which interference was, in fact, observed for linearly polarized incident light (but not for circularly polarized light) in the arrangement shown in figure 3.2. The incident light in these experiments was from a laser and could be described by a coherent state of the field, i.e. by an eigenket of the positive-frequency part of the electric field operator. In this case, the second and third terms in (3.149) do not necessarily vanish [see also [119]].

A heuristic semiclassical argument based on photon number–phase complementarity can be invoked to explain the presence or absence of interference: the interferometer in figure 3.2 cannot distinguish between circularly and linearly polarized photons because interference depends on the phase of the field, which is indefinite in the case of single photons. In the case of a coherent state, however, the phase is not indefinite and interference is not precluded.

Note that, aside from the specific scheme of figure 3.2, it must, in general, be impossible to have an apparatus that could *locally* determine the polarization state of a photon, without making measurements on a (non-local) EPR partner. (More generally, it must be impossible to determine the eigenstate of a single particle locally.) Similarly, it must be impossible to have an apparatus that faithfully preserves the polarization of a photon entering with (arbitrary) linear polarization while reversing the helicity of a photon entering with circular polarization [120]. Otherwise, as examples such as these suggest, superluminal communication would be possible.

Finally, we note that, in the case of an ordinary Michelson interferometer, the expression (3.149) is replaced by

$$\langle \hat{E}_{i}^{(-)}\hat{E}_{i}^{(+)}\rangle = \sum_{j=1}^{2}\sum_{k=1}^{2}[\alpha_{ij}\alpha_{jk}^{*} + \alpha_{ik}^{*}\beta_{ij} + \alpha_{ij}\beta_{ik}^{*} + \beta_{ik}^{*}\beta_{ij}]\langle \hat{A}_{k}^{(-)}\hat{A}_{j}^{(+)}\rangle \qquad (3.150)$$

and a single photon can 'interfere with itself', independently of polarization.

3.7 Mirror switching in cavity QED

The rate of spontaneous emission of an atom near a mirror or inside a cavity varies with the position of the atom. This aspect of 'cavity QED' is well established experimentally [121, 122]. Suppose we have a single excited atom at a position where spontaneous emission is completely suppressed and that, at a time T, one of the mirrors is suddenly replaced by a photodetector. Can a photon be counted immediately at time T or is the photon count ideally zero until a time $T' = T + T_R$, where T_R is a retardation time determined by the distance between the atom and the detector that replaced the mirror?

There are two plausible predictions. According to one argument, the atom cannot 'know' the mirror has been removed until the time $t = T + D/c$, where D is the atom–mirror distance, and the atom can have a non-vanishing probability of emitting a photon only after this time. Since the propagation time to the detector is D/c, a photon can be detected only after a time $t + D/c = T + 2D/c$, i.e. after a time $2D/c$ following the mirror 'switchout'.

Alternatively, it can be argued that, as in the case of a classical dipole radiator in a cavity, there are always fields (or, more precisely, probability amplitudes) propagating from the atom to the removable mirror and back to the atom and that the suppression of spontaneous emission implies a destructive interference of the two counter-propagating fields. The sudden removal of the mirror allows the part of the field propagating toward the mirror to escape from the cavity, so that a photon can be counted immediately after the switchout of the mirror.

We will show that the second prediction is the correct one [123]. Because the analysis of any real experiment will involve some complications that are irrelevant to the question of interest, we will consider an idealized model. This model consists of a two-state atom in the presence of a single plane mirror and an electric dipole atom–field interaction restricted to polarized field modes propagating only in the two directions normal to the mirror (figure 3.3) [124].

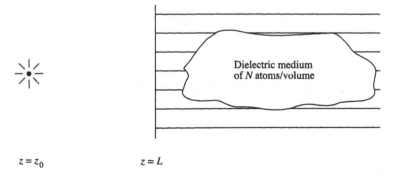

Dielectric medium
of N atoms/volume

$z = z_0$ $z = L$

Figure 3.3. Two-state atom at $z = z_0$ near a plane mirror at $z = L$. The field is restricted to modes with $\mathbf{k}$ vectors parallel to the z-axis.

The Heisenberg-picture electric field operator is $\hat{E}(z,t) = \hat{E}_0(z,t) + \hat{E}_s(z,t)$, where $\hat{E}_0(z,t)$ is the free field in the absence of any sources and

$$\hat{E}_s(z,t) = \frac{id}{2\epsilon_0} \int_0^t dt'\, [\hat{\sigma}(t') + \hat{\sigma}^\dagger(t')] \sum_k \omega_k U_k(z) U_k(z_0) e^{i\omega_k(t'-t)} + \text{h.c.}$$

(3.151)

is the source field due to the atom. Here $\hat{\sigma}$ and $\hat{\sigma}^\dagger$ are again two-state lowering and raising operators, respectively, d is the electric dipole matrix element for the two-level atom of transition frequency ω_0, and $U_k(z)$ is a mode function normalized in a volume of cross-sectional area S and length L. In our model,

$$U_k(z) = (2/SL)^{1/2} \sin k(L-z)$$

(3.152)

so that $U_k(L) = 0$ at the (perfectly conducting) mirror. In the limit $L \to \infty$, $\sum_k \to (L/\pi) \int dk = (L/\pi c) \int d\omega$ and

$$\hat{E}_s(z,t) \to \frac{d}{\epsilon_0 c S} \int_0^t dt'\, [\hat{\sigma}(t') + \hat{\sigma}^\dagger(t')] \frac{\partial}{\partial t'}$$
$$\times \int_{-\infty}^{\infty} d\omega\, \sin \frac{\omega}{c}(L-z) \sin \frac{\omega}{c}(L-z_0) e^{i\omega(t'-t)}. \quad (3.153)$$

We let $z, z_0, L \to \infty$ in such a way that $z - z_0$, $L - z_0$, and $L - z$ remain finite and positive. These limits are those appropriate for an atom at a distance $D = L - z_0$ from a single mirror. In this limit[12],

$$\hat{E}_s(z,t) = -\frac{d}{2\epsilon_0 c S} \left[\dot{\hat{\sigma}}_x \left(t - \frac{z-z_0}{c} \right) - \dot{\hat{\sigma}}_x \left(t - \frac{2L-z-z_0}{c} \right) \right] \quad (z > z_0)$$

(3.154)

where $\hat{\sigma}_x = \hat{\sigma} + \hat{\sigma}^\dagger$. It is important to note that *no approximations have been made in the derivation of this result from the Hamiltonian for our model.*

In the approximation $\dot{\hat{\sigma}}(t) \cong -i\omega_0 \hat{\sigma}(t)$, we obtain

$$\hat{E}_s(z,t) \cong \frac{id\omega_0}{2\epsilon_0 c S} \left[\hat{\sigma}\left(t - \frac{z-z_0}{c} \right) - \hat{\sigma}\left(t - \frac{2L-z-z_0}{c} \right) \right]$$
$$- \frac{id\omega_0}{2\epsilon_0 c S} \left[\hat{\sigma}^\dagger\left(t - \frac{z-z_0}{c} \right) - \hat{\sigma}^\dagger\left(t - \frac{2L-z-z_0}{c} \right) \right]$$
$$\equiv \hat{E}_s^{(+)}(z,t) + \hat{E}_s^{(-)}(z,t)$$

(3.155)

where the positive- and negative-frequency parts of the field are given (approximately) by

$$\hat{E}_s^{(+)}(z,t) \cong \frac{id\omega_0}{2\epsilon_0 c S} \left[\hat{\sigma}\left(t - \frac{z-z_0}{c} \right) - \hat{\sigma}\left(t - \frac{2L-z-z_0}{c} \right) \right] \quad (3.156)$$

[12] Unit step functions $\theta(t - (z-z_0)/c)$ and $\theta(t - (2L-z-z_0)/c)$ are implicit in the first and second terms, respectively. To simplify the equations, we omit the step functions except where they are crucial to our discussion. Terms involving reflection off the mirror at $z = 0$ are absent in equation (3.154) as a consequence of our assumption $L \to \infty$, i.e. that the mirror at $z = 0$ is infinitely far from the atom in our model.

and $\hat{E}_s^{(-)}(z, t) = \hat{E}_s^{(+)}(z, t)^\dagger$. In particular, at the position $z = z_0$ of the atom, we obtain from (3.156) the radiation reaction field

$$\hat{F}_s^{(+)}(z_0, t) = \frac{id\omega_0}{2\epsilon_0 cS}\left[\hat{\sigma}(t) - \hat{\sigma}\left(t - \frac{2D}{c}\right)\right] \cong \frac{id\omega_0}{2\epsilon_0 cS}\left[1 - e^{2i\omega_0 D/c}\right]\hat{\sigma}(t) = 0$$

(3.157)

for $e^{2i\omega_0 D/c} = 1$, where $D = L - z_0$ is the distance of the atom from the mirror. In other words, if $e^{2i\omega_0 D/c} = 1$, the radiation reaction responsible for spontaneous emission vanishes and spontaneous emission is inhibited.

The first term in brackets in equation (3.156) is a retarded field propagating from z_0 to z. The second term involves propagation from z_0 to the mirror and then to z. These terms, therefore, correspond to fields propagating in the positive and negative z directions, respectively, as can also be seen from a plane-wave expansion of the field. The 'forward'-propagating field

$$\hat{E}_{s,F}^{(+)}(L, T) \cong \frac{id\omega_0}{2\epsilon_0 cS}\hat{\sigma}\left(T - \frac{D}{c}\right)$$

(3.158)

for $z = L$ *can be measured instantaneously*: the photon-counting rate at an ideal broadband photodetector replacing the mirror at $z = L$ and $t = T$ is proportional to the normally ordered correlation function

$$\langle\hat{E}_{s,F}^{(-)}(L, T)\hat{E}_{s,F}^{(+)}(L, T)\rangle \cong \left(\frac{d\omega_0}{2\epsilon_0 cS}\right)^2\left\langle\hat{\sigma}^\dagger\left(T - \frac{D}{c}\right)\hat{\sigma}\left(T - \frac{D}{c}\right)\right\rangle$$

$$= \left(\frac{d\omega_0}{2\epsilon_0 cS}\right)^2 P\left(T - \frac{D}{c}\right)$$

(3.159)

where $P(t)$ is the probability at time t that the atom is in the excited state. There is, thus, an immediately non-vanishing photon-counting rate at $z = L$ when the detector replaces the mirror at time $t = T$.

Experiments have been performed to test the prediction that a photon can be detected immediately following the sudden replacement of a mirror by a photodetector. Branning *et al* [125] used the process of spontaneous parametric down-conversion in which, unlike the case of an atom in a cavity, inhibition can be obtained with cavities much longer than the emission wavelength [47, 126] and with the retardation time $2D/c$ much larger than in the case of inhibited spontaneous emission from an atom. Using a Pockels cell effectively to switch a cavity mirror on or off in a time of several ns—shorter than $2D/c$—they found that a photon can be counted immediately.

In an experiment based on second-harmonic generation from a Langmuir–Blodgett film, Kauranen *et al* [127] observed that there is a non-vanishing field between the emitter and a mirror even when the film is positioned such that emission is suppressed. This corroborates the theoretical analysis, which we now present in a bit more detail.

The rate at which field energy is lost from the cavity when the mirror is switched out at $t = T$ is

$$R(T) = \epsilon_0 c S \langle \hat{E}_{s,F}^{(-)}(L, T) \hat{E}_{s,F}^{(+)}(L, T) \rangle = \epsilon_0 c S \left(\frac{d\omega_0}{2\epsilon_0 c S} \right)^2 P \left(T - \frac{D}{c} \right)$$

$$= \beta \hbar \omega_0 P \left(T - \frac{D}{c} \right). \tag{3.160}$$

Here $\beta \equiv d^2 \omega_0 / 4 \hbar \epsilon_0 c S$ is half the spontaneous emission rate in the 'free-space' limit $D \to \infty$ in our model [124]. Both (3.158) and (3.160) are applicable at any time $t > D/c$ replacing T.

The probability that the atom at time t is in the excited state is given by [124]

$$P(t) = \left| \sum_{n=0}^{\infty} \frac{\beta^n}{n!} \left(t - \frac{2nD}{c} \right)^2 e^{-\beta(t - 2nD/c)} \theta \left(t - \frac{2nD}{c} \right) \right|^2 \tag{3.161}$$

for the case $e^{2i\omega_0 D/c} = 1$ of interest here. For times t sufficiently large compared with the 'photon bounce time' $2D/c$, $P(t)$ reaches a steady-state value; for $2\beta D/c \ll 1$, which is the case in standard cavity QED experiments, this steady-state value is $P_s \cong e^{-2\beta D/c}$. Assuming $T \gg 2D/c$, therefore, which must be the case if the mirror is switched out after a time when spontaneous emission is inhibited ($P > 0, \dot{P} = 0$), we have

$$R(T) = \beta \hbar \omega_0 P_s. \tag{3.162}$$

This has the following interpretation. The steady-state inhibition of spontaneous emission, i.e. the fact that the atom is not losing any net energy to the field by spontaneous emission, implies that any radiation emitted toward the mirror is exactly balanced by radiation reflected back from the mirror, a result consistent with (3.156). In particular, the rate at which radiant energy is transported toward the mirror is half the spontaneous emission rate, βP_s, times $\hbar \omega_0$, which is just (3.162). (The half is required because we are considering only the energy in the one-sided region $z_0 < z < L$.)

After the mirror is switched out at $t = T$, the atom cannot begin to lose energy to spontaneous emission until a time D/c later. The fact that we can, nevertheless, count a photon *before* $t = T + D/c$ might, therefore, appear to violate energy conservation. Consider, however, the (cycle-averaged) energy $W_F(T)$ associated with forward-propagating radiation in the space $z_0 < z < L$ at time T:

$$W_F(T) = \epsilon_0 S \int_{z_0}^{L} dz \, \langle \hat{E}_{s,F}^{(-)}(z, T) \hat{E}_{s,F}^{(+)}(z, T) \rangle$$

$$= \epsilon_0 S \left(\frac{d\omega_0}{2\epsilon_0 c S} \right)^2 \int_{z_0}^{L} dz \, P \left(T - \frac{z - z_0}{c} \right)$$

$$= \beta\hbar\omega_0 \int_{T-D/c}^{T} \mathrm{d}t'\, P(t') = \int_{T-D/c}^{T} \mathrm{d}t'\, R(t') = \frac{D}{c}R(T) = \frac{\beta D}{c}\hbar\omega_0 P_{\mathrm{s}}.$$

(3.163)

There is also non-vanishing energy associated with backward-propagating radiation and an interference between forward- and backward-propagating radiation. It follows that the non-vanishing photon-counting rate at $t = T$ occurs not at the expense of the atom but rather as a depletion of *field* energy, i.e. a depletion of the energy associated with the backward-propagating field and the interference of the counter-propagating fields. We now take up this point in more detail.

Our analysis thus far has relied on the intuitive idea that the Poynting vector associated with the forward-propagating radiation alone gives the rate of energy depletion from the cavity when the mirror is suddenly removed. For a better appreciation of where the energy comes from to register a count at the detector, we consider now the time dependence of the cavity energy after the mirror is removed. The mirror switchout at $t = T$ will affect the backward-propagating field in such a way that the field (3.156) is replaced by

$$\hat{E}_{\mathrm{s}}^{(+)}(z, t) \cong \frac{id\omega_0}{2\epsilon_0 cS}\left[\hat{\sigma}\left(t - \frac{z-z_0}{c}\right)\theta\left(t - \frac{z-z_0}{c}\right) - \hat{\sigma}\left(t - \frac{2L-z-z_0}{c}\right)\right.$$
$$\left.\times\, \theta\left(t - \frac{2L-z-z_0}{c}\right)\theta\left(T - t + \frac{L-z}{c}\right)\right]$$

(3.164)

where now we have explicitly included all appropriate step functions. The last step function accounts for the fact that backward-propagating waves persist at point z at times $t > T$ only if $L - z > c(t - T)$, i.e. if the information that the mirror is gone at $t = T$ has not yet propagated to z. Based on this expression, we calculate, in a manner analogous to (3.162), the cavity energy associated with the backward-propagating waves plus the interference of the forward- and backward-propagating waves in the region $z_0 < z < L$:

$$W_{\mathrm{B,BFI}} \cong \frac{1}{c}\beta\hbar\omega_0[D - c(t - T)]P_{\mathrm{s}}$$

(3.165)

for $T < t < T + D/c$. Thus

$$\frac{\mathrm{d}}{\mathrm{d}t}W_{\mathrm{B,BFI}}(t) = -\beta\hbar\omega_0 P_{\mathrm{s}} \qquad (T < t < T + D/c)$$

(3.166)

which is just $-R(T)$ [equation (3.162)], i.e. the rate at which energy associated with the *forward*-propagating radiation will escape from the cavity when the mirror is switched out.

This confirms the assertion that the immediate detection of a photon, in spite of inhibited spontaneous emission, occurs at the expense of cavity field energy or, actually the change in field energy associated with the backward-propagating radiation and its interference with forward-propagating radiation when the mirror is switched out. It is precisely this change, according to (3.162) and (3.166), that propagates *out* of the cavity and that can produce a photon count. We emphasize again that this occurs in spite of the fact that, back at the atom, there is still destructive interference and inhibited spontaneous emission, and a constant upper-state probability P_s, until time $T + D/c$. The *total* field energy has a constant expectation value up until this time. After $t = T + D/c$, of course, the atom radiates as it does in free space.

It is perhaps worth noting why, as a consequence of retardation, there will always be some field energy in the cavity before the mirror is removed. For a time $t = 2D/c$ after the atom is excited at $t = 0$, say, it will radiate uninhibitedly as if in free space. For times $0 < t < 2D/c$, therefore, the energy in the field is

$$W(t) = \hbar\omega_0[1 - P(t)] = \hbar\omega_0[1 - e^{-2\beta t}] \tag{3.167}$$

and the rate at which the field energy grows is

$$\dot{W}(t) = 2\beta\hbar\omega_0 e^{-2\beta t}. \tag{3.168}$$

At time $t = D/c$, when the radiated field reaches the mirror,

$$\dot{W}\left(\frac{D}{c}\right) = 2\beta\hbar\omega_0 e^{-2\beta D/c} \tag{3.169}$$

which, for $2\beta D/c \ll 1$, is approximately $2\beta\hbar\omega_0 P_s$, as noted earlier. The rate at which the energy of the *forward*-propagating radiation grows is half this, i.e. $R(D/c) \cong \beta\hbar\omega_0 P_s = R(T)$ as given by equation (3.162). As the atom quickly attains a steady state for the case $2\beta D/c \ll 1$ under consideration, the rate at which energy is put into the forward-propagating field will quickly equilibrate to the value $R(T)$. It is precisely this power that can be registered at the detector replacing the mirror. There is no contradiction with the fact of inhibited spontaneous emission because, although $\dot{P} = 0$, the steady-state probability P_s is always less than unity.

It also seems worth noting that, in the Schrödinger picture, the atom–field system at time $t < T$ is described to an excellent approximation by the state vector

$$|\psi(t)\rangle = a(t)|\text{atom excited}\rangle|\text{no photons}\rangle$$
$$+ \sum_k a_k(t)|\text{atom in lower state}\rangle|\text{one photon in mode } k\rangle \tag{3.170}$$

where $P(t) = |a(t)|^2$ is the probability that the atom is excited at time t. In the steady state of inhibited spontaneous emission, $P = P_s < 1$, i.e. $a, a_k < 1$ and the atom–field system is in an 'entangled' state.

3.8 Précis

The quantum theory of fields is formulated in such a way as to be consistent with causality (e.g. field commutators vanish for spacelike separations). In specific examples, however, an explicit proof of causality may not be an entirely trivial matter. In the Fermi problem of two atoms in vacuum, for example, causality can be shown to hold only when we go beyond the rotating-wave approximation in the Heisenberg picture or, equivalently, when we include energy-non-conserving processes in the Schrödinger or interaction pictures. In order to make causality explicit in the theory of photodetection, similarly, we should make the rotating-wave approximation leading to normal ordering only after we have solved for the (causal) electric field operator.

In like manner, it is not always a trivial matter to show where specific proposals for superluminal communication break down, even though we are certain that they must. In the process of disproving them, we can gain new insights: the no-cloning theorem, for instance, was motivated by a superluminal communication scheme based on quantum correlations. In fact, the highest possible cloning fidelity allowed by quantum theory is exactly that which forbids superluminal signalling. We have discussed these 'spooky' correlations and argued that they cannot be used to transmit *information* instantaneously.

Examples such as the 'superluminal quantum Morse telegraph' lead to the conclusion that, if superluminal communication is to be impossible, it must be impossible to determine the eigenstate of a single particle *locally*. It must be impossible to devise *any* apparatus that preserves the polarization of an incident photon with arbitrary linear polarization while reversing the helicity of an incident photon with circular polarization.

We have described a simple extension of Fermi's model in which the two atoms are embedded in a dielectric medium, and have discussed the implications of superluminal group velocity in this quantum-mechanical model. We showed that it is possible, without violating causality, for the initially unexcited atom to become excited, with high probability, as if a photon propagated superluminally from one atom to the other. We also discussed the quantum analog of the Sommerfeld–Brillouin theorem that the front velocity is equal to the speed of light in vacuum.

3.9 Appendix: On Einstein and hidden variables

> It seemed to me as if you had erected
> some dummy Einstein for yourself, which you then
> knocked down with great pomp.
> > W Pauli, letter to M Born, 31 March 1954 [100]

Do Bell's theorem and the experiments that have upheld the predictions of quantum theory *vis-à-vis* local hidden variable theories mean that Einstein was wrong in his objections to quantum theory? The widely held view that this is the case seems to be based on the following reasoning: Einstein argued that quantum theory is incomplete because it does not allow for an objective reality. Bell's theorem shows that no (local) hidden variable theory based on such an objective reality can be in full accord with the statistical predictions of quantum theory and, in particular, that there are certain inequalities that are predicted by hidden variable theories but not by quantum theory. Since experimental results violate these inequalities, there is no objective reality in the EPR sense and, therefore, Einstein's arguments against quantum theory, which stemmed from the presumption of such a reality, were incorrect.

But the case can be made that Bell's theorem and the experiments it stimulated are almost irrelevant to the question of whether Einstein was 'right' or 'wrong' in his objections to quantum theory, that Einstein *presupposed* the sort of correlations confirmed by the experiments, and that the experiments only rule out a class of theories he did not consider worthy of serious attention.

Recall what Einstein meant by 'completeness' and 'reality' in the context of a physical theory. In the famous EPR paper he, Podolsky, and Rosen carefully defined these terms [128]. According to EPR:

> If, without in any way disturbing a system, we can predict with certainty (i.e. with probability equal to unity) the value of a physical quantity, then there exists an element of physical reality corresponding to this physical quantity.

As for 'completeness of a physical theory,' EPR require as a necessary condition that *every element of the physical reality must have a counterpart in the physical theory.*

To illustrate their ideas, and to clarify their definitions of 'elements of physical reality' and 'completeness', EPR considered a two-particle system in which there are correlations between the coordinates and the momenta of the particles. In the modern literature, it is customary to deal instead with correlated two-state systems, such as a system of two spin-$\frac{1}{2}$ particles described by the singlet spin state

$$|\psi\rangle = \frac{1}{\sqrt{2}}[|+n\rangle_1|-n\rangle_2 - |-n\rangle_1|+n\rangle_2] \qquad (3.A.171)$$

$|\pm n\rangle_i$ is the state for which particle $i(=1, 2)$ has spin-up $(+)$ or -down $(-)$ in the direction of the (arbitrary) unit vector n. The particles are assumed to be separated by a large distance and not to interact. Let us imagine that the particles are moving in opposite directions along the z-axis, particle 1 along z and particle 2 along $-z$.

If we measure the spin component of particle 1 along the x-axis and find spin-'up', then, according to (3.A.171), we must find spin-'down' for particle 2. Likewise if we find that particle 1 has spin-down, then particle 2 must have spin-up. The spin components of the particles are correlated in this way.

Consider now the implications of this system in the context of EPR. In particular, consider what measurements on particle 1 tell us about particle 2. Since the particles are arbitrarily far apart and non-interacting, measurements made on particle 1 do not disturb particle 2. By making measurements of the x component of particle 1, we can predict, with certitude, the x component of the spin of particle 2. The x component of the spin of particle 2 is, therefore, an 'element of physical reality' in the EPR sense. We can choose instead to measure the y component of the spin of particle 1 and thereby conclude that the y component of the spin of particle 2 is likewise an element of physical reality.

The fact that *either* the x or y spin component of particle 2 can be predicted with absolute certainty, without ever making a measurement directly on particle 2, suggests that these elements of reality exist 'out there', independent of observation. But the x and y components of the spin correspond to non-commuting operators, and quantum theory states that they cannot have simultaneously precise values. In other words, according to EPR, there are simultaneous elements of physical reality in this system that are not accounted for by quantum theory, which therefore does not provide a 'complete' description of the system[13].

EPR did not argue from the perspective of any deterministic philosophy. Determinism plays essentially no role in their argument—*they even allow for the possibility of a non-deterministic theory when they use the phrase 'with probability equal to unity' in their definition of an element of reality.* No less a physicist than Max Born misunderstood for many years Einstein's primary objections to quantum theory. Pauli, in a letter to Born [100, p 221], pointed out that

> ...Einstein does not consider the concept of 'determinism' to be as fundamental as it is frequently held to be (as he told me emphatically many times), and he denied energetically that he had ever put up a postulate such as (your letter, para. 3): 'the sequence of such conditions must also be objective and real, that is, automatic, machine-like, deterministic'. In the same way, he *disputes* that he uses as criterion for the inadmissability of a theory the question: 'Is it rigorously deterministic?'

> Einstein's point of departure is 'realistic' rather than 'deterministic,' which means that his philosophical prejudice is a different one ...

Born later acknowledged that he 'had failed to understand what mattered to [Einstein]' [100, p 227].

Bell [130] noted

> It is remarkably difficult to get this point across, that determinism is not a *presupposition* of the [EPR] analysis. There is a widespread

---

[13] The word 'paradox' does not appear in the EPR paper but Einstein did refer to 'the paradox recently demonstrated by myself and two collaborators' in his 1936 essay on 'Physics and Reality' [129].

and erroneous conviction that for Einstein determinism was always *the* sacred principle. The quotability of his famous 'God does not play dice' has not helped in this regard.

Einstein's 'philosophical prejudice' in favour of 'realism', i.e, of a reality independent of observation, is almost certainly part of the mindset of most scientists. Feynman, for instance, in a lecture 'On the Philosophical Problems in Quantizing Macroscopic Objects' expressed exactly the same misgivings as Einstein about the standard, observer-dependent interpretation of quantum theory [131]:

> This is all very confusing, especially when we consider that even though we may consistently consider ourselves always to be the outside observer when we look at the rest of the world, the rest of the world is at the same time observing us, and that often we agree on what we see in each other. Does this then mean that my observations become real only when I observe an observer observing something as it happens? This is a horrible viewpoint. Do you seriously entertain the idea that without the observer there is no reality? Which observer? Any observer? Is a fly an observer? Is a star an observer? Was there no reality in the universe before 10^9 B.C. when life began? Or are *you* the observer? Then there is no reality to the world after you are dead? I know a number of otherwise respectable physicists who have bought life insurance.

Could our inability to make definite predictions in every instance mean that we are simply ignorant of 'hidden variables' whose values, if known, would uniquely determine the outcome of any measurement? Could the wavefunction in quantum theory only reflect aspects of a statistical distribution of such hidden variables, which could allow for an underlying objective reality in the EPR sense?

Conjectures about hidden variables go back as far as de Broglie's early work. In the early 1950s, Bohm raised again the possibility of hidden variables [132]. He elucidated various features of quantum theory in the context of the simple correlated two-state system of the preceding section. In particular, Bohm showed that a hidden variable theory could produce the same predictions of quantum theory *if it admitted a non-local interaction between separated systems*, a feature later noted by Bell [133]. Bell proved that any hidden variable theory in full agreement with the statistical predictions of quantum theory must have this non-local character. We will now briefly review these ideas, as well as a proof of the equivalent (contrapositive) statement of Bell's theorem: *no local hidden variable theory can reproduce all the statistical predictions of quantum mechanics.* We will begin by quickly reviewing, for the benefit of the reader not familiar with the territory, what is meant by hidden variables and locality in the context of Bohm's correlated spin system.

Suppose that all the spin components of each particle in the *Gedanken experiment* considered earlier have simultaneous elements of reality not accounted for by quantum theory. We imagine they are 'out there' all along, even though quantum theory does not allow us to predict them all simultaneously. We imagine furthermore that there are 'hidden' variables not included within quantum theory and that if only we knew the values of these variables, we could make definite predictions for all 'elements of reality' in the EPR sense. There might, for instance, be a hidden variable λ_1 whose value fully determines the x component of the spin of particle 1.

We assume that the spin component of particle 1 along any unit vector $\boldsymbol{a}$ is a function of λ_1 and $\boldsymbol{a}$ and denote this function by $A(\boldsymbol{a}, \lambda_1)$. We assume that the predictions of quantum theory are correct as far as they go and so we try to bring the hidden variable theory into as close an agreement as possible with quantum theory. Thus, we assume more specifically that $A(\boldsymbol{a}, \lambda_1)$ has only two possible values:

$$A(\boldsymbol{a}, \lambda_1) = \pm\tfrac{1}{2}. \tag{3.A.172}$$

Similarly, we assume a hidden variable λ_2 such that the measured value of the spin of particle 2 along any direction $\boldsymbol{b}$ is a function $B(\boldsymbol{b}, \lambda_2)$ with the two possible values

$$B(\boldsymbol{b}, \lambda_2) = \pm\tfrac{1}{2}. \tag{3.A.173}$$

In writing (3.A.172) and (3.A.173), we have implicitly made an important additional assumption: $A(\boldsymbol{a}, \lambda_1)$ is independent of $\boldsymbol{b}$ and $B(\boldsymbol{b}, \lambda_2)$ is independent of $\boldsymbol{a}$. In other words, we are assuming that a measurement of the spin component of particle 1 along $\boldsymbol{a}$ is independent of the direction $\boldsymbol{b}$ along which the spin of particle 2 is measured. This seems reasonable when we remember that the two particles can be arbitrarily far apart when we choose to measure their spin components and that we do not want to allow for any instantaneous action at a distance that might lead to a dependence of $A(\boldsymbol{a}, \lambda_1)$ on $\boldsymbol{b}$ or, for that matter, λ_2. This assumption defines what is called a *local* hidden variable theory.

The restrictions (3.A.172) and (3.A.173) on the possible values of the functions A and B will bring our hidden variable theory into agreement with quantum theory to the extent that the measured value of the component of the spin of each particle along any direction must be either $+\tfrac{1}{2}$ or $-\tfrac{1}{2}$. Since we want the theory to agree as fully as possible with the predictions of quantum theory, we will add to (3.A.172) and (3.A.173) the condition that, when $\boldsymbol{a} = \boldsymbol{b}$, the measured spins must always be opposite. Thus, for all possible values of λ_1 and λ_2 and for any direction $\boldsymbol{a}$, we require

$$A(\boldsymbol{a}, \lambda_1) = -B(\boldsymbol{a}, \lambda_2). \tag{3.A.174}$$

It is now easy to show (thanks to Bell) that the assumptions (3.A.172)–(3.A.174), however reasonable they might seem, are sufficient to rule out full agreement of the predictions of any local hidden variable theory with the predictions of quantum theory. Let $P(\lambda_1, \lambda_2)$ be the joint probability distribution for λ_1 and λ_2, so that

$$E(a, b) = \int d\lambda_1 \int d\lambda_2\, P(\lambda_1, \lambda_2) A(a, \lambda_1) B(b, \lambda_2) \tag{3.A.175}$$

is the expectation value for the product of the spin component of particle 1 in the a direction and the spin component of particle 2 in the b direction. Bell [133] considers now the difference $E(a, b) - E(a, c)$. From (3.A.172)–(3.A.175), it follows that[14]

$$
\begin{aligned}
E(a, &b) - E(a, c) \\
&= \int d\lambda_1 \int d\lambda_2\, P(\lambda_1, \lambda_2)[A(a, \lambda_1) B(b, \lambda_2) - A(a, \lambda_1) B(c, \lambda_2)] \\
&= \int d\lambda_1 \int d\lambda_2\, P(\lambda_1, \lambda_2) A(a, \lambda_1) B(b, \lambda_2)[1 - 4B(b, \lambda_2) B(c, \lambda_2)] \\
&= \int d\lambda_1 \int d\lambda_2\, P(\lambda_1, \lambda_2) A(a, \lambda_1) B(b, \lambda_2)[1 + 4A(b, \lambda_1) B(c, \lambda_2)].
\end{aligned}
$$
$$\tag{3.A.176}$$

Since the quantity in brackets is positive-definite and $|A(a, \lambda_1) B(b, \lambda_2)| = \frac{1}{4}$, we obtain the Bell inequality

$$|E(a, b) - E(a, c)| \leq \tfrac{1}{4} + E(b, c). \tag{3.A.177}$$

A local hidden variable theory constructed in accordance with the postulates (3.A.172)–(3.A.175) must, therefore, predict this inequality for the spin correlation experiment under consideration. However, it is easy to see that this inequality is *not* predicted by quantum mechanics, according to which $E(a, b) = -\frac{1}{4}a \cdot b$. That is, no local hidden variable theory can be in full accord with all the statistical predictions of quantum theory. This is Bell's theorem and, since Bell's original work, it has been proven in various other ways.

Einstein was well aware of Bohm's work on hidden variable theories—*local* versions of which were shown by Bell to be incompatible in general with the predictions of quantum theory. Although such theories are 'complete' (as well as deterministic) in the EPR sense, it is clear in the following, from a letter to Born, that Einstein had little enthusiasm for such theories [100, p 192]:

> Have you noticed that Bohm believes (as de Broglie did, by the way, 25 years ago) that he is able to interpret the quantum theory in deterministic terms? That way seems too cheap to me. But you, of course, can judge this better than I.

EPR did not propose any 'complete' and 'realistic' alternative to quantum theory but it is clear from the passage just cited that Einstein had no interest (or

---

[14] In the second equality, we use the fact that $B^2(b, \lambda_2) = \frac{1}{4}$ and, in the third, the fact that $B(b, \lambda_2) = -A(b, \lambda_1)$

had long since lost interest) in the 'cheap' sort of theory to which Bell's theorem applies and against which there is now ample experimental evidence[15]. Indeed, Einstein would almost certainly have *predicted* the outcome of the experiments testing the predictions of quantum theory against deterministic local hidden variable theories[16]. He obviously knew what quantum theory predicts in such experiments and he knew full well the great predictive success of quantum theory. According to Pauli, Einstein's objections had the flavor of ' the ancient question of how many angels are able to sit on the point of a needle ... Einstein's questions are ultimately always of this kind' [100, p 223].

And as Jammer has written [135],

> Although the Einstein–Podolsky–Rosen incompleteness argument was undoubtedly one of the major incentives for the modern development of hidden variable theories, it would be misleading to regard Einstein, as some recent authors do, as a proponent of or even as 'the most profound advocate' of hidden variables. True, Einstein was sympathetically inclined toward any efforts to explore alternatives, and as such also the ideas of de Broglie and of Bohm, but he never endorsed any hidden variable theory.

[15] For a review of these experiments, see [134].
[16] I have discussed this with five distinguished physicists who have opinions on the matter. Two of them agree with me.

Chapter 4

Fast light and signal velocity

Superluminal group velocities do not violate Einstein causality because they are not signal (information) velocities. In this chapter, we take up in more detail the question of what defines a signal and describe experiments aimed at clarifying the meaning of signal velocity. We will discuss the role of noise in the field, the medium, or the detector, with emphasis on the fact that *quantum noise limits the degree to which a pulse with superluminal group velocity can be measurably advanced.*

4.1 Experiments on signal velocities

In chapter 2, we described some of the first experiments that demonstrated that optical pulses can propagate with superluminal group velocities and can do so without any significant distortion of the pulse shape. We now briefly describe two experiments that further explore the difference between superluminal group velocity and the velocity of a *signal*, i.e. the difference between group velocity and the velocity at which *information* can be transmitted.

As noted in section 2.5, the signal velocity is sometimes defined as the velocity at which the half-the-peak-intensity point of an optical pulse propagates[1]. But we cannot regard this as a satisfactory definition of a signal because, for a Gaussian pulse, for example, the half-intensity point does not convey information that is not already present in the pulse's leading edge.

We argued that the propagation of new information, or a 'signal', requires a discontinuity in a waveform or one of its derivatives. A signal so defined has the satisfying property that it cannot be propagated with a velocity exceeding c: there can be no violation of Einstein causality. However, our discussion was completely classical and did not take any effects of detection or noise into account. It did not address the question of what is actually measured by a photodetector.

---

[1] Such a rather arbitrary definition obviously makes the measured signal velocity dependent on the sensitivity of the detector. Recall the remarks of Brillouin quoted near the end of section 2.5.

In section 4.3, we give an operational definition of signal velocity v_s based on the fact that the integrated photocurrent at the detector must exceed some threshold level, depending on noise and the allowed error rate, before it can be asserted that a *signal* has been received.

An experiment relating to this definition of v_s has been reported by Centini *et al* [136]. In their experiments, the peaks of chirped optical pulses passing through a GaAs cavity of thickness 450 μm were observed to have transit times that were positive, zero, or negative, i.e. the group velocity v_g could be positive, infinite, or negative, depending on the particular wavelength near 1550 nm. Pulses propagated in vacuum produced a detector response level of 71 mV, whereas the pulses transmitted by the cavity produced levels ranging from 22 to 71 mV, depending on the wavelength. Centini *et al* defined the operational v_s using a threshold detector level of 2 mV and their measurements showed, among other things, that v_s was always less than the phase velocity c/n, even when the group velocity was superluminal. The reader is referred to [136] and references therein for other conclusions of this work and for details of the experiments and the theory used to analyse them.

Another, particularly significant experiment bearing on the definition of signal velocity was reported by Stenner *et al* [137]. As in the experiments of Wang *et al* [49], anomalous dispersion in the spectral region between two Raman gain lines was used to obtain a superluminal group velocity, in this case employing the atomic coherence produced by two laser pulses in potassium vapour and using two cells, each of length $L/2 = 20$ cm, in order to avoid a parametric instability that could occur if a single cell of length L were used [138]. This resulted in greater gain and, consequently, a larger pulse advance than in the experiments of Wang *et al*: the 'group index' $n_g = n + \omega \, dn/d\omega$ ($v_g = c/n_g$) was found to be -19.6 ± 0.8, and the relative pulse advancement was about 10%.

The purpose of the experiments was to measure, evidently for the first time, the velocity with which information encoded on a superluminal pulse is transmitted. The pulses were shaped by a waveform generator driving an acousto-optic modulator. Near the peak of a Gaussian pulse the pulse amplitude was switched to a high amplitude (1) or a low amplitude (0) for the remaining duration of the pulse; and this was done at the same point in time on a pulse for either symbol 0 or 1. If this switching could be done instantaneously, we would have, in effect, a sharp front which, according to our earlier discussions, should propagate with the velocity c of light in vacuum. Of course, the finite electronic response time has the effect of smoothing out the switch in amplitude but the interesting question remains as to the velocity with which the information that we have a 0 or a 1 can be transmitted.

Before the point on a pulse at which the switching is done, there is of course, no information as to whether the pulse carries a 0 or a 1 and so the bit error rate (BER) is $1/2^2$. After the switching point, the information as to whether there

---

[2] The BER characterizes the measurement error probability per bit of information. If there is a

is a 0 or a 1 will increase from zero and the BER will decrease. Detection of a 0 or a 1 is assumed to occur when the BER falls below some predetermined threshold level. Note that the time at which information is actually detected in this sense will be greater than the time at which information actually appears at the detector, i.e. there is a 'detection latency' Δt [137] that depends on the details of the switching, the detection scheme, noise, and other factors. Note also that there will be a finite detection latency for vacuum-propagated pulses as well as for the advanced pulses.

The 0s and 1s were encoded both on pulses transmitted through the potassium cells and on pulses transmitted through vacuum and, from the calculated BERs and for a detection threshold set at BER $= 0.1$, the detection times T_{vac} and T_{adv} were determined. It was found that the time it takes to distinguish between 0s and 1s was *greater* for the advanced pulses than for the vacuum-propagated pulses, i.e. 'the information detection time for pulses propagating through the fast-light medium is longer than the detection time for the same information propagating through vacuum, even though the group velocity is in the highly superluminal regime for the fast-light medium' [137].

The time difference $T_i = T_{adv} - T_{vac}$ can be expressed as

$$T_i = (L/v_{i,adv} - L/v_{i,vac}) + (\Delta t_{adv} - \Delta t_{vac}) \qquad (4.1)$$

where the first term is associated with (possibly) different information velocities and the second with different detection latencies. Stenner *et al* [137] cite results of a theoretical model in which the first term vanishes, as would be expected in the case of a point of non-analyticity, due to the switching, in the temporal variation of a pulse: the point of non-analyticity should theoretically propagate with the velocity c, just like a Sommerfeld–Brillouin front. Their model gives results in qualitative agreement with the experimental data and yields $v_{i,adv} = 0.4(0.7 - 0.2)c$.

These experiments demonstrate that *the measured signal velocity depends on the detection process*, including quantum noise. In each case, the signal velocity for pulses with superluminal group velocities is found to be *subluminal* and consistent with Einstein causality. We now turn to some theoretical considerations of signal velocity relating specifically to quantum noise.

4.2   Can the advance of a weak pulse exceed the pulsewidth?

If a pulse with a superluminal group velocity could be advanced after propagation in a medium by a time that is large compared with the pulsewidth, and if this could be done at the single-photon level, we would have something very much

probability p_1 that a 0 will be erroneously detected as a 1, and a probability p_0 that a 1 will be erroneously detected as a 0, then the BER is defined as $\frac{1}{2}p_0 + \frac{1}{2}p_1$ (assuming 0 and 1 have equal transmission probabilities). The procedure for defining BER in the work of Stenner *et al* is rather technical and is described in the caption of figure 3 in their paper.

like a superluminally propagating particle (but satisfying Einstein causality). As discussed briefly in section 2.3, Chiao *et al* (CKK) [39] have suggested that an 'optical tachyon' might be realized in an optical amplifier with sufficiently long relaxation times. The group velocity in their model is given by equation (2.46) and indicates that an off-resonant pulse can propagate with a superluminal group velocity in an amplifier.

However, Aharonov, Reznik and Stern (ARS) [139] have presented general arguments, based on the unitary evolution of the state vector, that 'strongly questions the possibility that these systems may have tachyonlike quasiparticle excitations made up of a small number of photons'. They also consider a particular model as an analog of the CKK system.

In this section, we address the question of superluminal propagation at the one- or few-photon level and, in particular, the role played by quantum noise in the propagation of such extremely weak pulses. We begin with some physical considerations about the observation of superluminal propagation and we briefly compare the ARS and CKK models.

The quantum noise of interest here is associated with spontaneous emission and it could invalidate the CKK results in two ways. First, CKK assume that the atoms stay in their excited states as the pulse propagates through the amplifier. Radiative decay of the excited state will modify their dispersion relation and, if the decay is rapid enough, can lead to a subluminal rather than superluminal group velocity, since w in equation (2.46) can become negative. This can be avoided by using a sufficiently short pulse, during which radiative decay is negligible. Second, spontaneously emitted radiation might interfere with the *measurement* of the superluminal group velocity by introducing substantial noise. ARS address the latter possibility.

Although the ARS arguments are certainly compelling, they are based in part on an *analog* of an optical amplifier rather than a theory involving the interaction of the electromagnetic field with an atomic medium. In particular, their model is that of a single quantum field rather than coupled atomic and electromagnetic quantum fields. The dispersion relation associated with this model, and the criteria assumed by ARS for the observability of superluminal propagation, lead to the conclusion, by analogy to an optical amplifier, that spontaneous emission noise cannot be avoided no matter how short the pulse or the transit time through the amplifier. Specifically, the unstable modes appearing in their model—which 'are analogous to spontaneous emission in the optical model of an inverted medium of two-level systems' [139]—will preclude the observation of superluminal group velocity when the pulse is made up of a small number of photons; the quantum noise will be larger than the signal.

ARS state two *necessary* conditions for the observability of superluminal propagation [$c = 1$ in their units]:

(1) $v_g T \gg 1/\delta k$, where v_g is the group velocity, T is the time at which the wavepacket is observed, and δk is the spectral width of their initial pulse.

(2) $(v_g - 1)T \gg 1/\delta k$.

The first condition ensures that 'the point of observation [is] far outside the initial spread of the wavepacket'. The second allows us to 'distinguish between superluminal propagation and propagation at the speed of light'.

In the ARS model, the field ϕ satisfies

$$\frac{\partial^2 \phi}{\partial t^2} - \frac{\partial^2 \phi}{\partial z^2} - m^2 \phi = 0 \tag{4.2}$$

and the group velocity is[3]

$$v_g = \frac{k_0}{\sqrt{k_0^2 - m^2}} \tag{4.3}$$

where k_0 is the central value of the spatial frequency k for the initial pulse. For $m < k_0$, we can approximate v_g by $1 + m^2/2k_0^2$, so that conditions 1 and 2 are satisfied if

$$m^2 T \gg k_0^2/\delta k \gg k_0. \tag{4.4}$$

$k_0 \gg 1/T$—the condition that the observation time should be much larger than the central frequency of the pulse—then implies

$$mT \gg 1. \tag{4.5}$$

For $mT \gg 1$, the amplified quantum noise grows exponentially, as shown later [equation (4.47)]. ARS, therefore, conclude that the 'signal amplitude should be exponentially large' in order to distinguish it from noise. Thus, according to ARS, the observability of superluminality for an input pulse consisting of only a few photons would be clouded by spontaneous emission noise.

Let us now turn to the implications of conditions (1) and (2) for the actual system of interest, namely a very short optical pulse in an inverted medium [140]. Can we satisfy these conditions for observation times *short* compared with the radiative lifetime?

For a short optical pulse of central frequency ω propagating in an inverted medium ($w = 1$) with resonance frequency ω_0, the refractive index is [equation (2.40)]

$$n(\omega) \cong 1 - \frac{Ne^2 f}{4m\epsilon_0\omega_0} \frac{1}{\omega_0 - \omega} \equiv 1 - \frac{\omega_p^2}{4\omega_0\Delta} \tag{4.6}$$

for $\omega_p^2/(4\omega_0) \ll |\omega_0 - \omega| \equiv |\Delta|$. We are assuming that $|\Delta|$ is large compared with the absorption width, which, in our case, is the radiative decay rate. Equation (4.6) implies

$$\frac{v_g}{c} = \left(\frac{\mathrm{d}}{\mathrm{d}\omega}[\omega n(\omega)] \right)^{-1} = \frac{1}{1 - \omega_p^2/4\Delta^2} \tag{4.7}$$

---

[3] This follows from the dispersion relation $\omega^2 = k^2 - m^2$ implied by (4.2).

and

$$\frac{v_g}{c} - 1 = \frac{\omega_p^2/4\Delta^2}{1 - \omega_p^2/4\Delta^2} = \frac{\omega_p^2}{4\Delta^2}\frac{v_g}{c}. \tag{4.8}$$

Then conditions (1) and (2) of ARS become, respectively,

$$\frac{T}{1 - \omega_p^2/4\Delta^2} \gg \frac{1}{c\delta k} \sim \tau_p \tag{4.9}$$

$$\frac{(\omega_p^2/4\Delta^2)T}{1 - \omega_p^2/4\Delta^2} \gg \frac{1}{c\delta k} \sim \tau_p \tag{4.10}$$

with τ_p the pulse duration. Both conditions can be satisfied if, for instance, $T \gg \tau_p$ and $\omega_p^2/4\Delta^2$ is not too small. To avoid spontaneous emission during the observation time T, take $T \ll \tau_{RAD}$, where τ_{RAD} is the radiative lifetime of a single inverted atom. Then the ARS conditions require that

$$\tau_{RAD} \gg T \gg \tau_p. \tag{4.11}$$

There is another aspect of an inverted atomic medium that must be addressed, namely superfluorescence (SF), a collective phenomenon of the sample as a whole. We will denote by N_T, S, and L the number of atoms, the cross-sectional area, and the length of the sample, respectively, so that the density of atoms is given by $N = N_T/SL$. If collisional and other dephasing mechanisms are sufficiently weak, an inverted medium of N_T atoms can emit SF radiation at the rate

$$\tau_R = \tau_{RAD}/N_T \tag{4.12}$$

i.e. the radiative decay time can, in effect, be smaller by a factor of N_T than the single-atom τ_{RAD} assumed in the discussion thus far. The peak of the SF pulse occurs at a time [141–143]

$$\tau_D \sim \tau_R[\tfrac{1}{4}\ln(2\pi N_T)]^2 \tag{4.13}$$

following the excitation of the atoms. Evidently the quantum noise associated with SF will be small if

$$\tau_p, L/c < \tau_R < \tau_D. \tag{4.14}$$

We note for later purposes that

$$\omega_p^2 = \frac{Ne^2f}{m\epsilon_0} = \frac{4}{\tau_{RAD}}\frac{N_T}{SL}Sc = \frac{4}{\tau_R}\frac{c}{L} \tag{4.15}$$

where we have used the equation $1/\tau_{RAD} = e^2f/4m\epsilon_0 Sc$ for the single-atom radiative lifetime[4].

---

[4] This differs from the Einstein A coefficient $e^2\omega_0^2 f/2\pi\epsilon_0 mc^3$ for spontaneous emission because it is the rate into modes propagating *unidirectionally* with a *single polarization*. See Segev *et al* [140, appendix A]. This result is implicit in the definition of β in equation (3.160).

This brief summary lends support to the CKK suggestion but obviously a more quantitative analysis is called for. We now formulate, in the Heisenberg picture, a quantum theory of pulse propagation in an amplifier [140].

We begin with the Hamiltonian for N_T two-level atoms (TLAs) interacting with the quantized electromagnetic field via electric dipole transitions [cf (3.109)]:

$$\hat{H} = \tfrac{1}{2}\hbar\omega_0 \sum_{j=1}^{N_T} \hat{\sigma}_{zj} - d \sum_{j=1}^{N_T} \hat{\sigma}_{xj}\hat{E}(z_j) + \sum_k \hbar\omega_k \hat{a}_k^\dagger \hat{a}_k \tag{4.16}$$

where ω_0 and d have the same meaning as before and z_j is the z-coordinate of atom j. We consider a one-dimensional model in which the atoms occupy the region from $z = 0$ to $z = L$ and the field is a superposition of plane waves propagating in the z direction. The electric field operator is given by $\hat{E}(z) = \hat{E}^{(+)}(z) + \hat{E}^{(-)}(z)$, where

$$\hat{E}^{(+)}(z) = i \sum_k \left(\frac{\hbar\omega_k}{2\epsilon_0 S\ell}\right)^{1/2} \hat{a}_k e^{ikz} \qquad (k = \omega_k/c) \tag{4.17}$$

and $\hat{E}^{(-)}(z) = \hat{E}^{(+)}(z)^\dagger$. $S\ell$, where S, as before, is a cross-sectional area and ℓ a length, is the quantization volume. For simplicity, we consider only a single field polarization—linear polarization along the direction of the transition dipole moment of the TLAs.

We will work in the Heisenberg picture, in which the time-dependent electric field operator satisfies

$$\left(\frac{\partial^2}{\partial z^2} - \frac{1}{c^2}\frac{\partial^2}{\partial t^2}\right)\hat{E} = \frac{1}{\epsilon_0 c^2}\frac{\partial^2 \hat{P}}{\partial t^2} = \frac{d}{\epsilon_0 c^2 S}\sum_{j=1}^{N_T}\frac{\partial^2\hat{\sigma}_{xj}}{\partial t^2}\delta(z-z_j)$$

$$\rightarrow \frac{1}{\epsilon_0 c^2}Nd\frac{\partial^2}{\partial t^2}\hat{\sigma}_x(z,t) \tag{4.18}$$

where, in the last step, we have made the continuum approximation for the polarization density $\hat{P}$, assuming a uniform atomic density N. We now write

$$\hat{E}^{(+)}(z,t) = \hat{F}(z,t)e^{-i\omega(t-z/c)} \tag{4.19}$$

and assume $\hat{F}(z,t)$ is slowly varying in z and t compared with $\exp[-i\omega(t-z/c)]$. In this approximation,

$$2i\frac{\omega}{c}\left(\frac{\partial\hat{F}}{\partial z} + \frac{1}{c}\frac{\partial\hat{F}}{\partial t}\right) = \frac{Nd}{\epsilon_0 c^2}\frac{\partial^2\hat{\sigma}_x}{\partial t^2}e^{i\omega(t-z/c)}. \tag{4.20}$$

It will be convenient, as in chapter 3, to employ the atomic lowering and raising operators $\hat{\sigma} = \tfrac{1}{2}(\hat{\sigma}_x - i\hat{\sigma}_y)$ and $\hat{\sigma}^\dagger = \tfrac{1}{2}(\hat{\sigma}_x + i\hat{\sigma}_y)$, respectively, such that $[\hat{\sigma}, \hat{\sigma}^\dagger] = -\hat{\sigma}_z$, and to write

$$\hat{\sigma}(z,t) = \hat{s}(z,t)e^{-i\omega(t-z/c)} \tag{4.21}$$

where the operator $\hat{s}(z, t)$ is assumed to be slowly varying in the same sense as $\hat{F}(z, t)$. Then, in the rotating-wave approximation, we can replace (4.20) with

$$\frac{\partial \hat{F}}{\partial z} + \frac{1}{c}\frac{\partial \hat{F}}{\partial t} = \left(\frac{iNd\omega_0}{2\epsilon_0 c}\right)\hat{s} \qquad (4.22)$$

where, on the right-hand side, we have approximated ω by ω_0. This equation and the TLA Heisenberg equations [140]

$$\frac{\partial \hat{s}}{\partial t} = -i(\Delta - i\beta)\hat{s} - \frac{id}{\hbar}\hat{\sigma}_z\hat{F} \qquad (2\beta = 1/\tau_{RAD}) \qquad (4.23)$$

$$\frac{\partial \hat{\sigma}_z}{\partial t} = -2\beta(1 + \hat{\sigma}_z) - \frac{2id}{\hbar}(\hat{F}^\dagger \hat{s} - \hat{s}^\dagger \hat{F}) \qquad (4.24)$$

which follow straightforwardly from the Hamiltonian, provide a basis for a quantum theory of propagation in either amplifying or absorbing media.

In the semiclasical approximation in which the atom and field operators are replaced by their expectation values, equations (4.22)–(4.24) reduce to well-known Maxwell–Bloch equations.

Three limits are of particular interest:

(1) The limit $\beta \rightarrow 0$, $\Delta = 0$, and $\hat{\sigma}_z \rightarrow 1$ considered later gives equations (4.26)–(4.29) describing superfluorescence when the initial state of the field is the vacuum.
(2) The limit $\omega \gg \omega_0$ gives the ARS field equation, as discussed in section 4.2.1.
(3) The CKK case of large detuning, $\hat{\sigma}_z \rightarrow 1$, and a very short incoming pulse, is discussed in section 4.2.2.

If the field's central frequency ω is assumed to match exactly the atomic resonance frequency ω_0, so that $\Delta = 0$, and if we restrict ourselves to times short compared with the single-atom radiative lifetime τ_{RAD} and assume that the atoms remain with probability $\simeq 1$ in their excited states over times of interest, we can ignore (4.24) and replace $\hat{\sigma}_z(z, t)$ by 1 and equation (4.23) by

$$\frac{\partial \hat{s}}{\partial t} = -\frac{id}{\hbar}\hat{F}. \qquad (4.25)$$

In terms of the independent variables $\zeta = t - z/c$ and $\eta = z$,

$$\frac{\partial \hat{s}}{\partial \zeta} = -\frac{id}{\hbar}\hat{F} \qquad (4.26)$$

$$\frac{\partial \hat{F}}{\partial \eta} = \left(\frac{iNd\omega_0}{2\epsilon_0 c}\right)\hat{s} \qquad (4.27)$$

implying

$$\frac{\partial^2 \hat{s}}{\partial \eta \partial \zeta} = \left(\frac{\omega_p^2}{4c}\right)\hat{s} \qquad (4.28)$$

$$\frac{\partial^2 \hat{F}}{\partial \eta \partial \zeta} = \left(\frac{\omega_p^2}{4c}\right)\hat{F}. \tag{4.29}$$

Equations (4.26)–(4.29) have been used in studies of the build-up of superfluorescent radiation [141–143]. It will be useful for our purposes briefly to rederive here one of the most important results of those studies.

Equation (4.22) has the formal solution

$$\hat{F}(z,t) = \hat{F}_0(z,t) + \left(\frac{iNd\omega_0}{2\epsilon_0 c}\right)\int_0^z dz'\, \hat{s}\left(z', t - \frac{z-z'}{c}\right)\theta\left(t - \frac{z-z'}{c}\right)$$

$$= \hat{F}_0(z,t) + \left(\frac{iNd\omega_0}{2\epsilon_0 c}\right)\int_0^z dz'\, \hat{s}\left(z - z', t - \frac{z'}{c}\right)\theta\left(t - \frac{z'}{c}\right) \tag{4.30}$$

where we have chosen the retarded Green function over the advanced Green function in order to ensure causality. Here θ is the unit step function and $\hat{F}_0(z,t)$ is a solution of the homogeneous equation. We are interested in the expectation value $\langle \hat{F}^\dagger(L,t)\hat{F}(L,t)\rangle$ at the end ($z = L$) of the medium. For superfluorescence, the expectation value is taken over the vacuum state of the field, in which case the first term on the right-hand side of (4.30) does not contribute to expectation values. We may, therefore, ignore this term for our purposes. Defining $y = 2\sqrt{\zeta\eta}$, we find from (4.28) and (4.29) that $\hat{s}$ satisfies the differential equation for $I_0(y)$, the modified Bessel function of order zero [144]. The solution of interest for $\hat{F}(L,t)$ is then [141–143]

$$\hat{F}(L,t) = \left(\frac{iNd\omega_0}{2\epsilon_0 c}\right)\int_0^L dz'\, \hat{s}(L - z', 0)I_0\left(\omega_p\sqrt{(z'/c)(t - z'/c)}\right)\theta(t - z'/c). \tag{4.31}$$

In order to calculate $\langle \hat{F}^\dagger(L,t)\hat{F}(L,t)\rangle$, we require $\langle \hat{s}^\dagger(z',0)\hat{s}(z,0)\rangle$, which we evaluate later. We obtain [141–143]

$$\langle \hat{F}^\dagger(L,t)\hat{F}(L,t)\rangle = \left(\frac{d\omega_0}{2\epsilon_0 c}\right)^2 \frac{N}{S}\int_0^L dx\, \theta(t - x/c)I_0^2\left(\omega_p\sqrt{(x/c)(t - x/c)}\right). \tag{4.32}$$

For times large enough that I_0 may be replaced by its asymptotic form,

$$\langle \hat{F}^\dagger(L,t)\hat{F}(L,t)\rangle \sim \frac{\hbar\omega_0}{16\pi\epsilon_0 Sct}e^{4\sqrt{t/\tau_R}} \tag{4.33}$$

where we have used the equation $\omega_p^2 = Ne^2 f/m\epsilon_0 = 2Nd^2\omega_0/\hbar\epsilon_0$. Equating the intensity expectation value $2c\epsilon_0\langle \hat{F}^\dagger(L,t)\hat{F}(L,t)\rangle$ to the maximum expected SF intensity $N_T\hbar\omega_0/S\tau_R$, we arrive at the expression (4.13) for the time at which the SF pulse reaches its peak intensity. In the short-time limit, however,

$$\langle \hat{F}^\dagger(L,t)\hat{F}(L,t)\rangle \sim \left(\frac{d\omega_0}{2\epsilon_0 c}\right)^2 \frac{N}{S}ct \tag{4.34}$$

a result to which we will return in section 4.2.3.

Finally, let us derive the expression for $\langle \hat{s}^\dagger(z', 0)\hat{s}(z, 0)\rangle$ that we have used to obtain (4.32). For the initial state in which all the TLAs are in the upper state, $\langle \hat{s}_i(0)\rangle = 0$ and $\langle \hat{s}_i^\dagger(0)\hat{s}_j(0)\rangle = \delta_{ij}$. Then the operator

$$\hat{S} = \sum_{i=1}^{N_T} \hat{s}_i(0) \tag{4.35}$$

satisfies

$$\langle \hat{S}\rangle = 0 \tag{4.36}$$

$$\langle \hat{S}^\dagger \hat{S}\rangle = \sum_{i=1}^{N_T}\sum_{j=1}^{N_T} \langle \hat{s}_i^\dagger(0)\hat{s}_j(0)\rangle = N_T. \tag{4.37}$$

In the continuum limit,

$$\hat{S} = \frac{N_T}{L} \int_0^L dz\, \hat{s}(z, 0) \tag{4.38}$$

$$\langle \hat{S}^\dagger \hat{S}\rangle = \frac{N_T^2}{L^2} \int_0^L dz' \int_0^L dz'' \, \langle \hat{s}^\dagger(z', 0)\hat{s}(z'', 0)\rangle \tag{4.39}$$

and we can satisfy (4.36) and (4.37) by taking

$$\langle \hat{s}(z, 0)\rangle = 0 \tag{4.40}$$

$$\langle \hat{s}^\dagger(z', 0)\hat{s}(z'', 0)\rangle = \frac{L}{N_T}\delta(z' - z''). \tag{4.41}$$

4.2.1 Approximation leading to the ARS field equation

Our considerations thus far presume that the field's central frequency lies in the vicinity of the atomic resonance in the sense that the detuning Δ is very small in magnitude compared with ω and ω_0. Let us now suppose instead that the field frequency ω is very large compared with ω_0. In this case, we must work with the atomic operators $\hat{\sigma}_x$, $\hat{\sigma}_y$ instead of the slowly varying $\hat{s}$. It follows from the Hamiltonian (4.16) that

$$\ddot{\hat{\sigma}}_x + \omega_0^2 \hat{\sigma}_x = -\frac{2d\omega_0}{\hbar}\hat{\sigma}_z \hat{E} \cong -\frac{2d\omega_0}{\hbar}\hat{E} \tag{4.42}$$

in the approximation $\hat{\sigma}_z \cong 1$. The assumption $\omega \gg \omega_0$ implies

$$\ddot{\hat{\sigma}}_x \cong -\frac{2d\omega_0}{\hbar}\hat{E} \tag{4.43}$$

so that, from equation (4.18),

$$\left(\frac{\partial^2}{\partial t^2} - c^2 \frac{\partial^2}{\partial z^2} - \omega_p^2 \right) \hat{E} = 0. \tag{4.44}$$

This is identical to the equation of motion for the quantum field in the ARS model when we equate ω_p^2 to their m^2. From this perspective, the ARS equation of motion describes the interaction of the electromagnetic field with N unbound electrons ($\omega \gg \omega_0$) per unit volume. However, the usual plasma dispersion formula $n^2 = 1 - \omega_p^2/\omega^2$ for the refractive index n is replaced, in this case, by

$$n^2 = 1 + \omega_p^2/\omega^2. \tag{4.45}$$

This is a consequence of the assumption $\hat{\sigma}_z \cong 1$: had we assumed $\hat{\sigma}_z \cong -1$, we would have obtained the familiar plasma dispersion formula.

To describe the growth of the quantum noise with time in this model, we write (4.44) in the form

$$\frac{\partial^2 \hat{E}}{\partial \tau_1 \partial \tau_2} - \frac{m^2}{4} \hat{E} = 0 \tag{4.46}$$

where $\tau_1 = t - z/c$, $\tau_2 = t + z/c$. In terms of the independent variable $y = m\sqrt{\tau_1 \tau_2}$, equation (4.46) has solutions that are linear combinations of the zero-order modified Bessel functions $I_0(y)$, $K_0(y)$. For large t, the vacuum expectation value

$$\langle \hat{E}^2(z, t) \rangle \propto I_0^2(y) \sim \frac{e^{2mt}}{2\pi m t} \tag{4.47}$$

so that the quantum noise grows exponentially in time from the initial vacuum fluctuations, the fluctuations present before the medium in the ARS model is 'inverted'. This is the result cited after equation (4.5).

4.2.2 Signal and noise

We wish to determine to what extent the observation of the superluminal group velocity considered by CKK will be affected by quantum noise. The system of interest is described by the Heisenberg equations of motion (4.22) and (4.23). We approximate $\hat{\sigma}_z$ by 1, assuming that pulse durations τ_p and transit times L/c are sufficiently small that de-excitation of the initially inverted atoms by radiation (or any other decay process) is negligible. The situation here is different from that describing the onset of SF in that (a) the detuning Δ is not zero but is instead large and (b) the initial state of the field is not the vacuum but corresponds to a short pulse of radiation from some external source.

The equation for $\hat{s}(z, t)$ in the present model is

$$\frac{\partial \hat{s}}{\partial t} = -\mathrm{i}(\Delta - \mathrm{i}\beta)\hat{s} - \frac{\mathrm{i}d}{\hbar} \hat{F} \tag{4.48}$$

or

$$\hat{s}(z, t) = \hat{s}(z, t_0)e^{-i(\Delta - i\beta)(t-t_0)} - \frac{id}{\hbar} \int_{t_0}^{t} dt' \, \hat{F}(z, t')e^{i(\Delta - i\beta)(t'-t)} \qquad (4.49)$$

t_0 is some initial time, before any pulse is injected into the medium. We take $\hat{F}(z, t_0) = 0$, although of course what this really means is that there is no non-vanishing field or intensity in the medium at t_0, so that for practical purposes (expectation values) we can, in effect, ignore the operator $\hat{F}(z, t_0)$ in the equation for $\hat{s}(z, t)$.

The pulse is assumed to have a central frequency ω and to have no significant frequency components near the resonance frequency ω_0: $|\Delta|\tau_p > 1$. We assume that $|\Delta|\tau_p$ is large enough that we can approximate (4.49) by integrating by parts and retaining only the leading terms:

$$\hat{s}(z, t) \cong \hat{s}(z, t_0)e^{-i(\Delta - i\beta)(t-t_0)} - \frac{d}{\hbar} \frac{\Delta + i\beta}{\Delta^2 + \beta^2} \hat{F}(z, t) - \frac{id}{\hbar \Delta^2} \frac{\partial \hat{F}}{\partial t}. \qquad (4.50)$$

As will be clear from the analysis that follows, this approximation implies undistorted propagation of the incident pulse at the group velocity v_g, as assumed by CKK.

From (4.22),

$$\frac{\partial \hat{F}}{\partial z} + \frac{1}{c} \frac{\partial \hat{F}}{\partial t} \cong \left(\frac{iNd\omega_0}{2\epsilon_0 c} \right) \hat{s}(z, t_0)e^{-i(\Delta - i\beta)(t-t_0)} + \frac{g}{2} \hat{F} + i[n(\omega) - 1]\frac{\omega}{c} \hat{F}$$

$$+ \left(\frac{1}{c} - \frac{1}{v_g} \right) \frac{\partial \hat{F}}{\partial t} \qquad (4.51)$$

where

$$g \equiv \frac{Nd^2\omega_0}{\epsilon_0 \hbar c} \frac{\beta}{\Delta^2 + \beta^2} \qquad (4.52)$$

is the gain coefficient for propagation of a field with frequency ω in the inverted medium. We have used equation (4.6) for the refractive index $n(\omega)$ and (4.8) for $v_g/c - 1$. Writing $\hat{F}(z, t) = \hat{F}'(z, t) \exp^{(i[n(\omega)-1]\omega z/c)}$ and $\hat{s}(z, t_0) = \hat{s}'(z, t_0) \exp^{(i[n(\omega)-1]\omega z/c)}$ yields an equation in terms of the primed variables in which the term $i[n(\omega) - 1](\omega/c)z$ associated with phase velocity is eliminated. Then, ignoring for practical purposes the difference between the primed and unprimed variables, we have

$$\frac{\partial \hat{F}}{\partial z} + \frac{1}{v_g} \frac{\partial \hat{F}}{\partial t} = \frac{g}{2} \hat{F} + \left(\frac{iNd\omega_0}{2\epsilon_0 c} \right) \hat{s}(z, t_0)e^{-i(\Delta - i\beta)(t-t_0)} \qquad (4.53)$$

and, therefore,

$$\hat{F}(z, t) = \hat{F}(0, t - z/v_g)e^{gz/2} + \left(\frac{iNd\omega_0}{2\epsilon_0 c} \right) \int_0^z dz' \, \hat{s}(z', t_0)e^{g(z-z')/2}$$

$$\times \, e^{-i(\Delta-i\beta)[t-t_0-(z-z')/v_g]}\theta(t-t_0-(z-z')/v_g)$$

$$\equiv \hat{F}_s(0, t - z/v_g)e^{gz/2} + \hat{F}_n(z, t) \tag{4.54}$$

where the subscripts 's' and 'n' denote signal and noise, respectively. Here

$$\hat{F}_n(z, t) = \left(\frac{iNd\omega_0}{2\epsilon_0 c}\right) \int_0^z dz' \, \hat{s}(z', t_0)e^{g(z-z')/2}e^{-i(\Delta-i\beta)[t-t_0-(z-z')/v_g]}$$

$$\times \, \theta(t - t_0 - (z - z')/v_g) \tag{4.55}$$

is a quantum noise field associated with the quantum fluctuations of the atomic dipoles.

To appreciate the significance of g as defined by equation (4.52), consider the gain coefficient g_R for a radiatively broadened transition of frequency ω_0 and radiative decay rate $1/\tau_{RAD} = 2\beta$. For light of frequency $\omega = \omega_0 - \Delta$,

$$g_R = \frac{NS}{\tau_{RAD}} \frac{2\beta}{\Delta^2 + \beta^2} = \frac{Nd^2\omega_0}{\epsilon_0 \hbar c} \frac{\beta}{\Delta^2 + \beta^2} \tag{4.56}$$

if we assume that all the N atoms per unit volume are in the upper state of the amplifying transition. Thus, $g_R = g$, i.e. g is just the gain coefficient for amplification by stimulated emission. We note also that, from equation (4.8),

$$g = 2\beta \left(\frac{1}{c} - \frac{1}{v_g}\right) \tag{4.57}$$

in the case under consideration where the amplifying transition is radiatively broadened and the detuning is large compared with the gain bandwidth.

The operator $\hat{s}(z, t_0)$ has the expectation-value properties described earlier. These properties imply $\langle \hat{F}_n(z, t) \rangle = \langle \hat{F}_n^\dagger(z, t) \rangle = 0$ and

$$\langle \hat{F}_n^\dagger(z, t)\hat{F}_n(z, t) \rangle = \left(\frac{Nd\omega_0}{2\epsilon_0 c}\right)^2 \frac{L}{N_T}$$

$$\times \, e^{-2\beta(t-t_0)} \int_{z-v_g(t-t_0)}^z dz' \, e^{g(z-z')}e^{2\beta(z-z')/v_g}$$

$$= \left(\frac{Nd\omega_0}{2\epsilon_0 c}\right)^2 \frac{N}{S}\frac{c}{2\beta}\left[e^{gv_g t} - e^{-2\beta t}\right] \tag{4.58}$$

where we have used the relations (4.57) and $N_T = NSL$ and, to simplify the notation, we have taken $t_0 = 0$.

Since the atoms and field are initially uncorrelated, i.e.

$$\langle \hat{F}^\dagger(0, t - z/v_g)\hat{s}_j(t_0) \rangle = \langle \hat{F}_n^\dagger(0, t - z/v_g) \rangle \langle \hat{s}_j(t_0) \rangle = 0 \tag{4.59}$$

we have, at the end of the amplifier,

$$\langle \hat{F}^\dagger(L, t)\hat{F}(L, t) \rangle = \langle \hat{F}_s^\dagger(0, t - L/v_g)\hat{F}_s(0, t - L/v_g) \rangle e^{gL} + \langle \hat{F}_n^\dagger(L, t)\hat{F}_n(L, t) \rangle. \tag{4.60}$$

We define a signal-to-noise ratio

$$
\begin{aligned}
\text{SNR}(L, t) &\equiv \frac{\langle \hat{F}_s^\dagger(0, t - L/v_g) \hat{F}_s(0, t - L/v_g)\rangle e^{gL}}{\langle \hat{F}_n^\dagger(L, t) \hat{F}_n(L, t)\rangle} \\
&= \frac{\langle \hat{F}_s^\dagger(0, t - L/v_g) \hat{F}_s(0, t - L/v_g)\rangle e^{gL}}{(d\omega_0/2\epsilon_0 c)^2 (Nc/2S\beta)[e^{gL} - e^{-2\beta L/v_g}]} \\
&= \frac{\langle \hat{F}_s^\dagger(0, t - L/v_g) \hat{F}_s(0, t - L/v_g)\rangle}{(d\omega_0/2\epsilon_0 c)^2 (Nc/2S\beta)[1 - e^{-2\beta L/c}]} \\
&\simeq \frac{\langle \hat{F}_s^\dagger(0, t - L/v_g) \hat{F}_s(0, t - L/v_g)\rangle}{(d\omega_0/2\epsilon_0 c)^2 NL/S}.
\end{aligned}
\tag{4.61}
$$

In the denominators, we have taken $t = L/v_g$ for the time over which the atoms radiate and have used the fact that $2\beta(1/c - 1/v_g)L = gL$, the difference of two numbers that themselves are small according to our assumption that propagation times are small compared with the single-atom radiative decay rate, is $\ll 1$.

The numerator in equation (4.61) can be related to the expectation value q of the number of photons in the incident signal pulse as follows. The expectation value of the incident signal intensity is

$$
I_s(0, t) = \epsilon_0 v_g \langle \hat{F}_s(0, t) \hat{F}_s(0, t)\rangle = I_0 e^{-t^2/\tau_p^2}
\tag{4.62}
$$

for a Gaussian pulse of duration τ_p. Requiring that the energy fluence, namely $\int_{-\infty}^{\infty} dt\, I_s(z, t)$, be $q\hbar\omega/S \cong q\hbar\omega_0/S$ implies $I_0 = q\hbar\omega_0/(S\tau_p\sqrt{\pi})$ and, therefore,

$$
\langle \hat{F}_s^\dagger(0, t - L/v_g) \hat{F}_s(0, t - L/v_g)\rangle = q\frac{\hbar\omega_0}{\epsilon_0 v_g S\tau_p\sqrt{\pi}} e^{-(t-L/v_g)^2/\tau_p^2}.
\tag{4.63}
$$

Thus,

$$
\begin{aligned}
\text{SNR}(L, t) &= \frac{q}{2\tau_p\sqrt{\pi}} \frac{c}{v_g} e^{-(t-L/v_g)^2/\tau_p^2} \left(\frac{4\epsilon_0 \hbar c}{Nd^2\omega_0 L}\right) \\
&= \frac{q}{\sqrt{\pi}} \left(\frac{4c}{\omega_p^2 L\tau_p}\right) \frac{c}{v_g} e^{-(t-L/v_g)^2/\tau_p^2} \\
&= \frac{q}{\sqrt{\pi}} \frac{\tau_R}{\tau_p} \frac{c}{v_g} e^{-(t-L/v_g)^2/\tau_p^2}
\end{aligned}
\tag{4.64}
$$

where we have used equation (4.15).

Among the criteria given by CKK for the observation of a superluminal pulse is that 'the probe-pulse duration $[\tau_p]$ must not exceed $\tau_R = 4c/L\omega_p^2$'. This criterion implies, from equation (4.64), that $\text{SNR}(L, t) \geq (q/\sqrt{\pi})c/v_g$ and, therefore, that it is possible, even for $q \sim 1$, to have superluminal propagation with $\mathrm{SNR}(L, t) > 1$ if the pulse duration is short enough: $\tau_\mathrm{p} < \tau_\mathrm{R} c / v_\mathrm{g}$.

In order to relate this conclusion to ARS, we use equation (4.8) to write (4.64) as

$$\mathrm{SNR}(L, t) = \frac{q}{\sqrt{\pi}} \frac{\tau_\mathrm{p}}{\left(\frac{v_\mathrm{g}}{c} - 1\right) \frac{L}{c} \Delta^2 \tau_\mathrm{p}^2} e^{-(t - L/v_\mathrm{g})^2 / \tau_\mathrm{p}^2}. \tag{4.65}$$

We see from this expression that, if we impose the ARS condition (2), i.e. $(v_\mathrm{g}/c - 1) L/c \gg \tau_\mathrm{p}$, then

$$\mathrm{SNR}(L, t) \ll \frac{q}{\sqrt{\pi}} \frac{1}{(\Delta \tau_\mathrm{p})^2} e^{-(t - L/v_\mathrm{g})^2 / \tau_\mathrm{p}^2} \tag{4.66}$$

so that, given also the condition on $|\Delta| \tau_\mathrm{p}$ discussed before equation (4.50), *the signal-to-noise ratio will be very small when the ARS condition for strong distinguishability of superluminal propagation from propagation at the speed c is satisfied.*

In fact, if $(v_\mathrm{g}/c - 1) L/c \gg \tau_\mathrm{p}$ and, therefore, $\mathrm{SNR}(L, t)$ is very small for $q \approx 1$, then

$$t/\tau_\mathrm{R} = \frac{L/c}{\tau_\mathrm{R}} = \frac{v_\mathrm{g}}{c} \frac{L/v_\mathrm{g}}{\tau_\mathrm{R}} \gtrsim \frac{v_\mathrm{g}}{c} \frac{\tau_\mathrm{p}}{\tau_\mathrm{R}} \tag{4.67}$$

which, from (4.64), must be large. Then the noise associated with SF must be exponentially large [equation (4.33)]. It follows that q must be exponentially large in order to maintain a signal-to-noise ratio greater than unity. This is consistent with the ARS conclusion that 'for the signal amplitude to be larger than the amplitude of the fluctuations at the observation time, the signal amplitude should be exponentially large' [139].

Our results are, therefore, in agreement with ARS in that, if we require the separation of the superluminal pulse and a twin vacuum-propagated pulse to be much larger than the pulse duration, the signal-to-noise ratio will be very small at the one- or few-photon level. However, the results are not inconsistent with CKK: even at the one-photon level, we can achieve a signal-to-noise ratio greater than unity if this separation ($[v_\mathrm{g}/c - 1] L/c$) is smaller than the pulse duration τ_p [equation (4.65)].

4.2.3 Physical origin of noise limiting the observability of superluminal group velocity

Note that, when we set the time t in equation (4.34) for the short-time SF noise intensity equal to the 'observation time' L/c, we obtain exactly the noise intensity appearing in the denominator of equation (4.61). *Thus, the quantum noise that imposes limitations on the observability of superluminal group velocity is attributable to the initiation of superfluorescence.* We also note that the SF noise propagates at the speed of light and is, therefore, delayed with respect to the signal. This is a manifestation of a general result obtained by Segev *et al* [140].

4.2.4   Operator ordering and relation to ARS approach

Less obvious is the relation between the quantum noise we have considered—which stems from the *atomic dipole fluctuations* characterized by equations (4.40) and (4.41)—and the quantum noise of ARS, which is attributed to the *quantum fluctuations of the field*.

To establish the relation to the ARS approach, we return to our calculation of the noise intensity, using now *anti*-normally ordered field operators instead of the normally ordered operators used before. Thus, we consider now the expectation value $\langle \hat{F}(z,t)\hat{F}^\dagger(z,t)\rangle$ instead of $\langle \hat{F}^\dagger(z,t)\hat{F}(z,t)\rangle$. In this approach, the atomic dipole fluctuations play no explicit role, as can be seen from equation (4.54) and the fact that

$$\langle \hat{s}(z',t_0)\hat{s}^\dagger(z'',t_0)\rangle = 0 \qquad (4.68)$$

for excited atoms. In this case, however, the initially unoccupied modes of the field make a non-vanishing contribution as a consequence of non-normal ordering:

$$\langle \hat{F}(0,t-L/v_g)\hat{F}^\dagger(0,t-L/v_g)\rangle = \sum_k \frac{\hbar\omega_k}{2\epsilon_0 S\ell}\langle \hat{a}_k(0)\hat{a}_k^\dagger(0)\rangle e^{g(\omega_k)L}$$

$$\cong \sum_k \frac{\hbar\omega_k}{2\epsilon_0 S\ell}[g(\omega_k)L+1] \qquad (4.69)$$

which follows from (4.17) and (4.19) and the approximation $gL \ll 1$ upon which (4.64) is based. The contribution from the term that does not vanish as $L \to 0$ can be ignored, as it corresponds to vacuum quantum noise (energy $\frac{1}{2}\hbar\omega_k$ per mode) that is present even in the absence of the amplifier. In other words, the quantum noise of the field in the presence of the amplifier is

$$\langle \hat{F}(0,t-z/v_g)\hat{F}^\dagger(0,t-z/v_g)\rangle_n \equiv \sum_k \frac{\hbar\omega_k}{2\epsilon_0 S\ell}g(\omega_k)L$$

$$\to \frac{\ell}{2\pi c}\int d\omega \, \frac{\hbar\omega}{2\epsilon_0 S\ell}g(\omega)L$$

$$\cong \pi\left(\frac{\omega_0 d}{2\epsilon_0 c}\right)^2 \frac{NL}{c}\int_0^\infty d\omega \, \frac{\beta}{\Delta^2+\beta^2} \qquad (4.70)$$

where we have gone to the mode continuum limit, approximated ω by ω_0 in the numerator of the integrand, and used equation (4.52) for the gain coefficient. Performing the integration, we obtain exactly the noise term appearing in the denominator in the last line of (4.61). *But now the noise is attributable to the amplification of vacuum field fluctuations.*

Thus, we can attribute the quantum noise that limits the observation of superluminal group velocity to either the quantum fluctuations of the field in the inverted medium, as ARS do, or to the quantum fluctuations of the inverted atoms, as in our derivation of the signal-to-noise ratio. The situation here is similar to that in the theory of the initiation of SF, as discussed by Polder *et al* [142], or, as noted by those authors, to the theory of spontaneous emission by a single atom [8].

4.2.5   Limit of very small transition frequency

Since the origin of noise in the optical amplifier is associated ultimately with spontaneous emission, the question arises as to whether the signal-to-noise ratio might be increased by employing a transition having a very small transition frequency ω_0 and, therefore, a very large radiative lifetime. Indeed, since $\omega_p^2 \propto \omega_0$, the second line of equation (4.64) suggests, at first glance, that SNR $\to \infty$ in the limit $\omega_0 \to 0$. However, equation (4.7) shows that $v_g \to c$ in this limit: the superluminal effect itself becomes weaker as the spontaneous emission rate is made smaller.

In this connection, we invoke once again the form (4.65) of the signal-to-noise ratio. If we assume $|\Delta|\tau_p > 1$ in order that the pulse does not undergo substantial distortion as a consequence of strong absorption, then

$$\mathrm{SNR}(L, t) < q \frac{c\tau_p}{(v_g - c)L/c}. \qquad (4.71)$$

In other words, the signal-to-noise ratio must be smaller than the number of photons in the incident pulse mulitiplied by a factor equal to the length of the vacuum-propagated pulse divided by the separation of the vacuum-propagated pulse and the pulse emerging from the amplifier, *independent of the the atomic transition frequency or the radiative lifetime*. At the one- or few-photon level, the signal-to-noise ratio must, therefore, be less than unity under the ARS criteria for the observation of a superluminal group velocity, regardless of the frequency or strength of the amplifying transition.

4.2.6   Remarks

It may be useful to summarize the conclusions of this section briefly. We have considered the effects of quantum noise on the propagation of a pulse with a superluminal group velocity in an amplifying medium. In the case considered by CKK [39], where an off-resonant short pulse of duration τ_p propagates with superluminal group velocity v_g in an optical amplifier, we calculated a signal-to-noise ratio (SNR) and found that, for an incident pulse consisting of a single photon, the SNR $\ll 1$ under the condition $(v_g/c - 1)L \gg \tau_p$ assumed by ARS [139] for strong discrimination between the pulse propagating in the amplifier and a twin pulse propagating the same distance in vacuum. This result is fully consistent with the conclusions of ARS based on general considerations and, in particular, the reconstruction of the superluminal pulse from a truncated portion of the initial wave packet. However, if we impose the weaker condition that $(v_g/c - 1)L \gtrsim \tau_p$, then our conclusion is that SNR > 1 is possible. In this case, a superluminal group velocity is observable in the arrival statistics of many photons, not per shot.

These conclusions are consistent with figure 2.7: if the pulse is advanced by a time large compared with the pulse duration, then the output pulse is completely determined by a very small leading tail of the incident pulse. If this leading tail corresponds to a photon number less than one, then *the entire output pulse is, in effect, determined by quantum noise* and, consequently, the SNR ratio will be small [139].

We showed that, in the case of the optical amplifier, the quantum noise is attributable to the onset of superfluorescence and could be associated either with the quantum fluctuations of the field, along the lines of the ARS considerations, or with the quantum fluctuations of the atomic dipoles.

4.3 Signal velocity and photodetection

As we now discuss, quantum noise associated with spontaneous emission also acts effectively to retard the measured pulse by producing a background level of irradiation that must be exceeded before it can be asserted that the pulse has arrived.

We have defined a signal as a discontinuity such as a step-function turn-on of the field (section 2.5). According to this definition, the signal velocity of light is the velocity at which such a discontinuity propagates. Further consideration raises the question of how to define a signal velocity for a smoothly varying field described by an analytic function of time properly, for a very small portion of a pulse's leading edge can contain all the information about the entire pulse. More to the point, this leading edge can extend infinitely back in time and make it impossible to define unambiguously a signal arrival time.

In any event, this definition of a signal is not immediately applicable in the laboratory, where it is impossible to realize the infinite bandwidth associated with a step function. Moreover, this definition relies on purely classical concepts: in practice, one cannot extend the arrival time of a signal to any time before the detection of the first photon in a light pulse.

We now consider a more practical, operational definition of the signal velocity based on standard photodetection theory [145]. We analyse this signal velocity specifically for a superluminal group velocity in the case of a gain doublet medium, as discussed in section 2.4 [49, 50]. For the level scheme shown in figure 2.6, with the common detuning Δ_0 of the Raman and probe fields from the excited state $|0\rangle$ much larger than any of the Rabi frequencies or decay rates involved, we can adiabatically eliminate all off-diagonal density-matrix terms involving state $|0\rangle$. Then we obtain the following expression for the linear susceptibility as a function of the probe frequency [49, 148]:

$$\chi(\omega) = \frac{M}{\omega - \Delta\omega + i\gamma} + \frac{M}{\omega + \Delta\omega + i\gamma} \qquad (4.72)$$

where $\gamma > 0$ and $M > 0$ is a two-photon matrix element whose detailed form and numerical value are not required for our purposes.

Let us start by considering the detection of a 'signal' carried by a light pulse. We assign a time window T centred about a pre-arranged time t_0 at the detector and monitor the photocurrent produced by the detector. We assume there is a background level of irradiation that causes a constant photocurrent i_0 when no light pulse is sent. We further assume that a larger photocurrent $i_1(t)$ is registered when a light pulse is received. If the detector's integrated photocurrent $\int \mathrm{d}t\, i_1(t)$ rises above the background level by a certain amount, we assert that a 'signal' has been received. The time when this preset level of confidence is reached in the detection may be *defined* as the time of signal arrival.

From this point of view, the observable used to define the arrival of the signal carried by a light pulse is the integrated photon number

$$\hat{S}(L, t) = \eta \int_{t_0-T/2}^{t} \mathrm{d}t_1 \, \hat{E}^{(-)}(L, t_1)\hat{E}^{(+)}(L, t_1). \tag{4.73}$$

$\hat{E}^{(+)}(L, t_1)$ and $\hat{E}^{(-)}(L, t_1)$ are, respectively, the positive- and negative-frequency parts of the electric field operator at the exit port ($z = L$) of the medium and $t_0 = T_c + L/c$, where T_c is the time corresponding to the pulse peak. $T/2$ is half the time window assigned to the pulse, typically a few times the pulsewidth. η is a constant involving the quantum efficiency, and will be taken to be unity. The expectation value $\langle \hat{S}(L, t)\rangle$ gives the number of photons that have arrived at the detector's surface at an arbitrary time t. If $\langle \hat{S}_1(L, t)\rangle$ and $\langle \hat{S}_0(L, t)\rangle$ are, respectively, the photon number expectation value with and without an input pulse being present, then the photocurrent difference for an ideal detector is $\langle \hat{S}_1(L, t)\rangle - \langle \hat{S}_0(L, t)\rangle$. The second-order variance of the integrated photon number, $\langle \Delta^2 \hat{S}(L, t)\rangle$, gives the noise power due to quantum fluctuations. We, therefore, define an *optical signal-to-noise ratio* SNR$_o$ [146]

$$\mathrm{SNR}(L, t) = \frac{(\langle S_1(L, t)\rangle - \langle S_0(L, t)\rangle)^2}{\langle \Delta^2 S(L, t)\rangle}. \tag{4.74}$$

We define the arrival time t_s of a signal as the time when SNR(L, t) reaches some preset threshold level determined by the allowed error rate.

The positive-frequency part of the electric field operator of interest can be written as

$$\hat{E}^{(+)}(z, t) = \frac{1}{\sqrt{2\pi}} e^{-i\omega_0(t-z/c)} \int_0^{\infty} \mathrm{d}\omega \hat{a}(\omega)e^{-i\omega(t-z/v_g)} \tag{4.75}$$

where ω_0 is the carrier frequency of the pulse and $[\hat{a}(\omega), \hat{a}^\dagger(\omega')] = \delta(\omega - \omega')$. We assume plane-wave propagation in the z direction and that the group velocity approximation is valid.

In the experiments of interest [49], the transparent anomalously dispersive medium is a phase-insensitive amplifier, so that the general input–output relation can be applied [93, 147]:

$$\hat{a}_{\text{out}}(\omega) = g(\omega)\hat{a}_{\text{in}}(\omega) + \sqrt{|g(\omega)|^2 - 1}\,\hat{b}^\dagger(\omega). \qquad (4.76)$$

Here $\hat{a}_{\text{in}}(\omega)$ and $\hat{a}_{\text{out}}(\omega)$ refer, respectively, to the input ($z = 0$) and output ($z = L$) ports of the amplifier and $\hat{b}(\omega)$ is a bosonic operator ($[\hat{b}(\omega), \hat{b}^\dagger(\omega')] = \delta(\omega - \omega')$) that commutes with $\hat{a}_{\text{in}}(\omega)$ and $\hat{a}_{\text{in}}^\dagger(\omega)$ and whose appearance in (4.76) ensures that the commutation relations for the field operators $\hat{a}_{\text{out}}$ and $\hat{a}_{\text{out}}^\dagger$ are preserved. $|g(\omega)|^2$ is the power gain factor obtained from the imaginary part of $\chi(\omega)$.

Consider, first, the case of propagation over the distance L in vacuum, where $g(\omega) = 1$. We assume that the initial state $|\psi\rangle$ of the field is a coherent state such that $\hat{a}(\omega)|\psi\rangle = \alpha(\omega)|\psi\rangle$ for all ω, where $\alpha(\omega)$ is a c number. For such a state,

$$\hat{E}^{(+)}(0, t)|\psi\rangle = \alpha(t)e^{-i\omega_0 t}|\psi\rangle \qquad (4.77)$$

where

$$\alpha(t) = \pi^{-1/4}(N_{\text{p}}/\tau)^{1/2}e^{-(t-T_0)^2/2\tau^2} \qquad (4.78)$$

and N_{p} is the average number of photons in the initial (Gaussian) pulse of duration τ. It follows that

$$\text{SNR}_{\text{vac}}(L, t) = \langle \hat{S}_1(L, t)\rangle_{\text{vac}} = \text{SNR}_{\text{vac}}(0, t - L/c). \qquad (4.79)$$

In other words, the point $\text{SNR}_{\text{vac}}(L, t) = $ constant propagates at the velocity c without excess noise.

Next we treat the case of pulse propagation over the distance L in the anomalously dispersive medium, using equation (4.76) with $g(\omega) \neq 1$ and the same initially coherent field. We obtain, in this case,

$$\langle \hat{S}_1(L, t)\rangle - \langle \hat{S}_0(L, t)\rangle = |g(0)|^2 \langle \hat{S}_1(0, t - L/v_{\text{g}})\rangle_{\text{vac}} \qquad (4.80)$$

where

$$\langle \hat{S}_0(L, t)\rangle = \frac{1}{2\pi}\int_{t_0-T/2}^{t} dt_1 \int_0^\infty d\omega\,[|g(\omega)|^2 - 1] \qquad (4.81)$$

is the photon number in the absence of any pulse input to the medium. The fact that $\langle \hat{S}_0(L, t)\rangle > 0$ is due to amplified spontaneous emission (ASE) [146]: in the experiment of interest, the ASE is due to spontaneous Raman scattering.

For a probe pulse with sufficiently small bandwidth, the gain factor obtained from (4.72) is

$$|g(0)|^2 = e^{4\pi ML\gamma/\lambda(\Delta\omega^2+\gamma^2)} \qquad (4.82)$$

and the effective signal $\langle \hat{S}_1(L, t)\rangle - \langle \hat{S}_0(L, t)\rangle$ is proportional to the input signal $\langle \hat{S}_1(0, t - L/v_{\text{g}})\rangle_{\text{vac}}$ with time delay L/v_{g} determined by the group velocity v_{g}.

In an anomalously dispersive medium, $v_g = c/(n + \omega\, dn/d\omega)$ can be greater than c or negative, resulting in a time delay

$$\frac{L}{v_g} = \left[1 - \omega_0 M \frac{\Delta\omega^2 - \gamma^2}{(\Delta\omega^2 + \gamma^2)^2} \right] \frac{L}{c} \tag{4.83}$$

which is shorter than the time delay the pulse would experience upon propagation through the same length in vacuum: it can also become negative. In other words, the effective signal intensity defined here can be reached sooner than in the case of propagation in vacuum.

In order to determine with confidence when a signal is received, however, we must evaluate the SNR. Again using the commutation relations for the field operators, we obtain for the fluctuating noise background

$$\langle \Delta^2 \hat{S}(L, t) \rangle \equiv \langle \hat{S}^2(L, t) \rangle - \langle \hat{S}(L, t) \rangle^2$$
$$= |g(0)|^2 \langle \hat{S}_1(0, t - L/v_g) \rangle_{\text{vac}} + \langle \hat{S}_0(L, t) \rangle$$
$$+ 2|g(0)|^2 \text{Re}\left[\int_{t_0 - T/2}^{t} dt_1 \int_{t_0 - T/2}^{t} dt_2 \right.$$
$$\left. \times \alpha^*(t_1 - L/v_g)\alpha(t_2 - L/v_g) F(t_1 - t_2) \right]$$
$$+ \int_{t_0 - T/2}^{t} dt_1 \int_{t_0 - T/2}^{t} dt_2 |F(t_1 - t_2)|^2. \tag{4.84}$$

Here

$$F(t) = \frac{1}{2\pi} \int_{-\infty}^{\infty} d\omega \, [|g(\omega)|^2 - 1]e^{-i\omega t} \tag{4.85}$$

is a correlation function for the ASE noise. The four terms in equation (4.84) can be attributed to amplified shot noise, spontaneous emission noise, beat noise, and ASE self-beat noise, respectively [149]. Figure 4.1 shows the evolution of these noise terms within the time window T. Amplified shot noise is seen to dominate when the input pulse is strong.

Using equations (4.80) and (4.84), we compute $\text{SNR}_{\text{med}}(L, t)$ for the propagation through the anomalously dispersive medium. In figure 4.2, we plot the results of such computations for $\text{SNR}_{\text{med}}(L, t)$ as a function of time on the output signal. For reference, we also show the SNR for the identical pulse propagating over the same length in vacuum. It is evident that the pulse propagating in vacuum always maintains a higher SNR. In other words, for the experiments of interest here [49, 148], the signal arrival time as defined here is delayed, even though the pulse itself is advanced compared with propagation over the same distance in vacuum.

To further examine the signal velocity, let us require that, at a time t', the SNR of a pulse propagating through the medium be equal to that of the same pulse propagating through vacuum at a time t:

$$\text{SNR}_{\text{med}}(L, t') = \text{SNR}_{\text{vac}}(L, t). \tag{4.86}$$

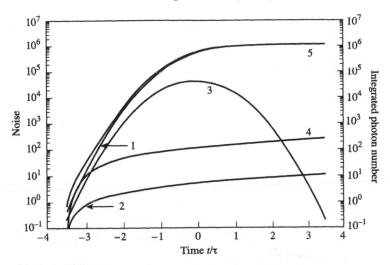

Figure 4.1. Evolution of quantum noise terms. Curves 1–5 indicate noise associated with terms 1–4 in equation (4.84) and the total noise, respectively. Parameters used in the figure are adopted from the experiments reported by Wang *et al* [49, 148]. There are 106 photons per pulse. Noise retards the detection of the signal by reducing the signal-to-noise ratio. From [145], with permission.

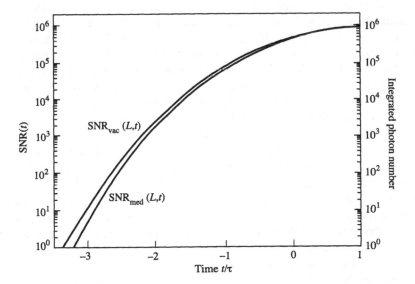

Figure 4.2. Signal-to-noise ratios for light pulses propagating through the gain medium and through vacuum, $SNR_{med}(L, t)$ and $SNR_{vac}(L, t)$. From [145], with permission.

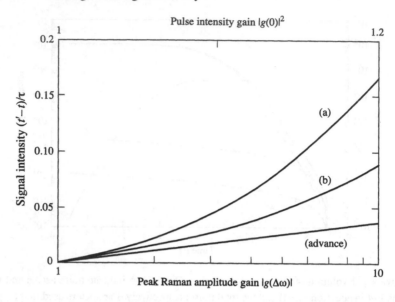

Figure 4.3. Delay in signal arrival time $\delta t = t' - t$ as a function of the gain coefficient. Curves (*a*) and (*b*) are for $(t - T_c)/\tau = -3$ and -1, respectively. Curve (*a*) is delayed more because at the early stage of the pulse, ASE self-beat noise produces noise much greater than the shot noise level. From [145], with permission.

Then we obtain a time difference $\delta t = t' - t$ that marks the retardation due to quantum noise. $\Delta t = t' - t + L/c$ gives the propagation time of the light signal and $L/\Delta t$ gives the signal velocity. In figure 4.3, we plot δt as a function of gain for $(t - T_c)/\tau = -3$ and -1. This corresponds to cases where the signal point is set at three and once times the pulsewidth on the leading edge of the pulse. We also plot for reference the pulse advance L/v_g. It is evident that the retardation in the SNR far exceeds the pulse advance. In other words, the *quantum noise* added in the process of advancing a signal effectively impedes the detection of the useful signal defined by the signal-to-noise ratio[5].

To summarize: we have presented an operational definition, based on photodetection, of the velocity of the signal carried by a light pulse. We have found for the experiments of Wang *et al* [49, 148] that, while the pulse and the effective signal are both advanced via propagation at a superluminal (or negative) group velocity, the signal velocity defined here is still bounded by c. The physical mechanism that limits the signal velocity is quantum fluctuation: because the transparent, anomalously dispersive medium is realized using closely placed gain lines, amplified quantum fluctuations introduce additional noise that effectively reduces the SNR in the detection of the signals carried by the light pulse.

[5] Zhu *et al* (2003) (S-Y Zhu, private communication) have obtained numerical results that differ slightly from those shown in [145], without affecting the conclusions therein.

Along similar lines, Wang *et al* [150] have found in numerical studies that the partial coherence of a pulse can change its group velocity from superluminal to subluminal.

4.4 Absorbers

The analyses in the two preceding sections applied to media with gain. In the case of absorbing media, a superluminal pulse will typically suffer a large decrease in amplitude, which adds to the difficulty of observing a large relative pulse advancement. Consider the example of a Lorentzian lineshape. From equations (1.33) and (1.34), and the expression $a(\omega) = 2n_I(\omega)\omega/c$ for the absorption coefficient, it follows that the group index

$$n_g(\omega) = n_R(\omega) + \omega \left(\frac{dn_R}{d\omega} \right)_\omega = 1 + \frac{c}{2\gamma} a(\omega) \left[\frac{\Delta^2 - \gamma^2}{\Delta^2 + \gamma^2} + \frac{\Delta}{\omega} \right]. \quad (4.87)$$

γ for an atomic vapour is typically $\sim 2\pi \times 10^9$ s^{-1}, in which case $ca(\omega)/2\gamma \sim 2.4a(\omega)$. Thus, a group index that differs greatly from unity (e.g. a large and negative value implying a superluminal group velocity) requires a large value of the absorption coefficient and, therefore, a strong attentuation of any superluminal pulse[6]. In fact, there appear to be no fast-light experiments—in either absorbers or amplifiers—where pulse advancements greater than about 25% of the pulse duration have been observed.

4.5 What is a signal?

It is not usually necessary to specify precisely what is meant by a signal[7]. A red traffic light tells a driver to stop and probably no one would argue that it thereby represents a signal. But a driver sitting at a red light is not learning anything *new* about whether he should remain stopped or drive on; that is, the red traffic light is not sending any *information*. Information might be said to be transmitted to the driver when the light turns from red to green. When we say that 'no signal can be transmitted faster than the speed of light in vacuum', we mean that no information can be transmitted faster than c. In other words, a signal in the sense used here is some measurable quantity *that carries information*, i.e. that transmits data that could not have been predicted with certainty beforehand. Thus, if our driver saw the traffic light turn from green to red as he approached the intersection, and knows that the light changes every 30 s, then he is not acquiring any information in this sense when the light turns back to green.

[6] The Garrett–McCumber theory of superluminal propagation of a Gaussian pulse in an absorber is reviewed in section 2.3.
[7] A dictionary defines a signal as 'a detectable physical quantity or impulse (as a voltage, current, or magnetic field strength) by which messages or information can be transmitted'.

Similarly, the arrival of the peak of a smooth superluminal pulse is not giving us any information that was not already contained in the leading edge of the pulse by analytic continuation (figure 2.7). New information appears only at points of non-analyticity on the waveform. Such points propagate at the velocity c, no matter how large the group velocity. Thus, as has been emphasized repeatedly, no signal (information) can be propagated with a velocity greater than c.

The principal message of this chapter is that the measurable advance in time (T_{adv}) of a superluminal pulse is reduced by noise arising from the field, the propagation medium, or the detector. We have analysed in some detail the effect of quantum noise in reducing T_{adv} in two specific examples.

The limiting effects of noise can be understood without detailed analyses. Figure 2.7 shows that, as the pulse advance is made larger, the advanced pulse is an analytic continuation of an increasingly small portion of the leading edge of the input pulse. Thus, for a large pulse advance relative to the pulsewidth, the truncated portion of the input pulse from which the advanced pulse is determined can correspond to an energy near or below the one-photon level, in which case the advanced pulse is determined primarily by quantum noise. In section 4.2.2, we obtained a signal-to-noise ratio in the case of an amplifying medium and confirmed, for this case, the general conclusion of Aharonov *et al* [139] that, for a very input weak pulse, the observed output will be dominated by noise. We showed that the quantum noise could be attributed either to the field or to the atoms constituting the medium. In other words, it should not be possible to observe an 'optical tachyon' except in the sense of an average over many shots, as in the single-photon tunnelling experiments of Chiao and Steinberg [44]. It should not be possible to observe a large relative pulse advance. (In the case of an absorbing medium, the advanced pulse will be even weaker than the small leading edge in the case of a large pulse advancement.)

The integrated photocurrent of a detector must exceed a certain threshold level, depending on the noise level and the allowed error rate, before it can register a pulse. Obviously, this has the effect of reducing the measured advance of a superluminal pulse. As discussed in section 4.3 for the experimental parameters of Wang *et al* [49], and as observed in the experiment of Centini *et al* [136], the measured advance implies that a superluminal pulse has a *signal* velocity less than c. The signal in this case is the detected arrival of the pulse.

Einstein causality requires that information encoded on a superluminal pulse cannot be transmitted with a velocity greater than c. This limit has been beautifully demonstrated in the experiments of Stenner *et al* [137]. Due to noise and detection latency, the measured signal (information) velocity is *less* than c. Theory based on Maxwell's equations without consideration of noise or the detection process predicts that the signal velocity should be exactly c, just as in the case of a Sommerfeld–Brillouin front.

The importance of noise in considerations of superluminal propagation and causality has also been emphasized by Wynne [151][8]. He considers the definition of information as a point of non-analyticity to be unsatisfactory because it ignores noise, whereas our point of view is that this definition is satisfactory for a description of the basic underlying (classical or quantum) physics, especially as it relates to Einstein causality. Wynne points out that, although an analytic pulse can, in principle, be 'extrapolated into the future', extrapolation to a significant degree (compared with the pulsewidth) can require a rather high-order Taylor expansion. Performing this Taylor expansion in practice will, therefore, require pulse samples at a correspondingly high number of points, and [151]

> [E]ach of these samples must contain a finite number of photons (in fact rather more than one photon) for this procedure to work. In addition, the sampling will have to be performed quickly with respect to the pulsewidth but slowly enough so that enough photons are detected to obtain meaningful samples. No matter how carefully the sampling and extrapolation are performed, the procedure will always result in a finite probability of making an error in predicting the future. This error can be made arbitrarily small by using a signal pulse with a peak amplitude that approaches infinity. However, to extrapolate a significant amount of time into the future (for example, ten pulsewidths) may require an unfeasibly large number of photons in the signal pulse. Extrapolation into the future only works for analytic and noiseless signals.

This bears on the question raised in section 4.2. But based on his definition of information (which is in accord with the definition used in communication theory), Wynne's conclusions are stronger than ours. He concludes that 'the only situation in which one might say that useful superluminal information exchange has taken place is if the superluminal advance is larger than the interval over which the input pulse is defined' [151]. The (unstated) conclusion seems to be that a measured time advance greater than the pulsewidth is impossible, regardless of the pulse intensity. We noted earlier that there appear to be no superluminal pulse propagation experiments in which an advance greater than about 25% of the full pulsewidth has been achieved.

4.6 Remarks

Before concluding the part of this book dealing mainly with fast light, I am impelled to repeat the apology in the preface to all the authors whose work I have not cited. The literature on the subject is simply too large for me to absorb

---

[8] Wynne's definition of a signal differs from ours. He defines a signal as 'any detectable waveform, which may contain information. Superluminal signal propagation does not necessarily imply superluminal transfer of information.' According to our definition, a signal *is* a carrier of information. This is also the definition adopted by Chiao and Steinberg [43], for instance. In reading the literature relating to superluminal pulses and causality, one should take care to note which definition is being used.

and to report sensibly on a sizable portion of it would require more than one book. It is appropriate, however, to briefly mention a few more experiments relating to some of the topics we have touched upon[9].

An experiment that certainly deserves mention is the early observation by Segard and Macke [153] of negative group velocity at millimetre wavelengths. In fact, these authors obtained a pulse advance on the order of 40% of the half-width-at-half-maximum pulsewidth, with modest pulse distortion. Tanaka *et al* [154] have observed comparable pulse advances and a group velocity of $-0.04c$ near an absorption line of rubidium vapour. Macke and Segard [155] have discussed some of the difficulties involved in realizing larger pulse advancements.

Superluminal pulse propagation has been observed in photonic grating media [156] and photonic crystals [157].

It should also be noted that negative group delays have been observed in electronic circuits. Mitchell and Chiao [158] have constructed an electronic bandpass amplifier with negative group delays at most frequencies for a modulated voltage pulse, and Kitano *et al* [159] have demonstrated negative group delays using an operational amplifier and an *RC* feedback circuit without any modulation of a carrier. These experiments can be analysed in the frequency domain using the linear input–output relations discussed at the beginning of section 1.3 [158].

Except for the discussion in chapter 2, we have not given much attention to the theoretical treatment of 'superluminal' behaviour in tunnelling. The arguments as to why there is no violation of Einstein causality are very much the same as in the theory of (non-evanescent) superluminal optical pulse propagation. We refer the reader to the review article by Chiao and Steinberg [43] and the many references therein, and to papers by Büttiker and Thomas [160] and Winful [161].

---

[9] A collection of papers with many references to the literature may be found in [152].

Chapter 5

Slow light

Group velocities can be *extremely small* compared with the speed of light (c) in vacuum. We begin this chapter by very briefly describing some early work in which pulses of light were slowed significantly compared with c. Then we discuss the phenomenon of electromagnetically induced transparency, which was the first physical mechanism employed to reduce the speed of a light pulse to virtually a snail's pace ('slow light'). Finally we describe other experiments and other physical processes that have been used to realize slow light.

5.1 Some antecedents

The observation of group velocities significantly less than c is not, in itself, new. We briefly recall here some experiments demonstrating group velocities smaller than c.

Near an absorption resonance the fact that $dn/d\omega < 0$ (anomalous dispersion) implies that the group velocity v_g can be superluminal. In section 2.4.1, we mentioned experiments in which the pulse repetition frequency of a mode-locked pulse train in a resonant absorber indicated group velocities very slightly greater than c. Near an amplification resonance, the sign of $dn/d\omega$ is reversed (cf section 2.2) and $v_g < c$. Casperson and Yariv [162] inferred group velocities $\sim c/2.5$ from the repetition frequency of mode-locked pulses in a 3.51-μm xenon discharge laser. The amplifying transition has high gain and a narrow Doppler resonance owing to the large mass of xenon. This results in a large value for $dn/d\omega$ and the small group velocities inferred.

Recall that, in the wings of an absorption resonance, the dispersion is normal ($dn/d\omega > 0$) and, therefore, $v_g < c$. For a strong transition with small linewidth, it is then possible to realize very small group velocities with a pulse central frequency well removed from the absorption resonance. Group velocities $\sim v_g/14$ in rubidium vapour have been demonstrated in this way [163]. The fact that the pulse frequencies are out in the wings of the absorption spectrum also results in relatively small attenuation of the slowed pulse.

Still smaller pulse propagation velocities are realized in self-induced transparency [164]. Pulse delays corresponding to peak velocities $\sim c/1000$ [165] and smaller have been observed.

We now turn our attention to some of the physics involved in realizing—and controlling—dramatically smaller group velocities.

5.2   Electromagnetically induced transparency

Extremely small group velocities have been obtained based on quantum interference effects that cause greatly reduced absorption and very rapid variation with frequency of the refractive index.

Let us begin with a very simple model for the sort of quantum interference that can lead to zero absorption. We consider a three-state 'atom' with equal transition dipole moments (μ) and equal applied electric field amplitudes ($\mathcal{E}$) at two equal transition frequencies (figure 5.1). The effect of an applied field on the atom in this model is described by the interaction Hamiltonian

$$H_{\text{Int}} = -\mu\mathcal{E}(|3\rangle\langle1| + |3\rangle\langle2|) + \text{h.c.} \tag{5.1}$$

Define two superposition states $|C\rangle = |1\rangle + |2\rangle$ and $|NC\rangle = |1\rangle - |2\rangle$ and note that $\langle3|H_{\text{Int}}|C\rangle = -2\mu\mathcal{E}$ and $\langle3|H_{\text{Int}}|NC\rangle = 0$.

The 'non-coupled' state $|NC\rangle$ is a *dark state*: an atom in such a superposition state does not interact with the applied field. When the applied field is turned on our three-state atom can be pumped into state 3, from which it can go by spontaneous emission into either the coupled state $|C\rangle$ or the non-coupled state $|NC\rangle$. Once in the latter state, it is trapped. Eventually the atom will find itself in the (non-absorbing) dark state by this process of 'coherent population trapping' [166].

Coherent population trapping was first observed by Alzetta *et al* [167] using a level scheme like that shown in figure 5.2. The states 1 and 2 in their experiment corresponded to hyperfine states of sodium which were driven into a coherent superposition state such that emission from state 3 was eliminated.

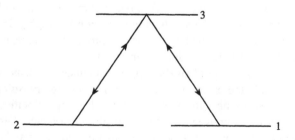

Figure 5.1. Three-state 'atom' in which the two allowed transitions $1 \leftrightarrow 3$ and $2 \leftrightarrow 3$ have the same transition frequency and electric dipole moment.

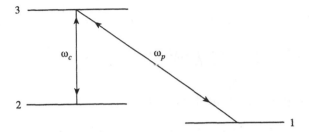

Figure 5.2. Three-level Λ scheme for realizing coherent population trapping and electromagnetically induced transparency. Transitions are allowed between states 1 and 3 and between states 2 and 3 but not between states 1 and 2. As a consequence of coherent population trapping in the presence of the coupling field (ω_c), a probe field at frequency ω_p can propagate without absorption.

Coherent population trapping in the presence of a coupling field can lead to electromagnetically induced transparency (EIT) [94, 168] for a probe field—i.e. the probe can propagate without absorption while the atoms remain unexcited. This is achieved when the probe and coupling frequencies differ by that of a non-allowed transition (figure 5.2). Associated with EIT is a rapidly varying refractive index and, consequently, a very small group velocity and, furthermore, the refractive index is unity and the group velocity dispersion is zero at line centre [169], as we will show.

The Hamiltonian for the three-level system of figure 5.2 is

$$\hat{H} = E_1\hat{\sigma}_{11} + E_2\hat{\sigma}_{22} + E_3\hat{\sigma}_{33} - \mu_{23}(\hat{\sigma}_{23} + \hat{\sigma}_{32})E(t) - \mu_{13}(\hat{\sigma}_{13} + \hat{\sigma}_{31})E(t) \quad (5.2)$$

where E_j is the energy eigenvalue associated with the state $|j\rangle$ in the absence of any atom–field coupling. We are using the $\hat{\sigma}$ operators employed in chapter 3, i.e. $\hat{\sigma}_{ij} = |i\rangle\langle j|$ at time $t = 0$, and we let these operators evolve in time according to the Heisenberg equation of motion

$$i\hbar\dot{\hat{\sigma}}_{ij} = [\hat{\sigma}_{ij}, \hat{H}]. \quad (5.3)$$

As in chapter 3, we denote operators by a caret ($\wedge$). μ_{ij} is the electric dipole matrix element between states i and j and we take it to be real without any loss of generality. We treat the applied electric field $E(t)$ classically. From the definition of the operators $\hat{\sigma}_{ij}$ at $t = 0$, and the orthogonality of the states $|1\rangle$, $|2\rangle$, and $|3\rangle$, it is easily seen that

$$[\hat{\sigma}_{ij}, \hat{\sigma}_{kl}] = \delta_{jk}\hat{\sigma}_{il} - \delta_{il}\hat{\sigma}_{kj}. \quad (5.4)$$

Since these commutation relations are satisfied at $t = 0$, they are satisfied at all times (by unitary time evolution). Using (5.3) and (5.4), therefore, we obtain the Heisenberg equations of motion

$$\dot{\hat{\sigma}}_{12} = -i\omega_{21}\hat{\sigma}_{12} + \frac{i}{\hbar}(\mu_{23}\hat{\sigma}_{13} - \mu_{13}\hat{\sigma}_{32})E(t) \quad (5.5)$$

$$\dot{\hat{\sigma}}_{23} = -i\omega_{32}\hat{\sigma}_{23} + \frac{i}{\hbar}\mu_{23}E(t)(\hat{\sigma}_{22}-\hat{\sigma}_{33}) + \frac{i}{\hbar}\mu_{13}E(t)\hat{\sigma}_{21} \qquad (5.6)$$

$$\dot{\hat{\sigma}}_{13} = -i\omega_{31}\hat{\sigma}_{13} + \frac{i}{\hbar}\mu_{23}E(t)\hat{\sigma}_{12} + \frac{i}{\hbar}\mu_{13}E(t)(\hat{\sigma}_{11}-\hat{\sigma}_{33}) \qquad (5.7)$$

where $\omega_{ij} = (E_i - E_j)/\hbar$.

For the applied electric field, we write[1]

$$E(t) = \tfrac{1}{2}\mathcal{E}_c[e^{-i\omega_c t} + e^{i\omega_c t}] + \tfrac{1}{2}\mathcal{E}_p[e^{-i\omega_p t} + e^{i\omega_p t}] \qquad (5.8)$$

where subscripts 'c' and 'p' refer to the 'coupling' field and the probe field, respectively (figure 5.2).

Since $\hat{\sigma}_{12}$, $\hat{\sigma}_{23}$, and $\hat{\sigma}_{13}$ oscillate most strongly at frequencies near ω_{21}, ω_{32}, and ω_{31}, respectively, we will make a rotating-wave approximation and replace equations (5.5)–(5.7) by

$$\dot{\hat{\sigma}}_{12} = -i\omega_{21}\hat{\sigma}_{12} + \frac{i}{2\hbar}\mu_{23}\mathcal{E}_c\hat{\sigma}_{13}e^{i\omega_c t} - \frac{i}{2\hbar}\mu_{13}\mathcal{E}_p\hat{\sigma}_{32}e^{-i\omega_p t} \qquad (5.9)$$

$$\dot{\hat{\sigma}}_{23} = -i\omega_{32}\hat{\sigma}_{23} + \frac{i}{2\hbar}\mu_{23}\mathcal{E}_c(\hat{\sigma}_{22}-\hat{\sigma}_{33})e^{-i\omega_c t} + \frac{i}{2\hbar}\mu_{13}\mathcal{E}_p\hat{\sigma}_{21}e^{-i\omega_p t} \qquad (5.10)$$

$$\dot{\hat{\sigma}}_{13} = -i\omega_{31}\hat{\sigma}_{13} + \frac{i}{2\hbar}\mu_{23}\mathcal{E}_c\hat{\sigma}_{12}e^{-i\omega_c t} + \frac{i}{2\hbar}\mu_{13}\mathcal{E}_p(\hat{\sigma}_{11}-\hat{\sigma}_{33})e^{-i\omega_p t}. \qquad (5.11)$$

We are interested in the expectation values of these operators and, in particular, in the expectation value $\langle\hat{\sigma}_{13}+\hat{\sigma}_{31}\rangle$ that determines the induced dipole moment and, therefore, the refractive index for the probe field. Since equations (5.9)–(5.11) are linear, we can effectively replace the operators by expectation values, i.e. $\hat{\sigma}_{ij} \to \langle\hat{\sigma}_{ij}\rangle \equiv \sigma_{ij}$. We can simplify further by assuming that the probe field is sufficiently weak that $\sigma_{11} \cong 1$, $\sigma_{22} = \sigma_{33} = \sigma_{23} = \sigma_{32} \cong 0$. Thus, for weak probe fields, we write

$$\dot{\sigma}_{12} = -i(\omega_{21}-i\gamma_{12})\sigma_{12} + \frac{i}{2\hbar}\mathcal{E}_c\mu_{23}\sigma_{13}e^{i\omega_{32}t} \qquad (5.12)$$

$$\dot{\sigma}_{13} = -i(\omega_{31}-i\gamma_{13})\sigma_{13} + \frac{i}{2\hbar}\mu_{23}\sigma_{12}\mathcal{E}_c e^{-i\omega_{32}t} + \frac{i}{2\hbar}\mu_{13}\mathcal{E}_p e^{-i\omega_p t}. \qquad (5.13)$$

We have made two additional modifications here of equations (5.9)–(5.11). First, we have assumed that the coupling field is exactly resonant with the $2 \leftrightarrow 3$ transition, i.e. $\omega_c = \omega_{32}$. Second, we have introduced the damping rates γ_{12} and γ_{13}, which are proportional to the homogeneous linewidths of the $1 \leftrightarrow 2$ and $1 \leftrightarrow 3$ transitions, respectively.

To solve these equations, it is convenient to introduce new variables S_{12} and S_{13} defined by writing

$$\sigma_{12} = S_{12}e^{-i(\omega_p-\omega_{32})t} \qquad \sigma_{13} = S_{13}e^{-i\omega_p t}. \qquad (5.14)$$

[1] We could allow $\mathcal{E}_p$ to be complex but this would have no real consequence.

Then (5.12) and (5.13) imply

$$\dot{S}_{12} = -i(\Delta - i\gamma_{12})S_{12} + \frac{i}{2\hbar}\mu_{23}\mathcal{E}_c S_{13} \tag{5.15}$$

$$\dot{S}_{13} = -i(\Delta - i\gamma_{13})S_{13} + \frac{i}{2\hbar}\mu_{13}\mathcal{E}_p + \frac{i}{2\hbar}\mu_{23}\mathcal{E}_c S_{12} \tag{5.16}$$

where $\Delta = \omega_{31} - \omega_p$ is the detuning of the probe field from the $1 \leftrightarrow 3$ transition. Equations (5.15) and (5.16) have the following steady-state solution for S_{13}:

$$S_{13} = \frac{(\mu_{13}\mathcal{E}_p/2\hbar)(\Delta - i\gamma_{12})}{(\Delta - i\gamma_{13})(\Delta - i\gamma_{12}) - (\mu_{23}\mathcal{E}_c/2\hbar)^2}. \tag{5.17}$$

The induced electric dipole moment at the probe frequency is

$$p = \mu_{13}(\sigma_{13} + \sigma_{31}) = \mu_{13}(S_{13}e^{-i\omega_p t} + S_{31}e^{i\omega_p t})$$
$$= \frac{1}{2}\alpha(\omega_p)\mathcal{E}_p e^{-i\omega_p t} + \frac{1}{2}\alpha^*(\omega_p)\mathcal{E}_p e^{i\omega_p t} \tag{5.18}$$

where the polarizability $\alpha(\omega_p)$ is given by

$$\alpha(\omega_p) = \alpha_R(\omega_p) + i\alpha_I(\omega_p) \tag{5.19}$$

with

$$\alpha_R(\omega_p) = \frac{\mu_{13}^2}{\hbar} \frac{\Delta(\Delta^2 - \frac{1}{4}\Omega_c^2 - \gamma_{12}\gamma_{13}) + \Delta\gamma_{12}(\gamma_{12} + \gamma_{13})}{[\Delta^2 - \gamma_{12}\gamma_{13} - \frac{1}{4}\Omega_c^2]^2 + \Delta^2(\gamma_{12} + \gamma_{13})^2} \tag{5.20}$$

$$\alpha_I(\omega_p) = \frac{\mu_{13}^2}{\hbar} \frac{\gamma_{12}(\gamma_{12}\gamma_{13} + \frac{1}{4}\Omega_c^2) + \Delta^2\gamma_{13}}{[\Delta^2 - \gamma_{12}\gamma_{13} - \frac{1}{4}\Omega_c^2]^2 + \Delta^2(\gamma_{12} + \gamma_{13})^2}. \tag{5.21}$$

In the approximation that the refractive index $n \cong 1$, $n = 1 + N\alpha/2\epsilon_0$, where N is the number density of atoms. This implies that the (real) refractive index and the (power) absorption coefficient are given by

$$n(\omega_p) = 1 + \frac{N}{2\epsilon_0}\alpha_R(\omega_p)$$
$$= 1 + \frac{N}{2\epsilon_0}\frac{\mu_{13}^2}{\hbar}\frac{\Delta(\Delta^2 - \frac{1}{4}\Omega_c^2 - \gamma_{12}\gamma_{13}) + \Delta\gamma_{12}(\gamma_{12} + \gamma_{13})}{[\Delta^2 - \gamma_{12}\gamma_{13} - \frac{1}{4}\Omega_c^2]^2 + \Delta^2(\gamma_{12} + \gamma_{13})^2} \tag{5.22}$$

$$a(\omega_p) = \frac{\omega_p}{c\epsilon_0}N\alpha_I(\omega_p)$$
$$= \frac{N\omega_p}{\epsilon_0 c}\frac{\mu_{13}^2}{\hbar}\frac{\gamma_{12}(\gamma_{12}\gamma_{13} + \frac{1}{4}\Omega_c^2) + \Delta^2\gamma_{13}}{[\Delta^2 - \gamma_{12}\gamma_{13} - \frac{1}{4}\Omega_c^2]^2 + \Delta^2(\gamma_{12} + \gamma_{13})^2} \tag{5.23}$$

respectively. Here, $\Omega_c = \mu_{23}\mathcal{E}_c/\hbar$ is the Rabi frequency characterizing the coupling field at the $2 \leftrightarrow 3$ transition. Note that if $\gamma_{12} = 0$, i.e. if there were no 'dephasing' of the $1 \leftrightarrow 2$ transition, the absorption coefficient would vanish (when $\omega_p = \omega_{31}$), corresponding to a perfect quantum interference effect.

Suppose that $\Delta = 0$, i.e. that the probe field is exactly resonant with the $1 \leftrightarrow 3$ transition. Then

$$n(\omega_p) = 1 \tag{5.24}$$

and

$$a(\omega_p) = \frac{2\pi N}{\epsilon_0 \lambda_p} \frac{\mu_{13}^2}{\hbar} \frac{\gamma_{12}}{\gamma_{12}\gamma_{13} + \frac{1}{4}\Omega_c^2} \tag{5.25}$$

where $\lambda_p = 2\pi c/\omega_p$ is the probe wavelength. Now the damping rate γ_{12}, being associated with the dipole *non-allowed* transition $1 \leftrightarrow 2$, is usually very small, so that the absorption at the probe frequency can be very small if $\Omega_c^2 \gg \gamma_{12}\gamma_{13}{}^2$. In other words, the effect of a strong coupling field at the $2 \leftrightarrow 3$ transition frequency is to produce EIT—very little absorption—at the probe frequency near the $1 \leftrightarrow 3$ transition. For large values of Ω_c, the absorption spectrum has two peaks separated by Ω_c (figure 5.3).

It is interesting to compare the absorption coefficient (5.25) with the absorption coefficient at the $1 \leftrightarrow 3$ transition frequency when there is no coupling field ($\Omega_c = 0$):

$$a(\omega_{31}) = \frac{2\pi N}{\epsilon_0 \lambda_p} \frac{\mu_{13}^2}{\hbar} \frac{1}{\gamma_{13}} \tag{5.26}$$

which is the well-known expression for the absorption coefficient for a homogeneously broadened transition at the centre of a Lorentzian profile. Comparison of (5.25) and (5.26) shows the crucial role for EIT of a strong coupling field in addition to the probe field.

We have assumed that $\omega_c = \omega_{32}$. More generally, we obtain for the polarizability

$$\alpha(\omega_p) = \frac{\mu_{13}^2}{\hbar} \frac{\Delta - \Delta_c - i\gamma_{12}}{(\Delta - i\gamma_{13})(\Delta - \Delta_c - i\gamma_{12}) - \frac{1}{4}\Omega_c^2} \tag{5.27}$$

where $\Delta_c = \omega_{32} - \omega_c$. This implies maximal transmission for $\Delta = (1 - \gamma_{12}/\gamma_{13})\Delta_c$. For high atomic number densities N, the transmission

$$T(\omega_p) = \exp(-N\alpha_I(\omega_p)kL) \tag{5.28}$$

($k = \omega_p/c$, L = propagation length) is significant only for frequencies very close to the maximal transmission frequency. The transmission can be estimated

---

[2] Such damping rates for forbidden transitions can be as small as a few Hz for atoms at rest. Larger damping rates are associated with transit-time broadening arising from the fact that the atoms spend a finite amount of time in the applied field. γ_{12} in this case can be reduced by introducing a noble buffer gas.

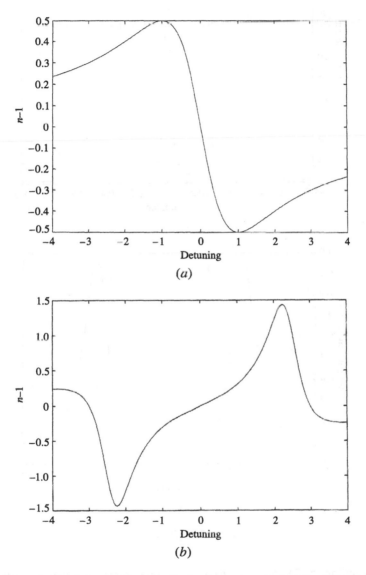

Figure 5.3. $n(\omega_p) - 1$ for (a) $\Omega_c = 0$, (b) $\Omega_c = 6\gamma_{13}$, and $a(\omega_p)$ for (c) $\Omega_c = 0$, (d) $\Omega_c = 6\gamma_{13}$. The detuning is $-\Delta/\gamma_{13} = [\omega_p - \omega_{31}]/\gamma_{13}$. $n(\omega_p) - 1$ is given in units of $N\mu_{13}^2/2\epsilon_0\hbar\gamma_{13}$, and $a(\omega_p)$ in units of $N\omega_p\mu_{13}^2/\epsilon_0 c\hbar\gamma_{13}$. In each case, $\gamma_{12}/\gamma_{13} = 0.02$.

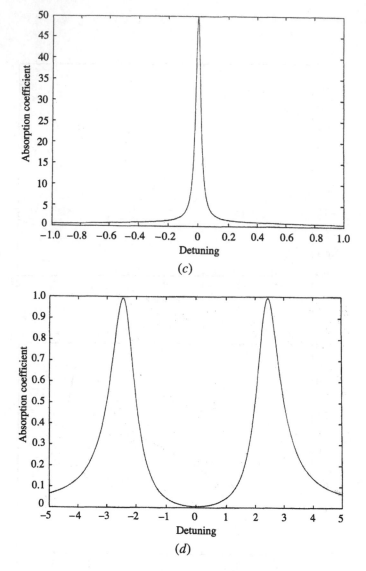

Figure 5.3. (Continued.)

by expanding (5.27) about this frequency: one obtains a transmission that is a Gaussian function of Δ with a half-width [170, 171]

$$\Delta\omega_{\text{tr}} \sim \frac{\Omega_c^2}{\sqrt{N\mu_{13}^2\omega_p\gamma_{13}L/\epsilon_0\hbar c}} \sim \frac{\Omega_c^2}{\gamma_{13}}\frac{1}{\sqrt{\eta kL}} \sim \frac{v_g}{L}\sqrt{\eta kL} \qquad (5.29)$$

where we have used $\gamma_{13} = \frac{1}{2}A_{31}$, $A_{31} = \mu_{13}^2\omega_p^3/3\pi\epsilon_0\hbar c^3$ being the rate of spontaneous emission at the $3 \rightarrow 1$ transition, and

$$\eta = 3N\lambda_\mathrm{p}^3/4\pi^2. \tag{5.30}$$

This result assumes $\gamma_{12} \ll \gamma_{13}$ and $\Omega_\mathrm{c}^2 \gg \gamma_{12}\gamma_{13}, |\Delta_\mathrm{c}|\gamma_{12}, \Delta_\mathrm{c}^2\gamma_{12}/\gamma_{23}$. Note that this transparency width *decreases* with increasing atomic density N. Using the fact that the absorption coefficient for a radiatively broadened transition is $a_0 = A_{31}\lambda^2 N/2\pi$ in the absence of EIT, we can write (5.29) equivalently as

$$\Delta\omega_\mathrm{tr} \sim \frac{v_\mathrm{g}}{L}\sqrt{a_0 L} = \frac{1}{\tau_\mathrm{d}}\sqrt{a_0 L} \tag{5.31}$$

where $\tau_\mathrm{d} = L/v_\mathrm{g}$ is the pulse delay in traversing a distance L.

Note also that the group velocity cannot be reduced to zero—i.e. the probe pulse cannot be stopped—for this would imply a vanishing window of transparency. In other words, if we attempt to make the group velocity go to zero, the transparency window becomes smaller than the spectral width of the pulse and there will be absorption rather than transparency. As discussed in the following chapter, however, the group velocity can go to zero if the Rabi frequency of the coupling field is made to vary appropriately in time.

Although the refractive index at $\omega = \omega_\mathrm{p}$ is unity, $dn/d\omega$ can be large and, therefore, the group velocity of a probe pulse can be very small. In the limit of large Ω_0, we obtain, from (5.22),

$$n(\omega_\mathrm{p}) \cong 1 - \frac{2N}{\epsilon_0} \frac{\mu_{13}^2}{\hbar\Omega_\mathrm{c}^2} \Delta \tag{5.32}$$

and

$$v_\mathrm{g}(\omega_\mathrm{p}) = \frac{c}{n(\omega_\mathrm{p}) + \omega_\mathrm{p}(dn/d\omega)_{\omega_\mathrm{p}}} \simeq \frac{\hbar c \epsilon_0 \Omega_\mathrm{c}^2}{2N\omega_\mathrm{p}\mu_{13}^2} \tag{5.33}$$

for $2N\omega_\mathrm{p}\mu_{13}^2/\hbar c\epsilon_0\Omega_\mathrm{c}^2 \gg 1$. A numerical example of the calculation of v_g from this formula is given later.

In figure 5.3, we plot the refractive index and the absorption coefficient [equations (5.22) and (5.23)] *versus* detuning Δ for a large value of Ω_c and for $\Omega_\mathrm{c} = 0$ and, in figure 5.4, we plot the group index n_g ($v_\mathrm{g} = c/n_\mathrm{g}$) for the same parameters. Note that the group index takes on large values near $\Delta = 0$, where the EIT effect is greatest.

EIT was first observed by Boller *et al* [172] in strontium vapour, the level scheme for which was of the Λ-type shown in figure 5.2 but with level 3 lying above the first ionization potential. The review articles by Harris [168] and Boyd and Gauthier [173] provide some historical background and many references relating to EIT.

As already noted, EIT can be interpreted as a consequence of coherent population trapping. The atoms are, in effect, pumped into the dark state, i.e. the transparency at the probe frequency is induced by driving the atoms with the

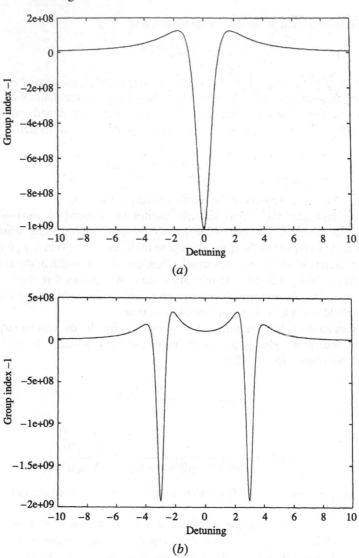

Figure 5.4. $n_g - 1$ *versus* frequency detuning $-\Delta/\gamma_{13} = [\omega_p - \omega_{31}]/\gamma_{13}$ for (a) $\Omega_c = 0$
and (b) $\Omega_c = 6\gamma_{13}$. $n_g - 1$ is in units of $N\mu_{13}^2/2\epsilon_0\hbar\gamma_{13}$. $\gamma_{12}/\gamma_{13} = 0.02$.

probe and coupling fields. Once the probe field is switched on, this pumping into
the dark state occurs on a time scale on the order of the radiative lifetime of the
excited state 3. If the probe field is turned on slowly (adiabatically) compared
with the Rabi frequency Ω_c, however, EIT occurs after a time $\sim \Omega_c^{-1}$, which can
be significantly shorter than the radiative lifetime.

It is interesting to note that the dark-state preparation can be done by applying the coupling field *before* the probe field is on. Because there is initially no population in the states 2 and 3 that are coupled by the coupling field, this circumstance is somewhat counterintuitive but it can be understood in terms of adiabatic preparation. Turning on the coupling field before the probe results in a dark eigenstate of the driven atom that is identical to the ground state. If the probe and coupling fields are turned on adiabatically, the S_{ij} follow the field variations in a quasi-steady-state way and the atom remains in the dark eigenstate [174]. In fact, the system remains in the dark state as long as the *difference* in the probe and coupling pulse envelopes varies slowly. In particular, if the temporal envelopes are identical, the atoms remain in a dark state even if the pulses vary rapidly in time. The pulses, in this case, are said to be 'matched'.

We have assumed that the probe pulse is weak and can be treated to first order in its effect on the atoms. If the probe is not weak, it can result in the propagation of a two-color pulse pair, called an 'adiabaton', that propagates with a single group velocity and can assume arbitrary pulse shapes, depending on the input pulses [175]. Non-adiabatic components of this pulse pair can make it unstable and, after a sufficiently large distance of propagation, can result in two matched pulses [176].

5.3   Slow light based on EIT

Associated with the narrow EIT spectral hole for the probe field is a rapid variation of the refractive index with frequency, a large value of the group index n_g (cf figure 5.4) and, therefore, a very small group velocity. Harris *et al* [169] estimated that for a ^{208}Pb vapour with a density of 7×10^{15} atoms cm^{-3}, a coupling laser wavelength of 405.9 nm and a Rabi frequency ~ 20 Ghz (laser intensity 283 kW cm^{-2}), an EIT probe pulse at 283 nm would have a group velocity $v_g = c/250$.

A group velocity $c/165$ in Pb vapour was subsequently observed by Kasapi *et al* [177] under EIT conditions with 55% transmission (absorption coefficient $\cong 600$ cm^{-1}) of the probe field in a 10-cm cell: without the coupling laser, the cell was 'nearly optically inpenetrable' (absorption coefficient $\cong 0.026$ cm^{-1}). They also observed that the transmitted field was nearly diffraction-limited.

An EIT pulse propagating without distortion at a group velocity $v_g \ll c$ undergoes substantial spatial compression compared with its free-space propagation, as noted by Harris *et al* [169]. Consider, for example, a Gaussian input pulse

$$\mathcal{E}_0(z = 0, t) = A e^{-t^2/\tau_p^2}. \tag{5.34}$$

Undistorted propagation at group velocity v_g implies that the field in the EIT medium is

$$\mathcal{E}(z, t) = \mathcal{E}_0(t - z/v_g) = A e^{-(t-z/v_g)^2/\tau_p^2} \tag{5.35}$$

compared with the field

$$\mathcal{E}(z,t) = \mathcal{E}_0(t - z/c) = Ae^{-(t-z/c)^2/\tau_p^2} \tag{5.36}$$

in free space. The EIT pulse is seen to undergo a spatial compression by the factor $c/v_g = n_g$. Note that the peak electric field remains the same as in free space if there is no absorption. The energy density of the field at frequency ω is [equation (2.61)]

$$u_\omega = \frac{n}{2\mu c v_g}|E_\omega|^2 \cong \frac{n_g}{2\mu_0 c^2}|E_\omega|^2 = \frac{1}{2}\epsilon_0 n_g|E_\omega|^2 \tag{5.37}$$

for $n \cong 1$ and $\mu \cong \mu_0$ (as is the case at optical frequencies). Thus, the electric field amplitude is the same as in free space while the energy density increases by the factor n_g and the intensity [equation (2.62)] $|S_\omega| = v_g u_\omega$ is the same as in free space. Harris and Hau [178] have discussed the fact that large nonlinear susceptibilities can accompany EIT. Note from the preceding remarks that any enhancement of nonlinear optical effects does not arise from an enhancement of the electric field strength, as also noted by Boyd and Gauthier [173].

5.3.1 Slow light in an ultracold gas

The first slowing down of light by many orders of magnitude—to 17 m s^{-1}—was observed by Hau *et al* [179] in experiments in which the EIT medium was an ultracold gas of sodium atoms. In these experiments, sodium atoms are cooled and trapped and optically pumped into the ($F = 1, M_F = -1$) state of the $^3S_{1/2}$ ground level (figure 5.5) and then put into a magnetic trap and evaporatively cooled to a temperature $T \sim 450$ nK with a number density $\sim 3 \times 10^{12}$ cm^{-3}. A linearly polarized coupling beam creates the quantum interference that allows a left-circularly polarized probe beam (figure 5.5), applied a few microseconds after the coupling beam and obtained from the same dye laser, to propagate with little absorption (EIT) at right angles to the coupling beam.

Doppler broadening is negligible and the width of the EIT spectral hole is much smaller than the natural linewidth of the 1 $\leftrightarrow$ 3 transition. This, of course, is responsible for the extremely strong variation of the refractive index with probe frequency and, therefore, the extremely small observed group velocities inferred from pulse delay measurements.

Using $\Omega_c = \mu_{23}\mathcal{E}_c/\hbar = (\mu_{23}/\hbar)\sqrt{2I_c/c\epsilon_0}$, we can write equation (5.33) as

$$v_g(\omega_p) = \frac{\mu_{23}^2}{\mu_{13}^2}\frac{I_c}{N\hbar\omega_p} \tag{5.38}$$

where I_c is the coupling field intensity. For a probe wavelength of 589 nm, a coupling field intensity of 12 mW cm^{-2}, and an atomic number density $N = 3.3 \times 10^{12}$ cm^{-3}, we obtain from this formula

$$v_g(\omega_p) = \frac{\mu_{23}^2}{\mu_{13}^2} \times 110 \text{ m s}^{-1} = 73 \text{ m s}^{-1} \tag{5.39}$$

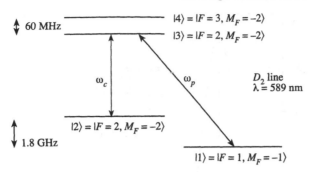

Figure 5.5. Transitions between hyperfine states of the sodium D_2 line in the experiments of Hau *et al.* From [179], with permission.

where we use $\mu_{23}^2/\mu_{13}^2 = 2/3$ based on the Wigner–Eckart theorem. For a coupling field intensity of 12 mW cm^{-2} and a peak atomic number density of 3.3×10^{12} cm^{-3}, the experimentally inferred group velocity was 32.5 m s^{-1}, corresponding to a time delay of several pulsewidths (figure 5.6). When the temperature was reduced sufficiently to obtain a Bose–Einstein condensate, the increase in the atomic density is predicted by equation (5.38) to result in a smaller group velocity and, in this case, a group velocity $v_g = 17$ m s^{-1} was observed. Pulse compression by the factor c/v_g implied that the ultraslow pulses had a spatial extent as small as 43 μm.

5.3.2 Slow light in a hot gas

Slow light is a consequence of a large value of $dn/d\omega$ associated with a sharp resonance. It might, therefore, be surmised that slow light cannot be realized in a hot gas, where atomic motion broadens the resonance (Doppler broadening). However, a very narrow EIT resonance can, in fact, be obtained in a hot gas if Doppler and other line-broadening mechanisms can be suppressed.

Defining the detuning $\delta = \omega_{21} - (\omega_p - \omega_c)$, which reduces to Δ when $\omega_c = \omega_{32}$, we can write the polarizability (5.27) as

$$\alpha(\omega_p) = 2\mu_{13}S_{13}/\mathcal{E}_p = \frac{3\epsilon_0\lambda_p^3}{8\pi^2}A_{31}\frac{i\gamma_{12} - \delta}{(\gamma_{12} + i\delta)(\gamma_{13} + i\Delta) + \frac{1}{4}\Omega_c^2}. \quad (5.40)$$

where we have used the equation $A_{31} = 8\pi^2\mu_{13}^2/3\epsilon_0\hbar\lambda_p^3$ for the radiative decay rate of the $3 \to 1$ transition.

This expression for the polarizability does not take atomic motion into account. An atom moving with velocity v along the direction of the probe field sees a field of frequency $\omega_p - k_p v$, where $k_p = n(\omega_p)\omega_p/c$. Likewise an atom with velocity v along the direction of the coupling field sees a field of frequency $\omega_c - k_c v$, where $k_c = n(\omega_c)\omega_c/c$. Assuming the coupling and probe fields are

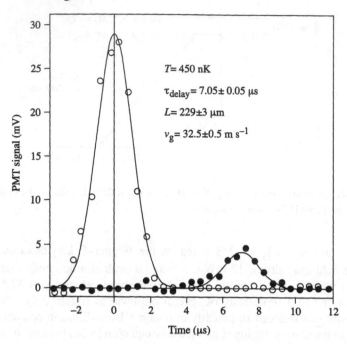

Figure 5.6. Experimental data of Hau *et al* showing the input pulse (open circles) and the pulse transmitted by an ultracold gas of length 229 ± 3 μm (filled circles). The transmitted pulse delay of 7.05 ± 0.05 μs corresponds to a group velocity of 32.5 ± 0.5 m s^{-1}. From [179], with permission.

propagating in the same direction, we can obtain the polarizability $\alpha(\omega_p, v)$ for an atom with velocity v along the propagation direction by replacing Δ in (5.40) by $\omega_{31} - (\omega_p - k_p v) = \Delta + k_p v$ and δ by $\omega_{21} - (\omega_p - k_p v - \omega_c + k_c v) = \delta + (k_p - k_c)v$:

$$\alpha(\omega_p, v) = \frac{3\epsilon_0 \lambda_p^3}{8\pi^2} A_{31} \frac{i\gamma_{12} - (\delta + (k_p - k_c)v}{[\gamma_{12} + i(\delta + (k_p - k_c)v)][\gamma_{13} + i(\Delta_p + k_p v)] + \frac{1}{4}\Omega_c^2}.$$
(5.41)

The polarizability $\alpha(\omega_p)$ for an atomic gas is then obtained by averaging $\alpha(\omega_p, v)$ over the (one-dimensional) Maxwell–Boltzmann velocity distribution $f(v)(\int_{-\infty}^{\infty} dv\, f(v) = 1)$ [180]:

$$\alpha(\omega_p) = \frac{3\epsilon_0 \lambda_p^3}{8\pi^2} A_{31}$$

$$\times \int_{-\infty}^{\infty} dv\, f(v) \frac{i\gamma_{12} - (\delta + (k_p - k_c)v}{[\gamma_{12} + i(\delta + (k_p - k_c)v)][\gamma_{13} + i(\Delta_p + k_p v)] + \frac{1}{4}\Omega_c^2}$$
(5.42)

The integration simplifies if we make the approximation $k_p = k_c = k$ and replace the (Gaussian) distribution $f(v)$ by the Lorentzian form [180]

$$f(kv) = \frac{\Delta\omega_D/\pi k}{(kv)^2 + \Delta\omega_D^2} \qquad (5.43)$$

where $\Delta\omega_D$ is the half-width-at-half-maximum (HWHM) of the distribution. Then (5.42) becomes [180]

$$\alpha(\omega_p) = \frac{3\epsilon_0\lambda_p^3}{8\pi^2}A_{31}\frac{i\gamma_{12} - \delta}{(i\Delta + \gamma_{13} + \Delta\omega_D)(i\delta + \gamma_{12}) + \frac{1}{4}\Omega_c^2}. \qquad (5.44)$$

Using the approximation $n \cong 1 + N\alpha/2\epsilon_0$ for $n \cong 1$, this implies the (power) absorption coefficient

$$
\begin{aligned}
a(\omega_p) &= \frac{2\omega_p}{c}n_I(\omega_p) = \frac{2\pi N}{\epsilon_0\lambda_p}\alpha_I(\omega_p) \\
&= \frac{3N\lambda_p^2}{4\pi}A_{31}\frac{\gamma_{12}}{\gamma_{12}(\gamma_{13} + \Delta\omega_D) + \frac{1}{4}\Omega_c^2} \qquad (5.45)
\end{aligned}
$$

for $\Delta = \delta = 0$, i.e. for $\omega_p = \omega_{31}$ and $\omega_c = \omega_{32}$. The group velocity $v_g = c/n_g$ follows similarly from the group index

$$n_g = \frac{3Nc\lambda_p^2}{8\pi}A_{31}\frac{\frac{1}{4}\Omega_c^2}{[\gamma_{12}(\gamma_{13} + \Delta\omega_D) + \frac{1}{4}\Omega_c^2]^2} \qquad (5.46)$$

for large Ω_c ($n_g \gg 1$). For $\frac{1}{4}\Omega_c^2 \gg \gamma_{12}(\gamma_{13} + \Delta\omega_D)$, the group velocity is

$$v_g = \frac{c}{n_g} = \frac{\hbar c\epsilon_0\Omega_c^2}{2N\omega_p\mu_{13}^2} \qquad (5.47)$$

which is identical to (5.33), which was obtained for *stationary* atoms. In other words, if the Rabi frequency Ω_c is large enough, it is possible to realize ultraslow group velocities in a hot gas as well as in a cold gas. Equation (5.45) shows that the absorption coefficient can likewise be very small.

The reason for this can be seen from (5.41). For co-propagating coupling and probe fields with $k_p = k_c$, as we have assumed, the only velocity dependence of $\alpha(\omega_p, v)$ is in the term $\gamma_{13} + i(\Delta_p + k_p v)$ in the denominator and, when Ω_c is much larger than $\gamma_{12}(\gamma_{13} + \Delta\Omega_D)$, the Doppler width $\Delta\omega_D$ has a negligible effect on the velocity-averaged polarizability (5.44). Unlike the case in cold gases, however, this 'Doppler-free' feature requires co-propagating coupling and probe fields.

The prediction of ultraslow group velocities based on EIT with co-propagating fields in a hot gas was verified in the experiments of Kash *et al* [180]. From pulse delay measurements, they inferred a group velocity $\sim$ 90 m s^{-1} in ^{87}Rb gas at 360 K. The ^{87}Rb density (2×10^{12} cm^{-3}) was comparable to the densities in the cold-gas experiments of Hau *et al* [179].

150   *Slow light*

It should be noted that, in these experiments, the dominant contribution to γ_{12} comes from the transit-time broadening associated with the time spent by the atoms in the 2-mm-diameter laser beam. γ_{12} was reduced to below 1 kHz using 30 Torr of He buffer gas. A larger beam diameter would reduce γ_{12} further and would presumably result in even smaller probe group velocities. A magnetic shield employing a high-permeability metal reduced the effect of stray magnetic fields that produce a Zeeman splitting of the hyperfine sublevels of sodium that are $(2F + 1)$-fold degenerate in the absence of a magnetic field.

5.4   Group velocity dispersion

In chapter 2 we noted that one of the arguments in the past against observing superluminal group velocities was that, when $v_g > c$, a pulse undergoes so much distortion in its shape that the concept of group velocity no longer makes sense. Experiments showed that this argument was incorrect, that pulses could, in fact, propagate with a superluminal group velocity and with little distortion. The slow pulses in the experiments described in the preceding section also propagate without significant distortion of their temporal profiles [cf equations (5.34) and (5.35)]. In this section, we briefly review the conditions necessary to realize distortionless linear pulse propagation [181, 182]. These considerations apply to both fast light and slow light.

Using equation (1.83), we write

$$\frac{\partial \mathcal{E}}{\partial z} + \frac{1}{v_g} \frac{\partial \mathcal{E}}{\partial t} + \frac{i}{2} \beta \frac{\partial^2 \mathcal{E}}{\partial t^2} = 0 \tag{5.48}$$

where we assume that the terms involving third and higher derivatives of $\mathcal{E}$ with respect to time are negligible, which is usually the case except for extremely short pulses. We have defined the group velocity dispersion (GVD) parameter

$$\beta = \frac{d^2 k}{d\omega^2} = \frac{2}{c} \frac{dn}{d\omega} + \frac{\omega}{c} \frac{d^2 n}{d\omega^2} \tag{5.49}$$

which is understood to be evaluated at the central frequency of the pulse [ω_L in equation (1.83)]. In terms of the new independent variables $\eta = z$ and $\tau = t - z/v_g$, we can rewrite (5.48) as

$$\frac{\partial \mathcal{E}}{d\eta} + \frac{i}{2} \beta \frac{\partial^2 \mathcal{E}}{\partial \tau^2} = 0. \tag{5.50}$$

If $\beta = 0$ (no group velocity dispersion), a pulse will propagate without distortion at the group velocity. GVD will cause the pulse to deviate from its initial shape. This is clear from the definition of β:

$$\beta = \frac{d}{d\omega}\left(\frac{dk}{d\omega}\right) = \frac{d}{d\omega}\left(\frac{1}{v_g}\right) = \frac{1}{c}\frac{dn_g}{d\omega} = -\frac{c}{v_g^2}\frac{dv_g}{d\omega}. \tag{5.51}$$

Thus, if $\beta > 0(< 0)$, the group velocity decreases (increases) with increasing frequency. Different frequency components of a pulse will propagate with different velocities and the pulse will not propagate without distortion at a well-defined group velocity.

One way to estimate the degree of pulse distortion is to compare equation (5.50) to the paraxial wave equation for a monochromatic field of wavelength $\lambda = 2\pi/k$ (see, for instance, [57]):

$$\frac{\partial \mathcal{E}}{\partial z} - \frac{i}{2} \frac{1}{k} \frac{\partial^2 \mathcal{E}}{\partial x^2} = 0 \qquad (5.52)$$

where we consider variations along only one direction (x) transverse to the direction (z) of beam propagation. If the beam at $z = 0$ has a width $\sim a$ in the transverse plane, then after a propagation distance z, it will have a width

$$\Delta x \sim z \frac{\lambda}{a}. \qquad (5.53)$$

Now equation (5.50) for the propagation of a plane wave in a dispersive medium has the same form as (5.52). The identifications $\eta \leftrightarrow z$, $\tau \leftrightarrow x$, and $\beta \leftrightarrow \lambda/2\pi$ imply from (5.53) that, if a pulse has an initial temporal width τ_p, it will have a width

$$\Delta t \sim L \frac{|\beta|}{t_p} \qquad (5.54)$$

after propagating a distance L in a dispersive medium. We define a characteristic propagation distance L_{GVD} at which the pulse spread becomes comparable to the initial pulse duration ($\Delta t \sim \tau_p$):

$$L_{GVD} \equiv \frac{\tau_p^2}{|\beta|}. \qquad (5.55)$$

Near the probe resonance, the refractive index $n(\omega)$ in EIT varies linearly with ω: $d^2 n/d\omega^2 = 0$ [equation (5.32)] and, therefore,

$$\beta = \frac{2}{c} \frac{dn}{d\omega}. \qquad (5.56)$$

Then

$$\frac{L}{L_{gvd}} = \frac{L|\beta|}{\tau_p^2} = \frac{2L}{c\tau_p^2} \left| \frac{dn}{d\omega} \right| = \frac{2}{\omega \tau_p} \frac{|L/v_g - L/c|}{\tau_p} \qquad (5.57)$$

where we have used the fact that $n = 1$ at the probe resonance. The second factor is the ratio of the pulse delay (compared with vacuum propagation) to the pulse duration, and is ~ 1 or less in all reported slow-light experiments. The first factor involves the ratio of the optical period to the pulse duration and is obviously very small. Thus,

$$\frac{L}{L_{gvd}} \ll 1 \qquad (5.58)$$

i.e. the distance of propagation in the experiments is much smaller than the distance at which GVD and pulse distortion are significant. For this reason, slow-light pulses in the EIT experiments described in the preceding section propagate with little distortion. The same considerations apply to the fast-light experiments described in chapter 2, where the pulse advance is smaller than the pulsewidth.

5.5   Slow light in solids

Solid materials would be preferable in possible applications of slow light. The difficulty with using solid-state media for realizing slow light with EIT is the large (compared with gases) damping rates that broaden spectral features and preclude the very large values of $dn/d\omega$ that can occur in ultracold gases with co-propagating (Doppler-free) coupling and probe fields.

Certain insulators doped with rare earths, however, exhibit narrow absorption features and spectral hole burning and, in particular, EIT [183] and slow light [184] have been observed using Pr-doped Y_2SiO_5 at 5 K. In the slow-light experiments, the coupling and probe fields propagate at right angles and the crystal becomes transparent as a consequence of optical pumping by the linearly polarized fields. The spectral hole in the absorption profile then prevents EIT but a third 're-pump' field at a different frequency results in a narrow absorption region, or 'anti-hole', centred at the spectral hole produced by the probe and coupling fields. Within such an anti-hole, EIT can occur at the probe frequency. Turukhin *et al* [184] have measured pulse delays $> 65\ \mu s$ in a 3-mm crystal, corresponding to a group velocity of 45 m s^{-1}.

It has also been possible to observe slow light in remarkably simple experiments using a *room-temperature* solid [185, 186]. The physical basis for slow light in this case—coherent population oscillations [187]—is entirely different from EIT and it is appropriate, therefore, to review it here.

5.5.1   Coherent population oscillations

Unlike EIT, the approach of Bigelow *et al* [185, 186] for slowing light can be understood using a *two*-level model for the propagation medium. The equations for a two-level atom in an applied electric field $E(t)$ can be obtained from the three-level model of section 5.2 by taking $\mu_{23} = 0$. Then equation (5.7) reduces to

$$\dot{\hat{\sigma}}_{13} = -i\omega_{31}\hat{\sigma}_{13} + \frac{i}{\hbar}\mu_{13}E(t)(\hat{\sigma}_{11} - \hat{\sigma}_{33}).\qquad(5.59)$$

For the operator $\hat{w}(t) \equiv \hat{\sigma}_{33}(t) - \hat{\sigma}_{11}(t)$, we obtain from the Hamiltonian (5.2) and the commutation rule (5.4) the Heisenberg equation of motion

$$\dot{\hat{w}} = -\frac{2i}{\hbar}\mu_{13}E(t)(\hat{\sigma}_{13} - \hat{\sigma}_{31}).\qquad(5.60)$$

As in section 5.2, we can replace the operators by their expectation values in the semiclassical approach in which the electric field $E(t)$ is treated classically. Denoting $\langle \hat{\sigma}_{13} \rangle$ by $\sigma(t)$, $\langle \hat{w}(t) \rangle$ by $w(t)$, μ_{13} by μ, and ω_{31} by ω_0, we write equations (5.59) and (5.60) more simply as

$$\dot{\sigma} = -i\omega_0 \sigma - \frac{i}{\hbar}\mu E(t)w(t) \tag{5.61}$$

and

$$\dot{w} = -\frac{2i}{\hbar}\mu E(t)(\sigma - \sigma^*). \tag{5.62}$$

We will be interested in the case in which the two-level atoms are driven by a bichromatic field

$$E(t) = \tfrac{1}{2}[E_d e^{-i\omega_d t} + E_p e^{-i\omega_p t}] + \text{c.c.} \tag{5.63}$$

In the rotating-wave approximation (RWA), we ignore the terms varying as $e^{i\omega_d t}$ and $e^{i\omega_p t}$ in (5.61) and replace that equation by

$$\dot{\sigma}(t) = -i\omega_0 \sigma - \frac{i}{2\hbar}\mu[E_d e^{-i\omega_d t} + E_p e^{-i\omega_p t}]w(t) \tag{5.64}$$

and, similarly, we replace (5.62) by

$$\dot{w}(t) = -\frac{i}{\hbar}\mu\sigma(t)[E_d^* e^{i\omega_d t} + E_p^* e^{i\omega_p t}] + \frac{i}{\hbar}\mu\sigma^*(t)[E_d e^{-i\omega_d t} + E_p e^{-i\omega_p t}] \tag{5.65}$$

i.e. we retain only the terms that are consistent with the slow variation of $w(t)$ compared with the frequencies ω_0, ω_d, and ω_p. As in section 5.2, we introduce, phenomenologically, damping rates that we denote here by γ_1 and γ_2:

$$\dot{\sigma}(t) = -i(\omega_0 - i\gamma_2)\sigma - \frac{i}{2\hbar}\mu[E_d e^{-i\omega_d t} + E_p e^{-i\omega_p t}]w(t) \tag{5.66}$$

$$\dot{w}(t) = -\gamma_1[w(t) - \overline{w}] - \frac{i}{\hbar}\mu\sigma(t)[E_d^* e^{i\omega_d t} + E_p^* e^{i\omega_p t}]$$

$$+ \frac{i}{\hbar}\mu\sigma^*(t)[E_d e^{-i\omega_d t} + E_p e^{-i\omega_p t}]. \tag{5.67}$$

$\gamma_2 = 1/T_2$ is proportional to the homogeneous linewidth of the transition, whereas $\gamma_1 = 1/T_1$ is the rate at which the population difference between the upper and lower levels relaxes to its equilibrium value $\overline{w}$ in the absence of any applied field.

If we assume that the probe field E_p is weak and treat it to first order, it can be seen by inspection of equations (5.66) and (5.67) that the solution for the population difference $w(t)$ will have the form

$$w(t) = w^{(0)} + w^{(\delta)}e^{-i\delta t} + w^{(-\delta)}e^{i\delta t} \tag{5.68}$$

where $\delta \equiv \omega_p - \omega_d$. In other words, the population difference of the two levels has components that oscillate at the beat frequency $\omega_d - \omega_p$ between the strong driving field E_d and the probe field E_p—*coherent population oscillation* at the beat frequency. These population oscillations are small if the population relaxation rate γ_1 is small compared with the beat frequency, for then the population cannot follow the intensity oscillations at the beat frequency. In other words, the beat frequency should be small compared with the population relaxation rate in order for coherent population oscillations to occur.

The steady-state solution of equations (5.66) and (5.67) for σ has frequency components at $n\omega_d + m\omega_p$, where n, m are integers. But if the probe field E_p is weak and we treat it only to first order in solving for σ, we find that σ oscillates primarily at the frequencies ω_d, ω_p, and $2\omega_d - \omega_p$, as can be seen by inspection of equations (5.66), (5.67), and (5.68). In this approximation, the steady-state solution for σ is found straightforwardly to be[3]

$$\sigma = \sigma(\omega_d)e^{-i\omega_d t} + \sigma(\omega_p)e^{-i\omega_p t} + \sigma(2\omega_d - \omega_p)e^{-i(2\omega_d - \omega_p)t} \tag{5.69}$$

where

$$\sigma(\omega_d) = \frac{\frac{1}{2}\Omega_d w}{\omega_d - \omega_0 + i\gamma_2} \tag{5.70}$$

$$\sigma(\omega_p) = \frac{\Omega_p w}{2D}\left[(\omega_p - \omega_d + i\gamma_1)(\omega_0 + \omega_p - 2\omega_d + i\gamma_2)\right.$$

$$\left. - \frac{\frac{1}{2}\Omega_d^2(\omega_p - \omega_d)}{\omega_d - \omega_0 - i\gamma_2}\right] \tag{5.71}$$

$$\sigma(2\omega_d - \omega_p) = \frac{\frac{1}{4}\Omega_d^2\Omega_p^* w(\omega_p - \omega_d + 2i\gamma_2)}{D(\omega_d - \omega_0 - i\gamma_2)}. \tag{5.72}$$

The Rabi frequencies of the drive and probe fields are defined by $\Omega_d = \mu E_d/\hbar$, $\Omega_p = \mu E_p/\hbar$ and we have also defined

$$w = \frac{[(\omega_d - \omega_0)^2 + \gamma_2^2]\overline{w}}{(\omega_d - \omega_0)^2 + \gamma_2^2 + \gamma_2\Omega_d^2/\gamma_1} \tag{5.73}$$

$$D = (\omega_p - \omega_d + i\gamma_1)(\omega_p - \omega_0 + i\gamma_2)(\omega_0 + \omega_p - 2\omega_d + i\gamma_2)$$

$$- \Omega_d^2(\omega_p - \omega_d + i\gamma_2). \tag{5.74}$$

If $\omega_d = \omega_0$, $\gamma_2 \gg \Omega_d, |\delta|$, and $\overline{w} = -1$, equation (5.71) simplifies to [185]

$$\sigma(\omega_p) = \frac{i\Omega_p}{\gamma_2}\left[\frac{1}{1 + \Omega_d^2/\gamma_1\gamma_2} - \Omega_d^2\frac{\gamma_1}{\gamma_2}\frac{1 + i\delta/\beta}{\delta^2 + \beta^2}\right] \tag{5.75}$$

---

[3] These solutions were obtained previously by a number of authors. See, for instance, Bloembergen and Shen [188] or Mollow [189].

where

$$\beta = \gamma_1[1 + \Omega_d^2/\gamma_1\gamma_2].\qquad(5.76)$$

Equation (5.75) implies the polarizability [cf equation (5.18)]

$$\alpha(\delta) = \frac{i\mu^2/\hbar\gamma_2}{1 + I_0}\left[1 - \frac{I_0(1 + I_0 + i\delta/\gamma_1)}{(\delta/\gamma_1)^2 + (1 + I_0)^2}\right]\qquad(5.77)$$

where [185]

$$I_0 \equiv \Omega_d^2/\gamma_1\gamma_2 = I_d/I_{\text{sat}}.\qquad(5.78)$$

I_d is the driving field intensity and I_{sat} is the saturation intensity of the two-level transition. The second term in brackets in equation (5.75) is attributable to coherent population oscillations.

5.5.2 Spectral hole due to coherent population oscillations

Under the assumptions made in deriving (5.77), the power absorption coefficient at the probe frequency is [190]

$$a(\delta) = \frac{2\omega_p}{c}n_I(\delta) = \frac{2\omega_p}{c}N\alpha_I(\delta)/2\epsilon_0 = \frac{a_0}{1 + I_0}\left[1 - \frac{I_0(1 + I_0)}{(\delta/\gamma_1)^2 + (1 + I_0)^2}\right]\qquad(5.79)$$

where

$$a_0 = \frac{N\omega_p\mu^2}{\epsilon_0\hbar c\gamma_2}\qquad(5.80)$$

is the (unsaturated) absorption coefficient at $\omega_p = \omega_d(\delta = 0)$. The second term in brackets gives rise to a spectral hole and is associated with coherent population oscillations .

An interesting feature of (5.79) is that it predicts a spectral hole in the absorption profile of a *homogeneously broadened* transition. Moreover, this hole is of width $\sim \gamma_1$, the population relaxation rate, and can be very small. In fact, holes as narrow as 37 Hz were observed by Hillman *et al* [191] in room-tempertaure ruby. In their experiment an argon ion laser at 514.5 nm was resonant with the transition from the ground state to the broad $^4F_2$ absorption band of ruby. Population excited in this band decays very rapidly (within a few ps) to the levels $2\bar{A}$ and $\bar{E}$, which decay (slowly, with a lifetime ~ 1 ms) to the ground state. The ruby crystal, thus, acts in effect as two-level system (figure 5.7). The laser was amplitude modulated with a modulation depth ~ 0.05 and the two weak sidebands served as the probe, with the carrier acting as the strong driving field. The driving and probe fields satisfy the equations

$$\frac{dI_0}{dz} = -\frac{a_0 I_0}{1 + I_0}\qquad(5.81)$$

and

$$\frac{dI_p}{dz} = -a(\delta)I_p\qquad(5.82)$$

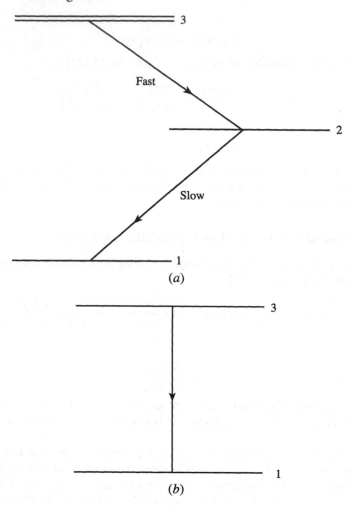

Figure 5.7. Level scheme in ruby in the experiments of Hillman *et al* [191]. (*a*) The broad level 3 excited by an Ar$^+$ laser decays very rapidly to an intermediate level 2. (*b*) Since level 2 decays slowly to the ground state, the ruby acts in effect as a two-level (1 ↔ 3) system with a small upper-level decay rate.

in the plane-wave approximation. Solution of these equations gave good agreement with the experimental results and, in particular, with the observed variation of the probe attenuation with the modulation frequency. At low laser powers (∼ 25 mW), the observed spectral hole had a width of 37 Hz, whereas at higher powers (∼ 1 W) approaching the saturation intensity (∼ 1.5 kW cm^{-2}), the hole was power broadened.

5.5.3   Slow light in room-temperature ruby

The refractive index that follows from (5.77) is

$$n(\delta) = 1 + N\alpha_R(\omega_p)/2\epsilon_0 = 1 + \frac{ca_0}{2\gamma_1\omega_p} \frac{I_0}{1+I_0} \frac{\delta}{(\delta/\gamma_1)^2 + (1+I_0)^2} \qquad (5.83)$$

from which the group index

$$n_g = n(\delta) + \omega_p \frac{dn}{d\delta} \qquad (5.84)$$

can be calculated. In experiments of Bigelow *et al* [185], the field input to the medium was amplitude modulated as in the experiments of Hillman *et al* [191]. In this case, the group index associated with the probe can be taken to be [185]

$$n_g^{(mod)} = n_d + \frac{\omega_p}{2\delta}[n(\delta) - n(-\delta)]$$

$$= n_d + \frac{ca_0}{2\gamma_1} \frac{I_0}{1+I_0} \left[\frac{1}{(1+I_0)^2 + (\delta/\gamma_1)^2} \right] \qquad (5.85)$$

where n_d is the refractive index at the drive frequency.

The experimental setup of Bigelow *et al* [185] is indicated in figure 5.8. The beam emerging from the ruby crystal and the input beam are stored on a digital oscilloscope and compared to determine the pulse delay and the pulse amplitudes. From the pulse delay, which was small in these experiments compared with the pulsewidths, it was inferred that the group velocity in the 7.25-cm ruby was as small as 57.5 ± 0.5 m s^{-1}. This value was obtained when the input drive power was 0.25 W, which is the power at which the spectral hole is deepest according to numerical simulations. These simulations were performed by integrating (5.81) and using the result in the model for the dispersion based on equation (5.85), letting a_0 and $\gamma_1 = 1/T_1$ be free parameters. The values $a_0 = 1.17$ cm^{-1} and $T_1 = 4.45$ ms, consistent with other measurements on ruby, gave good fits to

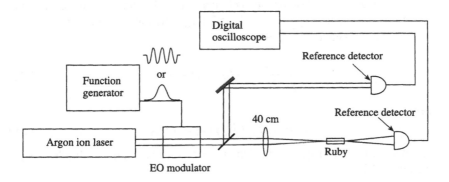

Figure 5.8. Experimental setup of Bigelow *et al*. From [185], with permission.

the measured pulse delays and the probe attenuation and, thus, confirmed that coherent population oscillations are the physical basis for the slow light and the spectral hole. The measured width of the spectral hole (half-width-at-half-maximum $1/2\pi T_1$) was 35.8 Hz at low laser powers, in good agreement with the measurements of Hillman *et al* [191].

Experiments were also performed in which single intense pulses without modulation were observed to propagate with little pulse distortion and with ultraslow group velocities. The longest pulses had the largest delays, with a 30 ms pulse delayed by 0.71 ms, corresponding to $v_g \sim 100$ m s^{-1}. The theory previously outlined, which assumes cw drive (pump) and probe fields, is not directly applicable but the fact that the longer pulses (containing smaller frequency components) have longer delays is consistent with expectations based on the theory: 'These relatively intense pulses can be thought of as producing their own pump field and are thus self-delayed' [185].

These single-pulse experiments are especially interesting because, as noted by the authors, they are thus far unique in that 'a separate pump beam is not required for generating ultraslow light'. Moreover, they can be explained using a simple two-level model. From equations (5.66) and (5.67) with an applied electric field

$$E(z,t) = \mathcal{E}(z,t)e^{-i\omega t} \tag{5.86}$$

$$\sigma(z,t) = S(z,t)e^{-i\omega t} \tag{5.87}$$

and $\omega = \omega_0$, we have

$$\frac{\partial S}{\partial t} = -\gamma_2 S - \frac{i\mu}{2\hbar}\mathcal{E}w \tag{5.88}$$

$$\frac{\partial w}{\partial t} = -\gamma_1(w+1) - \frac{2}{\gamma_2}\left(\frac{\mu|\mathcal{E}|}{\hbar}\right)^2 w \tag{5.89}$$

where we have taken $\overline{w} = -1$ for the case of an absorber. For ruby, the half-width-at-half-maximum homogeneous linewidth $\gamma_2/2\pi$ is much greater than the population decay rate γ_2 ($\gamma_2^{-1} \sim 4.45$ ms [185]). Then we can approximate S by $-i\mu\mathcal{E}w/2\hbar\gamma_2$ and equation (5.89) by

$$\frac{1}{\gamma_1}\frac{\partial w}{\partial t} = -(w+1) - Fw \tag{5.90}$$

where

$$F = (\mu|\mathcal{E}|/\hbar)^2/\gamma_1\gamma_2 = I/I_{\text{sat}}. \tag{5.91}$$

The equation for the propagation of the (slowly varying) field amplitude $\mathcal{E}$ is [cf equation (6.8)]

$$\frac{\partial \mathcal{E}}{\partial z} + \frac{1}{c}\frac{\partial \mathcal{E}}{\partial t} \simeq \frac{iN\mu\omega}{\epsilon_0}S \simeq \frac{N\mu^2\omega}{2\epsilon_0\hbar\gamma_2}\mathcal{E}w \tag{5.92}$$

or

$$\frac{\partial F}{\partial z} + \frac{1}{c}\frac{\partial F}{\partial t} = a_0 F w \qquad (5.93)$$

where $a_0 = N\mu^2\omega/\hbar c\epsilon_0\gamma_2$ is the small-signal (power) absorption coefficient.

Numerical integration of the simple equations (5.90) and (5.93) yields results in reasonable accord with experiment. In these computations, we have assumed an initial field intensity

$$F(z = 0, t) = F_0 e^{-t^2/\tau_p^2}. \qquad (5.94)$$

Figure 5.9 shows this initial intensity compared with the intensity obtained after a propagation distance of 7.25 cm, as in the experiments [185]. We have assumed an initial peak intensity equal to the saturation intensity, i.e. $F_0 = 1.0$. Note that, as observed experimentally, the pulse retains its Gaussian shape as it propagates through the absorber. Note also that the (ultraslow) group velocity, defined as the propagation distance divided by the time delay of the pulse peak compared with free-space propagation, varies with intensity and, therefore, with the propagation distance. The intensity remains Gaussian and retains essentially the same pulse duration as the initial pulse. Similar results are obtained for a range of initial intensities.

Such computations were performed independently by Agarwal and Dey [192] for both two- and three-level models. For initial peak intensities less than about three times I_{sat}, the two- and three-level models give comparable results for the group velocity.

Assuming $F_0 = 0.22$, we compute group velocities 306, 161, 112, and 96 m s^{-1} for pulse durations of 5, 10, 20, and 30 ms, respectively, compared with the experimental results 300, 159, 119, and 102 m s^{-1}. The computed transmission (figure 5.9) is consistent with the experimentally observed transmission $\sim 0.1\%$. Of course an accurate comparison of theory and experiment would have to take into account the focusing geometry and other factors but nevertheless, these results indicate that these single-pulse experiments demonstrating ultraslow group velocities are well described by a simple model that does not require atomic dipole coherence but which is based essentially on a simple rate equation (5.90) for the medium.

The propagation of optical pulses in resonant media has been studied in many experiments since the 1960s. The fact that the results of Bigelow *et al* can be explained by such simple and frequently employed models for a saturable medium strongly suggests that at least some of these experiments involved ultraslow light.

5.5.4 Fast light and slow light in a room-temperature solid

Bigelow *et al* [186] have also performed experiments in room-temperature alexandrite in which both fast light and slow light have been demonstrated,

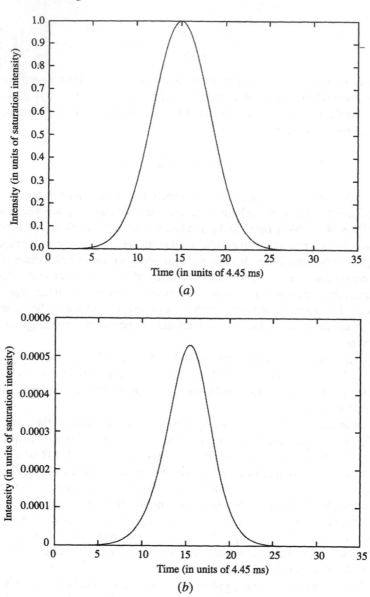

Figure 5.9. Results of numerical solution of equations (5.90) and (5.93) with $F_0 = 1.0$:
(*a*) initial intensity; (*b*) output intensity; (*c*) peak intensity *versus* distance of propagation;
and (*d*) group velocity *versus* distance of propagation.

depending on the laser wavelength. Like the slow-light experiments in room-
temperature ruby, these experiments are simpler than previous experiments on

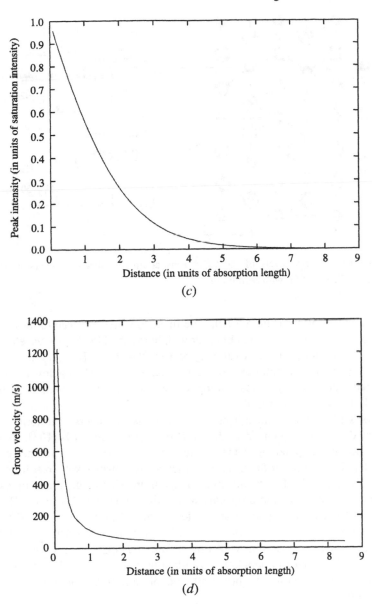

Figure 5.9. (Continued.)

fast light and slow light.

In alexandrite, the Cr^{+3} ions that replace the Al^{+3} ions in the host $BeAl_2O_4$ crystal are at sites with either mirror symmetry or inversion symmetry. Ions at the mirror sites have a T_1 time of 290 μs, whereas those at the inversion sites

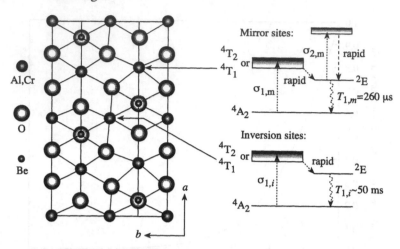

Figure 5.10. Energy levels of the Cr^{+3} ions at mirror sites and inversion sites in alexandrite. From [186], with permission.

have a T_1 of $\sim$ 50 ms (figure 5.10). The inversion sites are seen to have a level structure similar to that used to obtain slow light in ruby [185]. At a wavelength of 488 nm, these sites exhibit an extremely narrow (8.4 Hz, $\sim 1/2\pi T_1$) spectral hole and, using again the amplitude modulation technique [185], Bigelow *et al* [186] measured a probe delay corresponding to a group velocity as small as 91 m s^{-1} in the 4-cm alexandrite crystal.

For Cr^{+3} ions at the mirror sites, there is an inverse saturation effect associated with absorption from the *excited* level (figure 5.10) [193]. In other words, because population in the excited level can be lost to another excited level, the absorption can *increase* with increasing power levels, resulting in an absorption *anti-hole*. This results in a change in the sign of the group index and, therefore, the probe should propagate with a time *advance* (fast light). Bigelow *et al* [186] observed a spectral anti-hole of width 612 Hz ($= 1/2\pi T_1$) and a group index of -3.75×10^5 ($v_g = -800$ m s^{-1}) at a wavelength of 476 nm where the mirror sites dominate. As in the previous experiments with ruby, the theory based on coherent population oscillations gave very good agreement with the experimental data for the probe attenuation and group index for both slow light and fast light.

5.6 Remarks

There is much current interest in ultraslow light and our intention in this chapter has not been to give an up-to-date account of activity in this field but rather to describe some of the basic physics underlying some of the earliest experiments. As already noted, the interested reader may find other references as well as other perspectives in the reviews by Harris [168] and Boyd and Gauthier [173]. Among the references to early work, we should mention a paper by Tewari and Agarwal [194] in which the possibility of modifying the linear dispersion of a probe by applying a strong pump field was studied theoretically. Xiao *et al* [195] studied experimentally the dispersion associated with EIT in a hot Rb gas and, from the measured dispersion properties, it was inferred that the group velocity of the probe was $c/13$. It should also be noted that Budker *et al* [196] have reported a group velocity of 8 m s^{-1} in room-temperature ^{85}Rb vapour. The ultraslow group velocity in this case results from an extremely narrow nonlinear magneto-optic resonance.

What about applications? The study of slow light has mainly been done at the basic physics level but some potential applications have been noted, including the possibility of controllable optical delay lines in fibre-optic networks. Another possibility is discussed in the following chapter, where we discuss in some detail the degree to which group velocity can be controlled.

Chapter 6

Stopped, stored, and regenerated light

Light cannot only be dramatically slowed: it can be completely stopped, stored, and regenerated. In this chapter, we review the concept of dark-state polaritons, which provides an elegant way of understanding the stopping and controlling of light pulses that have been demonstrated with slow light based on electromagnetically induced transparency (EIT). We describe some of the first experiments on stopped light and compare the phenomenon with photon echoes. Finally, we briefly discuss the implications of stored and regenerated light for quantum memories and conclude with a summary of some related work.

6.1 Controlling group velocity

From the basic definition $v_g = c/[n + \omega \, dn/d\omega]$, it is obvious that group velocity can be changed simply by changing the central frequency of a light pulse. The theory of slow light in an EIT medium shows that the group velocity of a probe pulse can be controlled by varying the intensity of the coupling field [cf equation (5.38)]. In the case of slow light based on coherent population oscillations in a solid, the group velocity can be similarly controlled by varying the intensity of the driving field. It has also been predicted that the group velocity can be varied from subluminal to superluminal by applying an additional coupling field whose frequency is close to the non-allowed transition $(1 \leftrightarrow 2)$ in the Λ transition scheme for EIT in figure 5.2 [197]. With such an additional field, the solution for the expectation value $\langle \hat{\sigma}_{13}(t) \rangle$ (section 5.2) to first order in the probe field but to all orders in the other two fields is complicated but has been studied numerically. The results show that a probe pulse can propagate with a subluminal group velocity for small intensities of the additional field, a superluminal group velocity as the intensity is raised, and again a subluminal velocity as the intensity of the additional field is further increased. The additional field thus acts as a 'knob' for controlling the group velocity of a probe pulse [197].

It was noted in section 5.5.3 that group velocities $v_g \sim 100 \text{ m s}^{-1}$ have been observed in *single* pulses propagating in room-temperature ruby. The group velocity could be varied by changing the pulse duration. Talukder *et al* [198] have demonstrated that the group velocity of femtosecond pulses propagating in a dye could be varied from subluminal to superluminal by varying the dye concentration. Kim *et al* [199] have reported experiments in which the group velocity of pulses in cesium vapour changes from subluminal to superluminal as the pulse power is increased.

Light pulses can be delayed simply by a Fabry–Perot elaton. Near a transmission frequency determined by the imaginary part of the amplitude transmission factor for the etalon, there is also a steep variation of the real part, implying a strong reduction of the group velocity of a pulse in the case of high-finesse etalons. Photonic bandgap structures, similarly, can have dispersion curves such that the group velocity can be slowed substantially ('heavy photons') near a band edge. (See, for instance, [200].)

It has been demonstrated experimentally that the group velocity can be controlled to such an extent that a pulse of light can not only be slowed but completely stopped and then regenerated. The stopping can be associated with a temporary storage of light in collective metastable atomic states of the medium. In fact, it is possible in this way to store and then retrieve, after a relatively long period of time, the *quantum state* of the field [201–203]. A convenient and elegant way to understand this is based on the concept of *dark-state polaritons* [201], to which we now turn our attention.

6.2 Dark-state polaritons

Consider again the equations (5.15) and (5.16) that we used to describe EIT in the Λ transition scheme of figure 5.2. Assume that the probe field is resonant with the $1 \leftrightarrow 3$ transition, so that $\Delta = 0$ and

$$\dot{S}_{12} = -\gamma_{12} S_{12} + \frac{i}{2\hbar} \mu_{23} \mathcal{E}_c S_{13} \tag{6.1}$$

$$\dot{S}_{13} = -\gamma_{13} S_{13} + \frac{i}{2\hbar} \mu_{13} \mathcal{E}_p + \frac{i}{2\hbar} \mu_{23} \mathcal{E}_c S_{12}. \tag{6.2}$$

For times short compared with the (usually long) dephasing time γ_{12}^{-1} of the non-allowed transition $1 \leftrightarrow 2$, equation (6.1) implies

$$S_{13}(t) \cong -\frac{2i}{\Omega_c(t)} \frac{\partial S_{12}}{\partial t} \tag{6.3}$$

where the Rabi frequency $\Omega_c = \mu_{23} \mathcal{E}_c / \hbar$ associated with the coupling field is now explicitly allowed to be time-dependent. From (6.2), we have

$$\begin{aligned}
S_{12}(t) &= -\frac{\mu_{13} \mathcal{E}_p(t)}{\hbar \Omega_c(t)} - \frac{2i}{\Omega_c(t)} \left[\frac{\partial S_{13}}{\partial t} + \gamma_{13} S_{13} \right] \\
&\cong -\frac{\mu_{13} \mathcal{E}_p(t)}{\hbar \Omega_c(t)} - \frac{2i}{\Omega_c(t)} \left[\frac{\partial}{\partial t} + \gamma_{13} \right] \left(-\frac{2i}{\Omega_c(t)} \frac{\partial S_{12}}{\partial t} \right)
\end{aligned}$$

$$\cong -\frac{\mu_{13}\mathcal{E}_p(t)}{\hbar\Omega_c(t)} + \frac{4}{\Omega_c(t)}\left[\frac{\partial}{\partial t} + \gamma_{13}\right]\frac{1}{\Omega_c(t)}\frac{\partial}{\partial t}\left(\frac{\mu_{13}\mathcal{E}_p(t)}{\hbar\Omega_c(t)}\right)$$

$$\cong -\frac{\mu_{13}\mathcal{E}_p(t)}{\hbar\Omega_c(t)}. \tag{6.4}$$

This approximation is discussed later.

The probe field $E_p(z,t)$ in the plane-wave approximation propagates according to the wave equation

$$\left(\frac{\partial^2}{\partial z^2} - \frac{1}{c^2}\frac{\partial^2}{\partial t^2}\right)E_p(z,t) = \frac{N\mu_{13}}{\epsilon_0 c^2}\frac{\partial^2}{\partial t^2}(\sigma_{13} + \sigma_{31}) \tag{6.5}$$

where N is the density of atoms. As in chapter 5, we write $E_p(z,t)$ as a slowly varying envelope $\mathcal{E}_p(z,t)$ multiplying a carrier wave of frequency ω_p:

$$E_p(z,t) = \tfrac{1}{2}\mathcal{E}_p(z,t)e^{-i\omega_p(t-z/c)} + \text{c.c.} \tag{6.6}$$

and

$$\sigma_{13}(z,t) = S_{13}(z,t)e^{-i\omega_p(t-z/c)}. \tag{6.7}$$

Equation (6.5) yields

$$\frac{\partial\mathcal{E}_p}{\partial t} + c\frac{\partial\mathcal{E}_p}{\partial z} \cong \frac{iN\mu_{13}\omega_p}{\epsilon_0}S_{13} \cong \frac{iN\mu_{13}\omega_p}{\epsilon_0}\left[-\frac{2i}{\Omega_c}\frac{\partial S_{12}}{\partial t}\right]$$

$$\cong -\frac{\mathcal{N}g^2}{\Omega_c(t)}\frac{\partial}{\partial t}\left[\frac{\mathcal{E}_p(z,t)}{\Omega_c(t)}\right] \tag{6.8}$$

where $g^2 = 2\mu_{13}^2\omega_p/V\epsilon_0\hbar$ and $\mathcal{N} = NV$ is the number of atoms in the volume V. In deriving this expression, we have used the approximations (6.3) and (6.4).

If we take the coupling field intensity to be constant, (6.8) reduces to

$$\frac{\partial\mathcal{E}_p}{\partial z} + \frac{1}{v_g}\frac{\partial\mathcal{E}_p}{\partial z} = 0 \tag{6.9}$$

where the group velocity

$$v_g = \frac{\epsilon_0\hbar c\Omega_c^2}{2N\mu_{13}^2\omega_p} \tag{6.10}$$

in agreement with (5.33).

We have treated the probe and coupling fields classically. Since we have treated the (weak) probe field only to first order, equation (6.8) is linear in the probe field. If we continue to treat the (strong) coupling field as a prescribed classical field, the linearity of our equations with respect to $\mathcal{E}_p$ implies that we will formally obtain the same equation as (6.8) when we treat the probe field as a *quantum* field. In other words, the Heisenberg operator equation of motion for the probe field, subject to the same approximations used to arrive at (6.8), is formally identical to (6.8):

$$\frac{\partial \hat{\mathcal{E}}_p}{\partial t} + c\frac{\partial \hat{\mathcal{E}}_p}{\partial z} \cong -\frac{Ng^2}{\Omega_c(t)}\frac{\partial}{\partial t}\left[\frac{\hat{\mathcal{E}}_p(z,t)}{\Omega_c(t)}\right] \tag{6.11}$$

where $\hat{\mathcal{E}}_p$ is the Heisenberg-picture operator for the probe field. This is the equation for $\hat{\mathcal{E}}_p$ obtained by Fleischhauer and Lukin [201]. Similarly, the quantities σ_{ij} and S_{ij} can be replaced by Heisenberg operators $\hat{\sigma}_{ij}$ and $\hat{S}_{ij}$; and recall that σ_{ij} and S_{ij} have denoted the *expectation values* of the operators $\hat{\sigma}_{ij}$ and $\hat{S}_{ij}$.

To solve equation (6.11), Fleischhauer and Lukin [201] define a new quantum field $\hat{\Psi}(z,t)$ that is a superposition of atom and (probe) field lowering operators, as evidently first introduced for Raman transitions by Mazets and Matisov [204]:

$$\hat{\Psi}(z,t) = \cos\theta(t)\hat{F}_p(z,t) - \sin\theta(t)\sqrt{N}\hat{S}_{12}(z,t) \tag{6.12}$$

where

$$\cos\theta(t) = \frac{\Omega_c(t)}{\sqrt{\Omega_c^2(t) + Ng^2}} \tag{6.13}$$

$$\sin\theta(t) = \frac{\sqrt{N}g}{\sqrt{\Omega_c^2(t) + Ng^2}} \tag{6.14}$$

and

$$\hat{F}_p(z,t) = \frac{\mu_{13}}{\hbar g}\hat{\mathcal{E}}_p(z,t) \tag{6.15}$$

is a dimensionless electromagnetic field operator. Making again the approximation

$$\hat{S}_{12}(t) \cong -\frac{\mu_{13}\hat{\mathcal{E}}_p(t)}{\hbar\Omega_c(t)} \tag{6.16}$$

and using (6.11)–(6.14), one obtains, after straightforward algebra, the equation of motion [201]

$$\left[\frac{\partial}{\partial t} + c\cos^2\theta(t)\frac{\partial}{\partial z}\right]\hat{\Psi}(z,t) = 0. \tag{6.17}$$

Thus, the field $\hat{\Psi}(z,t)$ propagates with the *time-dependent* group velocity

$$v_g(t) = c\cos^2\theta(t) = \frac{c\Omega_c^2(t)}{\Omega_c^2(t) + Ng^2} \tag{6.18}$$

and equation (6.17) has the shape-preserving solution

$$\hat{\Psi}(z,t) = \hat{\Psi}\left(z - c\int_0^t d\tau\,\cos^2\theta(\tau), t = 0\right). \tag{6.19}$$

If $\hat{\Psi}(z, t)$ is written as the plane-wave expansion

$$\hat{\Psi}(z, t) = \sum_k \hat{\Psi}_k(t) e^{ikz} \qquad (6.20)$$

one obtains, from (6.12), the commutation relation

$$[\hat{\Psi}_k, \hat{\Psi}_{k'}^\dagger] \cong \delta_{kk'}[\cos^2\theta + \sin^2\theta] = \delta_{kk'}. \qquad (6.21)$$

In obtaining this result, the equal-time commutation relation $[\hat{S}_{12}, \hat{S}_{21}] = \hat{\sigma}_{11} - \hat{\sigma}_{22}$ is used, together with the approximation that the atoms remain in state 1 with high probability, so that we may replace $\hat{\sigma}_{11}$ by 1 and $\hat{\sigma}_{22}$ by 0. The result (6.21) states that the field $\hat{\Psi}$ is bosonic: the quasi-particles associated with $\hat{\Psi}$ are excitations of the combined atom–field system, i.e. they are *polaritons*.

Polaritons are quasi-particles belonging to mixed states of field and material excitations, e.g. mixed states of photons and electronic excitations (excitons) or mixed states of photons and phonons. The polaritons of the field $\hat{\Psi}(z, t)$ involve the atomic operator $\hat{\sigma}_{12}$ associated with the *non-allowed* transition $1 \leftrightarrow 2$, whereas the better known polaritons, if described similarly, would involve corresponding operators associated with *allowed* transitions. To see why the quasi-particles belonging to the field $\hat{\Psi}$ are called *dark-state* polaritons by Fleischhauer and Lukin, consider $\hat{\Psi}$ defined for a single atom as

$$\hat{\Psi} = \cos\theta \, \hat{F}_p - \sin\theta \, \hat{S}_{12} \qquad (6.22)$$

and assume also a single probe field mode, for which the slowly varying operator $\hat{\mathcal{E}}_p$ is given by

$$\hat{\mathcal{E}}_p = \left(\frac{2\hbar\omega_p}{\epsilon_0 V}\right)^{1/2} \hat{a}_p e^{i\omega_p t} = \frac{\hbar g}{\mu_{13}} \hat{a}_p e^{i\omega_p t} \qquad (6.23)$$

where $\hat{a}_p$ is the photon annihilation operator for the mode ($[\hat{a}_p, \hat{a}_p^\dagger] = 1$). Then

$$\hat{F}_p = \hat{a}_p e^{i\omega_p t} \equiv \hat{A}_p \qquad (6.24)$$

and

$$\hat{\Psi} = \cos\theta \, \hat{A}_p - \sin\theta \, \hat{S}_{12}. \qquad (6.25)$$

The excited states (quasi-particles) of $\hat{\Psi}$ are generated by applying the operator $\hat{\Psi}^\dagger$ to the ground state $|0\rangle|1\rangle$, where $|0\rangle$ is the vacuum state of the field and $|1\rangle$ is the ground state of the unperturbed atom (figure 5.2). Thus, for instance, the first excited state is

$$\begin{aligned}
\hat{\Psi}^\dagger |0\rangle|1\rangle &= [\cos\theta \, \hat{A}_p^\dagger - \sin\theta \, \hat{S}_{21}]|0\rangle|1\rangle \\
&= \cos\theta |1_p\rangle|1\rangle - \sin\theta |0\rangle|2\rangle \\
&= \frac{\Omega_c |1_p\rangle|1\rangle - g|0\rangle|2\rangle}{\sqrt{\Omega_c^2 + g^2}}
\end{aligned} \qquad (6.26)$$

where $|1_p\rangle$ is the one-photon state of the probe field.

The interaction Hamiltonian in this single-mode single-atom model is

$$\hat{H}_{int} = -\hbar\Omega_c(\hat{S}_{23} + \hat{S}_{32}) - \hbar g(\hat{A}_p^\dagger\hat{S}_{13} + \hat{S}_{31}\hat{A}_p). \tag{6.27}$$

It follows that

$$\hat{H}_{int}\hat{\Psi}^\dagger|0\rangle|1\rangle = \frac{-\hbar g\Omega_c|0\rangle|3\rangle + \hbar\Omega_c g|0\rangle|3\rangle}{\sqrt{\Omega_c^2 + g^2}} = 0. \tag{6.28}$$

In other words, the state $\hat{\Psi}^\dagger|0\rangle|1\rangle$ is a dark state. In fact, if we take $g = \Omega_c$ as in our simple model of section 5.2 [cf equation (5.1)], we see that $\hat{\Psi}^\dagger|0\rangle|1\rangle$ is just the non-coupled state $|NC\rangle$. More generally, the excited states of $\hat{\Psi}$ are generated by applying $\hat{\Psi}^n$ to the ground state $|0\rangle|1\rangle$ and these states, likewise, satisfy

$$\hat{H}_{int}(\hat{\Psi}^\dagger)^n|0\rangle|1\rangle = 0 \tag{6.29}$$

i.e. they are dark states.

In the $\mathcal{N}$-atom case, the excited states generated by applying $\hat{\Psi}_k^\dagger$ to the ground state $|1_1 1_2 \ldots 1_N\rangle$, where 1_j denotes the state 1 of atom j, is the symmetric state

$$\frac{1}{\sqrt{N}}\sum_{j=1}^{\mathcal{N}}|1_1 1_2 \ldots 2_j \ldots 1_N\rangle \tag{6.30}$$

and, likewise, the state $(\hat{\Psi}_k^\dagger)^n|1_1 1_2 \ldots 1_N\rangle$ is a symmetric state with n atoms in state $|2\rangle$. As in the single-atom model, these states are (orthogonal) eigenstates with eigenvalue 0 of the interaction Hamiltonian, i.e. they are dark states that do not interact with light.

Thus, the polaritons of the coupled atom–field system in EIT are called dark-state polaritons [201]. Equation (6.18) shows that the group velocity of the polariton field can be controlled by varying the strength of the coupling field. If θ is adiabatically varied from 0 ($\Omega_c^2 \gg \mathcal{N}g^2$) to $\pi/2$, for instance, the group velocity changes from c to 0, i.e. the probe pulse is *stopped*. According to (6.12), when this happens, the polariton excitations are entirely atomic:

$$\hat{\Psi}(z, t) = -\sqrt{\mathcal{N}}\hat{S}_{12}(z, t) \tag{6.31}$$

and a state with photons in the field and all the atoms in state 1 is changed to a state with no photons and atoms in state 2. But more interesting is the fact that the *pulse shape and quantum state* of the probe pulse are stored in a collective atomic state of the medium *and can be recovered* by changing the polariton mixing angle θ from $\pi/2$ to 0. Because the reduction of the group velocity to zero is basically linear in the probe field, the quantum state of the probe is preserved and stored in the atomic medium. This information is 'imprinted' in the atoms and retrieved when the mixing angle is brought back to its original value.

It should be noted that the shape-preserving solution (6.19) is not a soliton, since it does not require any special pulse shape dictated by a nonlinear coupling of the field to the medium.

Note also that, in the excitation transfer from the probe field to the atoms, the probe field energy is not all transferred to the atoms. Conservation of energy requires that some—in fact, most—of the probe field energy is transferred to the coupling field as the atoms undergo the Raman transition $1 \rightarrow 2$.

Let us now examine the approximation (6.4) that leads to (6.17). It is clear, first of all, that this approximation requires that the Rabi frequency $\Omega_c(t)$ not vary too rapidly. Non-adiabatic corrections to the equation of motion (6.17) are obtained using the approximation contained in the penultimate line. Retaining only the leading terms, one obtains [171]

$$\left[\frac{\partial}{\partial t} + c\cos^2\theta\frac{\partial}{\partial z}\right]\hat{\Psi} \sim -\gamma_{13}\left(\frac{d\theta}{dt}\right)^2\frac{\sin^2\theta}{\mathcal{N}g^2}\hat{\Psi} + c^2\gamma_{13}\frac{\cos^2\theta\sin^4\theta}{\mathcal{N}g^2}\frac{\partial^2}{\partial z^2}\hat{\Psi}$$

(6.32)

for times $t \gg \gamma_{13}^{-1}$. For the first term on the right-hand side to be negligible, its integral over time should be small:

$$\gamma_{13}\int_0^\infty dt \left(\frac{d\theta}{dt}\right)^2\frac{\sin^2\theta}{\mathcal{N}g^2} = \gamma_{13}\int_0^\infty dt \frac{(d\theta/dt)^2}{\Omega_c^2(t) + \mathcal{N}g^2} \ll 1 \qquad (6.33)$$

whereas for the second term to be negligible, we should have (for $\sin\theta \cong 1$)

$$\frac{c^2\gamma_{13}}{\mathcal{N}g^2}\frac{1}{L_p^2} \ll \frac{c}{L} \qquad (6.34)$$

or

$$\mathcal{N}\left(\frac{L_p}{L}\right)\frac{g^2}{\gamma_{13}}\tau_p \gg 1 \qquad (6.35)$$

where $L_p = c\tau_p$ is the pulse length. In terms of the absorption coefficient a_0 in the absence of EIT [cf equation (5.31)], we can write this condition equivalently as

$$L \ll L_p\sqrt{a_0 L}. \qquad (6.36)$$

We can write this condition in yet another, instructive way using the fact that the pulse is compressed in EIT by the factor c/v_g (section 5.3): $L_p = (v_g/c)c\tau_p = v_g/\Delta\omega_{pulse}(0)$, where $\Delta\omega_{pulse}(0)$ is the initial spectral width of the pulse. Then (6.36) becomes $L \ll v_g\sqrt{a_0 L}/\Delta\omega_{pulse}(0)$. But, from (5.31), $\Delta\omega_{tr}(0) = v_g\sqrt{a_0 L}/L$ for the initial spectral width of the EIT transparency window. Thus [171],

$$\Delta\omega_{pulse}(0) \ll \Delta\omega_{tr}(0). \qquad (6.37)$$

That is, *the spectrum of the initial probe pulse must be narrower than the EIT transparency window*: the pulse duration cannot be too short. Note that this condition is most easily satisfied for an optically dense medium ($a_0 L \gg 1$).

The condition (6.33) obviously requires that the coupling field not be changed too rapidly. Using (6.13), (6.14), and (6.18), and again the approximation $\sin \theta \cong 1$[1], we can write this condition as [171]

$$T \gg \frac{\ell_{abs}}{c} \frac{v_g^0}{c} \tag{6.38}$$

where the time

$$T = \frac{\Omega_c^2}{\int_0^\infty dt \, \dot{\Omega}_c^2(t)} \tag{6.39}$$

is a measure of the time scale over which $\Omega_c^2(t)$ varies, $v_g^0 = \Omega_c^2/\mathcal{N}g^2$ is the initial group velocity, before the coupling field is varied, and $\ell_{abs} = 1/a_0$ is the ordinary (non-EIT) absorption length. The adiabaticity condition (6.38) is easily met experimentally.

Numerical solutions of the Maxwell–Bloch equations [171, 205] support these rough estimates for adiabaticity. In fact, they indicate that the probe pulse can be nearly perfectly regenerated even when $T \to 0$ and the atomic dipole coherence does not adiabatically follow the coupling field. This is consistent with the observations of Liu *et al* [206] described in the following section.

In addition to the conditions (6.33) and (6.35) for adiabaticity, we have assumed [recall (6.3)] that the dephasing rate γ_{12} of the non-allowed transition $1 \leftrightarrow 2$ is negligible. This rate limits the time over which a light pulse can be stored in the medium before it can be faithfully restored. In fact, γ_{12} can be made quite small (γ_{12}^{-1} on the order of seconds) by using buffer gases and wall coatings to minimize the effects of collisions on the $1 \leftrightarrow 2$ coherence. If there are $\mathcal{N}_e$ atoms in state $|2\rangle$ during the storage of the pulse, the maximum storage time is on the order of $(\mathcal{N}_e\gamma_{12})^{-1}$.

Note that it has been assumed that the coupling field is a function of time but not distance z. This is a valid assumption if the coupling field propagates at right angles to the propagation direction of the probe field and a reasonable approximation if the group velocity of a coupling field co-propagating with the probe is much larger than the probe group velocity. More generally, assuming that the coupling field propagates as in free space [$\Omega_c(z, t) = \Omega_c(t - z/c)$], one obtains [201]

$$\left[\frac{\partial}{\partial t} + c \cos^2 \theta(z, t) \frac{\partial}{\partial z} \right] \frac{\hat{\mathcal{E}}_p}{\Omega_c} = 0. \tag{6.40}$$

In this case, the pulse shape is not preserved exactly, since the group velocity now depends on z.

[1] The approximation $\sin \theta \cong 1$, or $\mathcal{N}g^2/\Omega_c^2 = 2N\omega_p\mu_{13}^2/\hbar c\epsilon_0\Omega_c^2 \gg 1$, has already been used in our discussion of EIT [cf equation (5.33)]. It is roughly equivalent to the condition that the atomic density be much larger than the photon density.

6.3 Stopped and regenerated light

The first, beautiful experimental demonstration of the stopping and storage of light based on the preceding ideas was reported by Liu *et al* [206]. As in the first demonstration of ultraslow light [179], the propagation medium was an ultracold gas of sodium atoms. The peak density of the cloud of 1.1×10^7 sodium atoms was 1.1×10^{13} cm^{-3} and the temperature (0.9 μK) was slightly higher than the critical temperature for Bose–Einstein condensation. The cloud was 339 μm in length and 55 μm in width along the transverse direction. The three-level EIT transition scheme in these experiments was realized using hyperfine states of the sodium D$_1$ line at 589.6 nm (figure 6.1) and the coupling and probe fields in this case were *co-propagating*. The atoms were magnetically trapped in the state $|1\rangle = |3S, F = 1, M_F = -1\rangle$.

The coupling field was turned on a few ms before the probe pulse and the two fields had orthogonal linear polarizations before passing through a quarter-wave plate (figure 6.1) that resulted in two fields with orthogonal circular polarizations incident on the sodium cloud. After passage through the cloud, the fields passed through a second quarter-wave plate that restored their original linear polarizations, after which they were separated by a polarizing beam splitter. The fields that passed through the central portion of the cloud were incident on photomultiplier tubes (PMT) and their intensities were simultaneously recorded.

Figure 6.2(*a*) shows a reference pulse (open circles) and a Gaussian fit (dotted curve) obtained by averaging 100 probe pulses in the absence of the atomic cloud, together with the coupling field (broken curve) and the (normalized) probe pulse (filled circles) transmitted by the cloud. At 6.3 μs, indicated by the arrow, the probe pulse is spatially compressed and confined to the 339 μm length of the cloud. The delay of the probe pulse relative to the reference pulse is 11.8 μs, corresponding to a group velocity $v_g = (339 \ \mu\text{m})/(11.8 \ \mu\text{s}) = 29$ m s^{-1}. The spatial compression factor $c/v_g = 10^7$ (section 5.3) and the free-space probe pulse length of 3.4 km ($\tau_p = 11.3 \ \mu$s) implies that the spatial extent of the pulse in the cloud is $(3.4 \ \text{km})/(10^7) = 340 \ \mu$m, i.e. the pulse is spatially compressed to the cloud size.

Figure 6.2(*b*) shows the regeneration of the probe pulse after the coupling laser is turned off at 6.3 μs and then turned back on at 44.3 μs. The revived probe pulse has nearly the same shape as the original. Figure 6.2(*c*) shows the regeneration of the probe pulse when the coupling laser is turned back on after a much longer time (839.3 μs), and figure 6.2(*d*) is a plot of the probe pulse transmission *versus* the storage time. The fit to the data shown by the straight line implies that the dephasing time of the non-allowed transition ($1 \leftrightarrow 2$) is about 0.9 ms. This is comparable to the inverse of the elastic collision rate for the cold sodium atoms [206].

Experiment thus supports the predictions of the theory described in the preceding section: the probe pulse can be stopped and regenerated depending on whether the coupling field is off or on. Optical information can be coherently

(a)

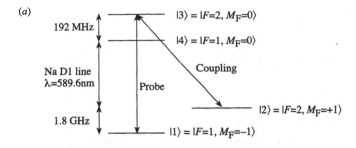

(b)

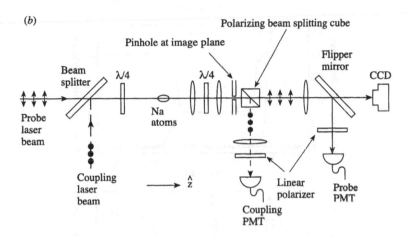

Figure 6.1. (a) Transitions between hyperfine states of the sodium D_1 line in the experiments of Liu *et al* and (b) experimental set-up for the observation of stored and regenerated light (see text). From [206], with permission.

stored in the atoms and then 'read out' by changing the coupling field. Memory of the wavevector, polarization, and shape of the probe pulse is preserved in the relative phases between different atoms in the medium over times shorter than the dephasing time γ_{12}^{-1}. Figures 6.2(c) and 6.2(d) show that, for storage times approaching γ_{12}^{-1}, the regenerated pulse is no longer a faithful reproduction of the original pulse and the probe transmission decreases.

The amplitude of the regenerated probe pulse increased when the coupling field was turned on with an intensity larger than that of the original coupling field, again consistent with theory. Experiments in which two or three short coupling pulses were turned on sequentially demonstrated multiple read-outs of the information stored in the atoms [206].

The ultracold atoms and the co-propagating coupling and probe fields in these experiments greatly reduce the effects of atomic motion and Doppler broadening. However, as in the realization of ultraslow group velocities, the

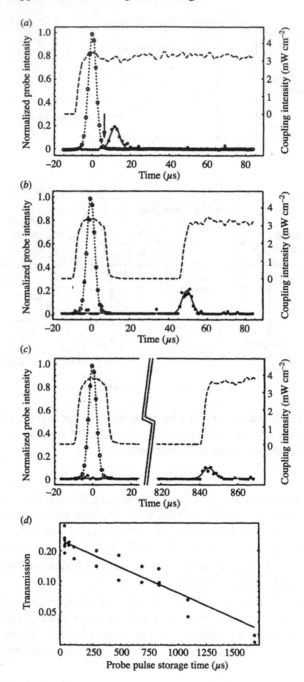

Figure 6.2. Experimental data of Liu *et al* on the storing and regeneration of a probe pulse (see text). From [206], with permission.

stopping, storing, and regeneration of light can also be achieved in hot vapours. Phillips *et al* [207] have demonstrated light storage in a 4-cm-long cell containing ^{87}Rb vapour at temperatures $\sim$ 70–90 °C and densities $\sim$ 10^{11}–10^{12} cm^{-3}. The three levels in their EIT Λ transition scheme are

$$|1\rangle = |5S_{1/2}, F = 2, M_F = 2\rangle$$
$$|2\rangle = |5S_{1/2}, F = 2, M_F = 0\rangle$$
$$|3\rangle = |5P_{1/2}, F = 1, M_F = 1\rangle. \tag{6.41}$$

The co-propagating probe and coupling fields had opposite circular polarizations. The probe was spatially compressed from its free-space length ($\sim$ 10–30 μs $\times$ 3×10^8 m s^{-1}) in the cell by more than five orders of magnitude as a result of the slow group velocity ($v_g \sim 1$ km s^{-1}, $c/v_g \sim 3 \times 10^5$). While the probe pulse was in the cell, the coupling field was turned off over a time interval of about 3 μs and turning it back on resulted in the regeneration of the probe pulse. Storage times roughly comparable to those demonstrated by Liu *et al* were realized.

Finally, we note that the stopping and storing of light was also observed in a cold solid in the experiments of Turukhin *et al* [184] mentioned in section 5.5.

6.4   Echoes

The stopping and regeneration of the probe field as the mixing angle is varied from 0 to $\pi/2$ to 0 results ideally in an exact replica or 'echo' of the probe field. The question is frequently raised as to whether this regeneration is related to photon echoes [208], the analog of the spin echoes known since 1950 in nuclear magnetic resonance [209].

Photon echoes can occur in an inhomogeneously broadened collection of two-level atoms, i.e. a collection of atoms in which different atoms have slightly different transition frequencies. A short, intense laser pulse that brings the atoms midway between their lower and upper states—a $\pi/2$ pulse—produces a macroscopic polarization that results in the emission of a pulse whose duration is on the order of the inverse of the inhomogeneous linewidth. Physically, this time duration is the time it takes for the macroscopic polarization to be washed out as the individual atomic dipole moments oscillate at their different transition frequencies. The application later of a π pulse causes the individual dipoles to begin *rephasing*, resulting again in a macroscopic polarization and, thence, the photon echo[2].

Among the echo phenomena possible in gases is the Raman echo [211], in which two fields interact with a three-level atom as in figure 5.2. In this case, two bichromatic pulses with central frequencies ω_p and ω_c are applied at times $t = 0$ and $t = \tau$, and the application of a third pulse of frequency ω_c at time 2τ results

---

[2] This terse description of photon echoes is only meant to serve as a reminder for the reader who has previously encountered the concept. For a proper discussion, see, for example, Allen and Eberly [210].

in the Raman echo of frequency ω_p at time 2τ. In the case of the Raman echo, both fields can be far off resonance from their respective transitions.

The stopping and regeneration of light based on EIT and a temporally varying coupling field is quite different from these echo phenomena, which result from a rephasing of induced dipole moments as in the case of a Doppler-broadened transition in a gas. For photon echoes, the gas should be optically thin and the 'storage' time is relatively short compared with the rather long storage times observed in the experiments described in section 6.3 and, for both photon and Raman echoes, only a small fraction of the incident light is stored. Moreover, they involve inherently dissipative processes and are, therefore, unlikely to be implementable, for instance, for quantum memories. Finally, the formation of the echoes does not involve ultraslow group velocities.

6.5 Memories

A frequently mentioned potential application[3] of stored and regenerated light is in 'quantum information' processing (QIP) and, in particular, for 'quantum memories' for photons.

QIP is extremely challenging for several reasons, not the least of which is the realization of quantum memories that can store information with little 'decoherence' and that can be accessed on short time scales. As is well recognized, quantum optical systems are attractive for this purpose because atom–field interactions can be made controllable and reversible and decoherence processes can be alleviated using long-lived hyperfine states, as in EIT (e.g. state 2 in figure 5.2).

Considerable progress in this direction has been made in the area of cavity quantum electrodynamics (cavity QED) [212]. Consider again the three-level Λ system shown in figure 5.2 and assume such an atom is in a cavity such that the $1 \leftrightarrow 3$ transition is resonant with a single quantized field mode while the $2 \leftrightarrow 3$ transition is resonant with a classically prescribed field with Rabi frequency $\Omega_c(t)$. The interaction Hamiltonian is

$$\hat{H}_{int} = -\hbar g(\hat{a}\hat{\sigma}_{31} + \hat{a}^\dagger \hat{\sigma}_{13}) - \hbar\Omega_c(t)(\hat{\sigma}_{23} + \hat{\sigma}_{32}) \qquad (6.42)$$

and the system has dark states

$$|\psi_n\rangle = \cos\theta(t)|1\rangle|n+1\rangle - \sin\theta(t)|2\rangle|n\rangle \qquad (6.43)$$

such that

$$\hat{H}_{int}|\psi_n\rangle = 0. \qquad (6.44)$$

Here $|n\rangle$ is the probe field state with n photons and

$$\tan\theta(t) = \frac{g\sqrt{n+1}}{\Omega_c(t)}. \qquad (6.45)$$

---

[3] See, for instance, [201] and [206] and references therein.

Thus, by adiabatically changing the coupling field Rabi frequency $\Omega_c(t)$, we can (reversibly) switch between the states $|1\rangle|n+1\rangle$ and $|2\rangle|n\rangle$, i.e. between atom and probe field excitations. Such stimulated Raman adiabatic passage (STIRAP) [213] is the basis for some proposals for QIP [214].

The adiabaticity condition for STIRAP is that the time T characterizing the state transfer should satisy

$$g^2 nT \gg \gamma \qquad (6.46)$$

where γ is the rate of spontaneous emission out of level 3. Obviously T should not be larger than the decoherence time $1/n\kappa$, where κ is the cavity photon loss rate. Thus, (6.46) implies the condition

$$g^2 \gg \kappa\gamma \qquad (6.47)$$

for adiabaticity. This condition defines the so-called *strong coupling regime* of cavity QED and it is difficult to satisfy in practice. The cavity QED approach to QIP also suffers from the related sensitivity of the coupling to the position of the atom in the cavity.

As noted by Fleischhauer and Lukin [201], the storing and regeneration of light offers certain advantages over cavity QED for QIP. For one thing, the adiabaticity condition (6.35) for coherent excitation transfer between the atoms and the field is more easily satisfied than the cavity QED adiabaticity condition (6.47) because it involves the number of atoms, $\mathcal{N}$. This is a reflection of the fact that the storing and regeneration is a *collective* atomic coherence phenomenon.

Consider first EIT with a *constant* coupling field as a technique for storing information. The pulse delay time $\tau_d = L/v_g$, which we can regard as the storage time, is related to the width of the EIT transparency window by [equation (5.31)] $\tau_d = \sqrt{a_0 L}/\Delta\omega_{tr}$. Thus, the figure of merit τ_d/τ_p for storage is limited by the opacity of the medium[4]:

$$\frac{\tau_d}{\tau_p} = \frac{\Delta\omega_p}{\Delta\omega_{tr}}\sqrt{a_0 L} \leq \sqrt{a_0 L}. \qquad (6.48)$$

An increase in τ_d requires a decrease in the group velocity but, as discussed in section 5.2, this implies that the EIT transparency window gets narrower and that there is a consequent increase in the probe absorption.

In the case of *stopped* light resulting from a *temporally varying* coupling field, however, the narrowing of the transparency window is accompanied by a simultaneous narrowing of the probe pulse spectral width. As the group velocity is adiabatically slowed, the spatial profile and length of the pulse are unaffected [equation (6.19)] while the electric field is reduced in amplitude and broadened in time. There is, therefore, narrowing of the pulse spectrum that goes along with the narrowing of the transparency window and, provided the initial pulse spectral

[4] Of course the condition $\tau_d < \gamma_{12}^{-1}$ must also be satisfied.

width satisfies (6.37), the pulse spectrum remains within the transparency window as the group velocity is adiabatically decreased [171].

For light stopping and storage to be useful as a quantum memory technique, it must also store and regenerate quantum information with high fidelity, i.e. it must be robust against decoherence effects even though it involves a many-particle entangled state. While further work in this direction is warranted, the fact that dark-state polaritons do not undergo spontaneous emission means that they are immune to at least one major source of decoherence.

6.6   Some related work

We have considered only situations where spatial dispersion can be ignored, i.e. where there is no dependence of the refractive index n on the wavevector k. If we assume that $n = n(\omega, k)$, $k = n\omega/c$, then the definition $v_g = d\omega/dk$ implies that

$$v_g = \frac{c - \omega\, dn/dk}{n + \omega\, dn/d\omega}. \tag{6.49}$$

This suggests that the group velocity can be made very small (or negative) not only by making the denominator large (or negative), as considered thus far, but also by making the numerator small (or negative) [215]. A simple example where spatial dispersion occurs is when we have an atomic medium moving with a velocity v. Then each atom sees a Doppler shift $-kv$ in a co-propagating field of frequency ω and, consequently, the refractive index of the moving medium depends on k as well as ω. Using $k \cong \tilde{k}$ and $\omega \cong \tilde{\omega} - \tilde{k}v$, where the tilde labels quantities in the frame moving with the atoms, we obtain

$$v_g = \frac{d\omega}{dk} = \frac{d\tilde{\omega}}{d\tilde{k}} - v = \tilde{v}_g - v \tag{6.50}$$

where $\tilde{v}_g$ is the group velocity in the frame moving with the atoms. In effect, there is a light 'dragging' reminiscent of the Fresnel drag (section 1.6). If the atoms are moving with the velocity $v = \tilde{v}_g$, then $v_g = 0$ and the light may be said to be 'frozen' in the laboratory frame. There do not appear to be any experimental demonstrations of frozen light or negative group velocities in moving media. An excellent review of light propagation in moving media is given by Leonhardt and Piwnicki [216].

In chapter 5, we mentioned self-induced transparency (SIT) as one of the effects in which small group velocities ($\sim 10^2 - 10^4$ m s^{-1}) are observed. Bullough and Gibbs [217] have discussed at some length the differences and similarities between SIT and the information storage and retrieval considered in this chapter. They emphasize the fact that the theory of SIT is intrinsically nonlinear, while that of EIT and the stopping and regeneration of light is essentially a linear refractive index theory and that, in quantum-mechanical terms, both theories are formulated at the level of expectation values.

A noteworthy point made by these authors is that the stopping of a light pulse and the recovery of its phase information was already predicted and observed many years ago in the context of *non-degenerate* SIT. In the case of the sodium D_1 $F = 2 \leftrightarrow 2$ transitions, for example, there are two relevant transition dipole moments of magnitude p and $\frac{1}{2}p$; thus, a 4π pulse for p is a 2π pulse for $\frac{1}{2}p$. The initial absorption $\propto p^2$ and $\propto \frac{1}{4}p^2$ is, of course, dominated by the larger dipole moment. An initial 4π pulse will break up into two 2π pulses as in non-degenerate SIT, with the more intense pulse travelling faster. This leading pulse, being a π pulse for the $\frac{1}{2}p$ dipole moment, is then strongly attenuated and *stopped—while retaining its phase information—*and the second pulse propagating in the medium excited by the first pulse grows in amplitude and speed and overtakes the first pulse at its stopping point. Then this pulse is attenuated and stopped and overtaken by the first pulse that is amplified and advanced by the excited medium ahead of it. In other words, there is a pulse 'leap-frogging' that was observed in the experiments [218]. If damping is negligible, the leap-frogging should proceed indefinitely.

Chapter 7

Left-handed light: basic theory

We have mainly been concerned thus far with 'abnormal' group velocities associated with strong variations of the refractive index with frequency. Now we turn our attention to abnormal values of the refractive index itself, specifically to the possibility that the refractive index can be *negative*. Although there are no known naturally occurring materials with a negative refractive index, there is no fundamental reason why the refractive index cannot be negative and, as we shall see in the following chapter, it is possible to fabricate structures that do, in fact, exhibit negative refraction.

In this chapter, we discuss some of the basic theory of negative refraction and its consequences. A negative refractive index is shown to be implied by simultaneously negative values of the permittivity ϵ and the permeability μ. In order to avoid certain inconsistencies, dispersion of negative-index materials must be taken into account. We quantize the field in a negative-index medium and show that the Doppler and Cerenkov effects are reversed in such a medium. After a discussion of the Fresnel formulas in the case that one of the media at a dielectric interface has a negative refractive index, we turn our attention to evanescent waves and the possibility that they can be amplified in a negative-index slab. Implications of negative refractive index for surface modes at a dielectric interface are discussed. We consider the 'perfect lens,' an aberration- and reflection-free *planar* lens capable of subwavelength resolution, and show that absorption must be kept very small if such a lens is to be realized.

7.1 Introduction

In section 1.2, we reviewed the derivation of expressions such as (1.25) for the refractive index, taking the magnetic permeability μ to have the value $\mu_0 = 4\pi \times 10^{-7}$ N A^{-2} as in free space. If the field frequency ω is sufficiently large compared with atomic transitions of significant oscillator strength [cf equation (1.27)], the permittivity $\epsilon(\omega)$ can be negative and the refractive index purely imaginary: $n = i n_{\mathrm{I}}$, $n_{\mathrm{I}} > 0$ for a passive material. In this case, $\exp[ikz] =$

$\exp[in(\omega)\omega z/c] = \exp[-n_I\omega z/c]$. This occurs, for instance, in the case of free electrons ($\omega_{ji} = 0$) when the field frequency is smaller than the plasma frequency ω_p: $\epsilon(\omega) = \epsilon_0(1 - \omega_p^2/\omega^2) < 0$.

Equation (1.33) shows that the real part of the permittivity ϵ can change sign and become negative (and large) as the field frequency is swept across a resonance. Similarly, the permeability μ can be negative near a resonance in ferromagnetic materials. The difficulty in realizing *simultaneously* negative values of ϵ and μ lies in the fact that the resonance frequencies for which $\epsilon < 0$ tend to be much larger than the resonance frequencies for which $\mu < 0$—high frequencies (typically optical or infrared) in the former case and much lower frequencies in the latter. Moreover, the resonance regions tend to be very narrow in either case.

For atomic media the magnetic susceptibility χ_m ($= \mu/\mu_0 - 1$) is of order v^2/c^2 at optical frequencies, where v is the electron velocity [219]. Thus

> ...there is certainly no meaning in using the magnetic susceptibility from optical frequencies onward, and in discussing such phenomena we must put $\mu = [\mu_0]$. To distinguish between *B* and *H* in this frequency range would be an over-refinement. Actually, the same is true for many phenomena even at frequencies well below the optical range.

Note that if either ϵ or μ is negative and the other positive, then the refractive index is purely imaginary and there is no propagating wave. This occurs, for instance, when a wave with frequency below the plasma frequency is incident on a plasma. The incident wave is reflected.

In this chapter, we take for granted that ϵ and μ can be simultaneously negative and explore some consequences of this double negativity. As mentioned in section 1.8, materials with this property are referred to as *doubly negative, left-handed*, or simply as *metamaterials*, the last term connoting the fact that the only materials known to have both $\epsilon < 0$ and $\mu < 0$ are purposefully fabricated rather than naturally occurring.

Veselago [28, 29] was evidently the first to consider seriously the possibility, and some consequences, of having a negative refractive index. In particular, he showed that a negative electric permittivity ϵ *and* a negative magnetic permeability μ, imply a negative refractive index and a phase velocity in a direction opposite to the direction of the group velocity and energy flow. As also noted in section 1.8, the possibility that the refractive index (or phase velocity) could be negative was briefly raised by Mandelstam in connection with Snell's law in 1944 [30, 31]. Lamb [220], in an article 'On group-velocity' in 1904, considered the possibility that waves on strings could have a phase velocity in the direction opposite to that of the group velocity and Schuster [221], in the same year, noted that the phase and group velocities of an electromagnetic wave could have opposite signs in a region of anomalous dispersion[1]. Experimental

---

[1] I learned of these and other early contributions from a website of Alexander Moroz (www.wavescattering.com/negative) that was brought to my attention by Gary D Doolen.

confirmations of negative group velocity were described in chapter 2.

7.2   Negative ϵ and μ imply negative index

At first thought, it would seem that having $\epsilon < 0$ and $\mu < 0$ should not affect the refractive index $n = \sqrt{\epsilon\mu/\epsilon_0\mu_0}$. However, ϵ, μ, and n are complex, and we can write

$$\epsilon/\epsilon_0 = r_\epsilon e^{i\theta} \qquad \mu/\mu_0 = r_\mu e^{i\phi} \tag{7.1}$$

and

$$n = \sqrt{\epsilon\mu/\epsilon_0\mu_0} = \sqrt{r_\epsilon r_\mu} e^{i(\theta+\phi)/2}. \tag{7.2}$$

The requirement that the imaginary part of n be positive for a passive (non-absorbing) medium implies that

$$0 \le \tfrac{1}{2}(\theta + \phi) < \pi. \tag{7.3}$$

If the real parts of ϵ and μ are both negative, i.e. $\cos\theta < 0$ and $\cos\phi < 0$, then

$$\frac{\pi}{2} < \frac{1}{2}(\theta + \phi) < \frac{3\pi}{2}. \tag{7.4}$$

To satisfy both (7.3) and (7.4), we must have

$$\frac{\pi}{2} < \frac{1}{2}(\theta + \phi) < \pi \tag{7.5}$$

and, therefore, a negative (real) refractive index[2]:

$$n_R \equiv \mathrm{Re}[n] = \sqrt{r_\epsilon r_\mu} \cos \tfrac{1}{2}(\theta + \phi) < 0. \tag{7.6}$$

Thus, the requirement that n has a positive imaginary part leads to the conclusion that, if ϵ and μ have negative real parts, the real part of n must also be negative. In a fundamental sense, this conclusion may be said to follow from causality, which, as discussed in section 1.3, demands that $\epsilon(\omega)$ and $n(\omega)$ (and similarly $\mu(\omega)$) be complex.

An immediate consequence of a negative refractive index[3] can be seen from Snell's law[4]. Consider a plane wave incident from a medium with $n > 0$ onto a medium with $n < 0$. In particular, consider a wave incident from vacuum onto

---

[2] The choice of a negative refractive index when ϵ and μ are both negative is discussed in more detail and justified by Ziolkowksi and Heyman [222].

[3] The 'refractive index' will, henceforth, mean the real part of the complex refractive index. This quantity, which determines the phase delay (or advance) of a propagating electromagnetic wave, will be denoted by n.

[4] Snell's law is easily shown to be applicable regardless of the signs of the refractive indices. See section 7.7.

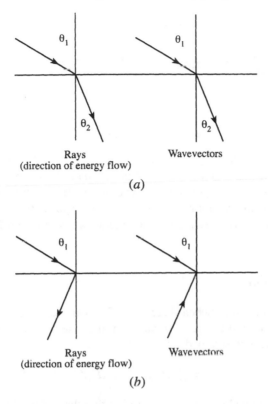

Figure 7.1. Rays and wavevectors when light is incident from vacuum onto a medium with a refractive index (a) $n > 0$ and (b) $n < 0$. In the latter case, the direction of the wavevector and, therefore, the phase velocity is *opposite* to the direction of energy flow.

a negative-index medium (NIM). If θ_1 and θ_2 are the angles of incidence and transmission, respectively, then according to Snell's law

$$\sin \theta_1 = n \sin \theta_2 \qquad (7.7)$$

$n < 0$ implies that $\sin \theta_2 < 0$: the transmitted rays make a *negative* angle with respect to the normal to the interface. In other words, the refracted rays are bent to the same side of the normal as the incident rays—the 'wrong' way compared with the usual case of positive-index media. This is shown in figure 7.1. Experimental evidence for this aspect of NIMs is discussed in the following chapter.

   Consider now the closely related example of a plane wave propagating in an NIM. Write

$$E = E_0 e^{i(k \cdot r - \omega t)} \qquad (7.8)$$

and, likewise, for D, B and H. From the Maxwell curl equations

$$\nabla \times E = - \frac{\partial B}{\partial t} \qquad (7.9)$$

$$\nabla \times \boldsymbol{H} = \frac{\partial \boldsymbol{D}}{\partial t} \tag{7.10}$$

and the constitutive relations $\boldsymbol{D} = \epsilon(\omega)\boldsymbol{E}$ and $\boldsymbol{B} = \mu(\omega)\boldsymbol{H}$ for an isotropic medium, we obtain

$$\boldsymbol{k} \times \boldsymbol{E}_0 = \omega\mu(\omega)\boldsymbol{H}_0 \tag{7.11}$$

and

$$\boldsymbol{k} \times \boldsymbol{H}_0 = -\omega\epsilon(\omega)\boldsymbol{E}_0. \tag{7.12}$$

Thus, if $\epsilon(\omega) > 0$ and $\mu(\omega) > 0$, the vectors $\boldsymbol{E}$, $\boldsymbol{H}$, and $\boldsymbol{k}$ form a right-handed triad and the Poynting vector $\boldsymbol{S} = \boldsymbol{E} \times \boldsymbol{H}$ points in the direction of the wavevector $\boldsymbol{k}$. But if $\epsilon(\omega) < 0$ and $\mu(\omega) < 0$, the vectors $\boldsymbol{E}$, $\boldsymbol{H}$, and $\boldsymbol{k}$ form a *left*-handed triad and the Poynting vector points in a direction *opposite* to the direction of the wavevector $\boldsymbol{k}$. For this reason, electromagnetic waves in an NIM may be called *left-handed*.

7.3 Dispersion

Veselago [29] noted that 'when there is no frequency dispersion nor absorption we cannot have $\epsilon < 0$ and $\mu < 0$, since in this case the total energy would be negative', i.e. the field energy density

$$u_\omega = \tfrac{1}{4}[\epsilon|\boldsymbol{E}_\omega|^2 + \mu|\boldsymbol{H}_\omega|^2] < 0 \tag{7.13}$$

where ϵ and μ are real under the assumption that there is no absorption. Taking dispersion into account, we have [equation (2.60)]

$$u_\omega = \frac{1}{4}\left[\frac{\mathrm{d}}{\mathrm{d}\omega}(\epsilon\omega)|\boldsymbol{E}_\omega|^2 + \frac{\mathrm{d}}{\mathrm{d}\omega}(\mu\omega)|\boldsymbol{H}_\omega|^2\right] \tag{7.14}$$

away from any absorption resonances. This is positive because

$$\frac{\partial(\epsilon\omega)}{\partial\omega} > 0 \quad \text{and} \quad \frac{\partial(\mu\omega)}{\partial\omega} > 0. \tag{7.15}$$

These conditions are another consequence of causality and may be derived as follows. From the dispersion relation between the real and imaginary parts of $\epsilon(\omega)$ [equation (1.48) with $g = \epsilon - 1$], namely

$$\epsilon_R(\omega) - 1 = \frac{2}{\pi}\mathrm{P}\int_0^\infty \frac{\omega'\epsilon_I(\omega')}{\omega'^2 - \omega^2}\,\mathrm{d}\omega' \tag{7.16}$$

it follows that

$$2\omega[\epsilon_R(\omega) - 1] + \omega^2\frac{\mathrm{d}\epsilon_R}{\mathrm{d}\omega} = \frac{4\omega}{\pi}\mathrm{P}\int_0^\infty \frac{\omega'^3\epsilon_I(\omega')}{(\omega'^2 - \omega^2)^2}. \tag{7.17}$$

For a passive medium, $\epsilon_I(\omega') \geq 0$ at all frequencies and the integral is positive. Thus [223],

$$\omega \frac{d\epsilon_R}{d\omega} > 2(1 - \epsilon_R) \tag{7.18}$$

and, therefore,

$$\frac{d}{d\omega}(\epsilon_R \omega) > 2 - \epsilon_R \tag{7.19}$$

and $d(\epsilon_R \omega)/d\omega > 0$ for $\epsilon_R < 0$. The second inequality in (7.15) is derived in the same way. Thus, (7.15) is satisfied for frequencies at which absorption is negligible.

It can also be shown that—away from any absorption resonances—the group index $n_g = n + \omega \, dn/d\omega$ is positive in an NIM. To see this, write $n = -\sqrt{|\epsilon||\mu|}/\sqrt{\epsilon_0 \mu_0}$, where $\epsilon = \epsilon_R$ and $\mu = \mu_R$, so that

$$n_g \sqrt{\epsilon_0 \mu_0} = -\sqrt{|\epsilon||\mu|} - \frac{\omega}{2}\sqrt{\frac{|\epsilon|}{|\mu|}} \frac{d|\mu|}{d\omega} - \frac{\omega}{2}\sqrt{\frac{|\mu|}{|\epsilon|}} \frac{d|\epsilon|}{d\omega}$$

$$= -\frac{1}{2}\sqrt{\frac{|\epsilon|}{|\mu|}}\left(|\mu| + \omega\frac{d|\mu|}{d\omega}\right) - \frac{1}{2}\sqrt{\frac{|\mu|}{|\epsilon|}}\left(|\epsilon| + \omega\frac{d|\epsilon|}{d\omega}\right)$$

$$= \frac{1}{2}\sqrt{\frac{|\epsilon|}{|\mu|}}\frac{d}{d\omega}(\mu\omega) + \frac{1}{2}\sqrt{\frac{|\mu|}{|\epsilon|}}\frac{d}{d\omega}(\epsilon\omega) > 0. \tag{7.20}$$

This result will be used in the following section.

7.4 Maxwell's equations and quantized field

Let us consider now the electric and magnetic fields in spectral regions removed from any absorption resonance in an NIM. For a monochromatic electric field, we write

$$E(r, t) = C\alpha(t)F(r) \tag{7.21}$$

where $\alpha(t) = \alpha(0)\exp(-i\omega t)$, $F(r)$ is a mode function, and C is a constant. Maxwell's equations

$$\nabla \cdot D = 0 \tag{7.22}$$

$$\nabla \cdot E = 0 \tag{7.23}$$

$$\nabla \times E = -\frac{\partial B}{\partial t} \tag{7.24}$$

$$\nabla \times H = \frac{\partial D}{\partial t} \tag{7.25}$$

for a medium without free charges and currents, together with the constitutive relations, imply from (7.21) that

$$D(r, t) = \epsilon C\alpha(t)F(r) \tag{7.26}$$

$$B(r, t) = -\frac{i}{\omega} C\alpha(t) \nabla \times F(r) \tag{7.27}$$

$$H(r, t) = -\frac{i}{\mu\omega} C\alpha(t) \nabla \times F(r). \tag{7.28}$$

It then follows from (7.14) that the field energy associated with frequency ω is

$$\begin{aligned} U_\omega &= \int d^3r \, u_\omega \\ &= \frac{1}{4}|C|^2|\alpha(t)|^2 \left[\frac{d}{d\omega}(\epsilon\omega) \int d^3r \, |F|^2 + \frac{1}{\mu^2\omega^2} \frac{d}{d\omega}(\mu\omega) \int d^3r \, |\nabla \times F|^2 \right]. \end{aligned} \tag{7.29}$$

We assume that the mode function $F(r)$ is normalized, $\int d^3r \, |F(r)|^2 = 1$. The fact that F must satisfy the Helmholtz equation

$$\nabla^2 F + k^2 F = 0 \qquad (k^2 = \epsilon\mu\omega^2 = n^2\omega^2/c^2) \tag{7.30}$$

as required by (7.22)–(7.25), together with standard identities of vector calculus[5], implies that

$$\int d^3r \, |\nabla \times F|^2 = k^2 \int d^3r \, |F|^2 = k^2 = n^2\omega^2/c^2 \tag{7.31}$$

and, therefore,

$$U_\omega = \frac{|C|^2}{4\mu}|\alpha(t)|^2 \left[\mu\frac{d}{d\omega}(\epsilon\omega) + \epsilon\frac{d}{d\omega}(\mu\omega) \right] = \frac{n|C|^2}{2\mu c^2}|\alpha(t)|^2 \frac{d}{d\omega}(n\omega) \tag{7.32}$$

where the last equality follows from the definition $n^2 = \epsilon\mu c^2$.

The constant C up to this point is arbitrary. Let us now set

$$C = (\mu c^2/nn_g)^{1/2} \tag{7.33}$$

where again we introduce the group index

$$n_g = d(n\omega)/d\omega. \tag{7.34}$$

Then,

$$U_\omega = \tfrac{1}{2}|\alpha(t)|^2. \tag{7.35}$$

Let us furthermore write

$$\alpha(t) = \alpha(0)\exp(-i\omega t) = p(t) - i\omega q(t) \tag{7.36}$$

[5] See, for instance, [8, appendix C].

which implies that

$$\dot{q} = p \tag{7.37}$$

$$\dot{p} = -\omega^2 q. \tag{7.38}$$

Thus, we have the Hamilton equations of motion for a simple harmonic oscillator. To quantize the field mode of frequency ω, we quantize this harmonic oscillator, replacing q and p by operators $\hat{q}$ and $\hat{p}$ satisfying

$$[\hat{q}, \hat{p}] = i\hbar. \tag{7.39}$$

The photon annihilation and creation operators are then $\hat{a} = (1/\sqrt{2\hbar\omega})(\hat{p} - i\omega\hat{q})$ and $\hat{a}^\dagger = (1/\sqrt{2\hbar\omega})(\hat{p} + i\omega\hat{q})$, with

$$[\hat{a}, \hat{a}^\dagger] = 1. \tag{7.40}$$

The electric field

$$E(r, t) = \tfrac{1}{2}[C\alpha(t)F(r) + C^*\alpha^*(t)F(r)^*] \tag{7.41}$$

is similarly replaced by the operator

$$\hat{E}(r, t) = \left(\frac{\mu c^2 \hbar\omega}{2nn_g}\right)^{1/2} [\hat{a}(t)F(r) + \hat{a}^\dagger(t)F(r)^*] \tag{7.42}$$

when we quantize. For plane-wave modes, we can take

$$F(r) = (i/\sqrt{V})e_k \exp(ik \cdot r) \tag{7.43}$$

where e_k is a unit polarization vector ($k \cdot e_k = 0$) and V is a quantization volume:

$$\hat{E}(r, t) = i\left(\frac{\mu c^2 \hbar\omega}{2nn_g V}\right)^{1/2} \left[\hat{a}(t)e^{ik\cdot r} - \hat{a}^\dagger(t)e^{-ik\cdot r}\right]e_k \tag{7.44}$$

and we have taken e_k to be real. Similarly, from equations (7.26)–(7.28), we write the quantized fields

$$\hat{B}(r, t) = i\left(\frac{\mu c^2 \hbar}{2nn_g\omega V}\right)^{1/2} [\hat{a}(t)e^{ik\cdot r} - \hat{a}^\dagger(t)e^{-ik\cdot r}]k \times e_k \tag{7.45}$$

$$\hat{D}(r, t) = i\left(\frac{n\epsilon\hbar\omega}{2n_g V}\right)^{1/2} [\hat{a}(t)e^{ik\cdot r} - \hat{a}^\dagger(t)e^{-ik\cdot r}]e_k \tag{7.46}$$

$$\hat{H}(r, t) = i\left(\frac{c^2\hbar}{2nn_g\mu\omega V}\right)^{1/2} [\hat{a}(t)e^{ik\cdot r} - \hat{a}^\dagger(t)e^{-ik\cdot r}]k \times e_k. \tag{7.47}$$

These expressions apply to a single-mode field in a dispersive dielectric, provided the mode frequency is far from any absorption resonance. In particular, they apply to the case of an NIM where $\epsilon(\omega)$, $\mu(\omega)$, and $n(\omega)$ are all negative. For the case of an NIM, note the importance in these equations of the fact that $n_g > 0$ [equation (7.20)].

The field energy (7.35) becomes, on quantization,

$$\hat{H}_{\text{field}} = \hbar\omega(\hat{a}^\dagger\hat{a} + 1/2). \tag{7.48}$$

The operator corresponding to the Poynting vector

$$\hat{S} = \hat{E} \times \hat{H} \tag{7.49}$$

similarly, is

$$\hat{S} = \frac{\hbar c^2}{nn_g V}(\hat{a}^\dagger\hat{a} + 1/2)e_k \times (k \times e_k) \tag{7.50}$$

when we cycle-average. Writing

$$k = \frac{n\omega}{c}z \tag{7.51}$$

where z is the unit vector pointing in the z direction, we have

$$\hat{S} = \frac{\hbar c\omega}{n_g V}(\hat{a}^\dagger\hat{a} + \tfrac{1}{2})z = \frac{\hbar\omega v_g}{V}(\hat{a}^\dagger\hat{a} + \tfrac{1}{2})z$$
$$= z v_g(\hat{H}_{\text{field}}/V) = v_g(\hat{H}_{\text{field}}/V) \tag{7.52}$$

where the group velocity $v_g = zc/n_g = zv_g$. Equations (7.51) and (7.52) show that, in an NIM, the Poynting vector and the k vector point in opposite directions: E, H, and k define a left-handed triad. Note also that the Poynting vector and the group velocity are in the same direction but that the group and phase velocities are in opposite directions in an NIM.

The multimode generalization of (7.42), for instance, is

$$\hat{E}(r, t) = i \sum_{k\lambda} \left(\frac{\mu c^2 \hbar\omega}{2nn_g V}\right)^{1/2} [\hat{a}_{k\lambda}(t)e^{ik\cdot r} - \hat{a}^\dagger_{k\lambda}(t)e^{-ik\cdot r}]e_{k\lambda} \tag{7.53}$$

λ labels the polarization of mode k, λ ($k \cdot e_{k\lambda} = 0$, $\lambda = 1, 2$). More generally, we can write such multimode expansions for situations where mode functions other than plane waves are more convenient.

7.4.1 Radiative rates in negative-index media

The possibility that the refractive index can be negative obviously invites a reconsideration of various processes. For example, it is well known that, in the absence of local field corrections, the spontaneous emission rate for an electric dipole transition of frequency ω_0 is $A' = n(\omega_0)A$, where A is the free-space Einstein A coefficient[6]. How is this formula to be understood if $n(\omega_0) < 0$?

The coupling constant for the field mode k, λ and the atomic transition with electric dipole matrix element d in the $-r \cdot E$ interaction is

$$V(\omega) = -i\left(\frac{\mu c^2 \hbar \omega}{2nn_g V}\right)^{1/2} d \cdot e_{k\lambda}. \tag{7.54}$$

Fermi's golden rule then implies the spontaneous emission rate

$$\frac{2\pi}{\hbar}|V(\omega_0)|^2 \rho_e(\omega_0) \tag{7.55}$$

where ω_0 is the transition frequency and ρ_e is the density (in energy) of final states:

$$\rho_e(\omega_0)\hbar\,d\omega = \frac{V}{(2\pi)^3}d^3k = \frac{V}{(2\pi)^3}k^2\,d\Omega_k\,dk$$
$$= \frac{V}{8\pi^3 c^3}n^2(\omega)\omega^2\frac{d}{d\omega}[n(\omega)\omega]\,d\omega\,d\Omega_k \tag{7.56}$$

where $d\Omega_k$ is the differential element of solid angle about k. The rate of spontaneous emission into all solid angles and polarizations is then, from (7.54)–(7.56) [225],

$$A' = \frac{2\pi}{\hbar}\frac{\mu(\omega_0)c^2\hbar\omega_0}{2nn_g V}\frac{Vn^2(\omega_0)\omega_0^2}{8\pi^3 c^3\hbar}n_g(\omega_0)\sum_\lambda \int d\Omega_k\,|d \cdot e_{k\lambda}|^2$$
$$= n(\omega_0)\frac{\mu(\omega_0)}{\mu_0}A \tag{7.57}$$

where $A = |d|^2\omega_0^3/3\pi\epsilon_0\hbar c^3$ is the free-space radiation rate. Equation (7.57) differs from the familiar result cited earlier by the factor $\mu(\omega_0)/\mu_0$, which ensures that $A' > 0$ in an NIM.

Absorption and stimulated emission are likewise affected. The Einstein B coefficient is calculated in the same manner as A' to be

$$B' = \frac{\mu(\omega_0)/\mu_0}{n_g(\omega_0)n(\omega_0)}B \tag{7.58}$$

where B is the coefficient for an atom in free space. Obviously $B' > 0$ for both positive- and negative-index media. This generalizes the expression $B' = B/n^2(\omega_0)$ that appears frequently in the literature [226]. The latter is seen to be applicable if dispersion is negligible $[n_g(\omega_0) \to n(\omega_0)]$ and $\mu(\omega_0) \approx \mu_0$. The expressions given here for A' and B' also generalize the results obtained in [227], where it was assumed that the host medium is non-magnetic ($\mu = \mu_0$).

[6] See, for instance, [224].

7.5   Reversal of the Doppler and Cerenkov effects

One of the weird things about NIMs is that the Doppler effect is reversed: if a source is moving towards a detector, the emitted radiation is observed to have a *smaller* frequency than when the source is stationary [29, 228].

Consider an excited two-level atom of mass m moving with velocity v in an NIM, and let ω_0 be its natural transition frequency. The atom emits a photon of wavevector k and frequency ω, after which it moves with velocity v'. Conservation of energy gives[7]

$$\tfrac{1}{2}mv'^2 + \hbar\omega = \tfrac{1}{2}mv^2 + \hbar\omega_0 \qquad (7.59)$$

while conservation of linear momentum implies

$$mv' = mv - \hbar k. \qquad (7.60)$$

Ignoring terms involving $1/c^2$ in this non-relativistic treatment—i.e. ignoring the recoil energy $\tfrac{1}{2}\hbar^2 k^2$ compared with mc^2—and using

$$k = \frac{n\hbar\omega}{c}s \qquad (7.61)$$

where s is the unit vector in the direction of the Poynting vector (and the group velocity), we obtain, from equations (7.59)–(7.61), the classical expression

$$\omega \cong \omega_0 1 + \frac{n}{c}v\cos\theta \qquad (7.62)$$

where θ is the angle between v and s. Thus, for $n < 0$, a photon detected in the direction of the atom's initial velocity will have a frequency smaller than ω_0; and a photon detected in the opposite direction will have a frequency larger than ω_0. Since the Poynting vector of the emitted field points in the direction opposite to k in an NIM (section 7.4), equation (7.60) implies that, upon emission, the atom will recoil in the *same* direction as the Poynting vector of the emitted field in an NIM.

Another effect that is 'reversed' in an NIM is the Cerenkov effect [29, 228, 229]. Consider an electron (for instance) with initial energy E moving with velocity v in a medium with refractive index n and let E' be its energy after emitting a photon of energy $\hbar\omega$. Similarly, let p and p' be the initial and final linear momenta of the electron and k the wavevector of the emitted photon. Then, equations (7.59) and (7.60) generalize to

$$E = E' + \hbar\omega \qquad (7.63)$$

$$p = p' + \hbar k \qquad (7.64)$$

---

[7] The kinematic treatment of the Doppler and Cerenkov effects in this section follows Fermi's treatment of the Doppler effect [69] in terms of the recoil of the emitter, as well as Ginzburg's related discussion of Cerenkov radiation [229].

where we now use the relativistic expressions

$$E = \sqrt{m^2c^4 + p^2c^2} \tag{7.65}$$

$$p = \frac{mv}{\sqrt{1 - v^2/c^2}} \tag{7.66}$$

and, likewise, for E' and p'. For the wavevector k and the linear momentum $\hbar k$ of the photon, we again use equation (7.61).

If $n = 1$, equations (7.63)–(7.65) and (7.61) can only hold if $\omega = 0$: a particle in uniform motion in vacuum cannot radiate[8].

If $n \neq 1$, however, (Cerenkov) radiation is possible even if v is constant. Equations (7.63)–(7.66) and (7.61) in this case imply [229]

$$\hbar\omega = \frac{2mc[n(\omega)v\cos\theta - c]}{[n^2(\omega) - 1]\sqrt{1 - v^2/c^2}} \tag{7.67}$$

and

$$\cos\theta = \frac{c}{n(\omega)v}\left(1 + \frac{\hbar\omega}{2mc^2}[n^2(\omega) - 1]\sqrt{1 - v^2/c^2}\right) \tag{7.68}$$

where θ is again the angle between v and s. In order to have $\omega > 0$ and $|\cos\theta| < 1$, it is necessary to have $|vn(\omega)/c| > 1$. In other words, radiation of frequency ω by a charge moving with constant velocity in a dielectric is only possible if the velocity exceeds the phase velocity of light at the frequency ω. If we let $\hbar \to 0$, we recover the classical expression

$$\cos\theta = \frac{c}{n(\omega)v} \tag{7.69}$$

for the emission cone of Cerenkov radiation. When $n(\omega) < 0$, the direction of Cerenkov radiation is reversed, i.e. the Poynting vector (and the group velocity) of the emitted radiation makes an obtuse angle with the velocity of the electron; and if the electron is incident on a medium with $n(\omega) < 0$, radiation of frequency ω will appear in the *backward* direction.

These effects follow from the fact that the phase and group velocities are in opposite directions. As already mentioned, the possibility that the phase and group velocities could be in opposite directions was known long before the current interest in negative refractive index. In fact, the reversal of the Doppler and Cerenkov effects when the phase and group velocities are opposite was noted by Pafomov [228] in 1959 and is discussed, for instance, in the monograph by Ginzburg [229]. Opposite phase and group velocities can occur in anisotropic media or when there is spatial dispersion [230] or when the refractive index is negative.

---

[8] We are, of course, assuming that the internal state of the particle does not change.

7.5.1   On photon momentum in a dielectric

The previous simple derivations rely on the assumption that the linear momentum
of a photon is given by

$$p = \hbar k \quad (k = n\omega/c). \tag{7.70}$$

The form of the linear momentum of the electromagnetic field in a dielectric
has been discussed and debated for many years; the review by Brevik [231] and
chapter 12 of Ginzburg [229] are particularly good references. The subject has
recently attracted renewed interest[9]. The two most frequently advocated forms of
the electromagnetic momentum density are the Abraham form[10]

$$g_A = \frac{1}{c^2}S = \frac{1}{c^2}E \times H \tag{7.71}$$

and the Minkowksi form

$$g_M = D \times B. \tag{7.72}$$

Equation (7.52) yields

$$g_A = \frac{z}{c^2}v_g(H_{field}/V) \tag{7.73}$$

which implies that the linear momentum of a single photon of energy $\hbar\omega$ is [233]

$$p_A = z\frac{\hbar\omega}{c}\frac{v_g}{c} = \frac{1}{n(\omega)}\frac{v_g}{c}\hbar k. \tag{7.74}$$

For the Minkowksi form, we obtain instead

$$g_M = \epsilon\mu z v_g(H_{field}/V) \tag{7.75}$$

which implies the photon linear momentum [233]

$$p_M = n(\omega)\frac{v_g}{c}\hbar k. \tag{7.76}$$

Note that

$$|p_A| = |\frac{v_g}{c}|\frac{\hbar\omega}{c} \tag{7.77}$$

and

$$|p_M| = n^2(\omega)|\frac{v_g}{c}|\frac{\hbar\omega}{c}. \tag{7.78}$$

[9] See, for instance, the papers by Loudon [232] and Garrison and Chiao [233] and references therein.
[10] Since the question of which form of the momentum is more appropriate in a given situation arises
in both classical and quantum theories, we do not bother to distinguish here between classical and
quantum fields.

If dispersion is ignored ($v_g/c \rightarrow 1/n$), these expressions for p_A and p_M reduce to those given by Ginzburg [229]:

$$|p_A| = \frac{\hbar\omega}{c|n|} \tag{7.79}$$

and

$$|p_M| = \frac{|n|\hbar\omega}{c}. \tag{7.80}$$

Our derivations here for the Doppler and Cerenkov effects, which dealt with a monochromatic field and did not require variations of the index with respect to frequency, suggest that the Minkowksi form for the field momentum is the correct one [equations (7.70) and (7.80)]. But this argument in favour of the Minkowksi form is superficial since, as discussed by Ginzburg [229] for the special case of a dispersionless and non-magnetic medium, the momentum conservation laws based on the Minkowksi and Abraham forms of the field momentum 'are identical—they differ merely in the different splitting up into terms of the same sum' [229, p 283]. Thus, in the formulation using the Abraham field momentum, there appears an additional momentum

$$p_m^A = \frac{n^2 - 1}{c^2} \int d^3r \int dt \, \frac{\partial}{\partial t}(E \times H) \tag{7.81}$$

imparted to the dielectric. For a single-photon plane wave in the assumed dispersionless and non-magnetic medium, it follows easily from this expression that

$$p_m^A = \frac{n - 1}{n} \frac{\hbar\omega}{c} \frac{k}{k} \tag{7.82}$$

and, therefore, that [229]

$$p_A + p_m^A = \frac{1}{n} \frac{\hbar\omega}{c} \frac{k}{k} + \frac{n - 1}{n} \frac{\hbar\omega}{c} \frac{k}{k} = n \frac{\hbar\omega}{c} \frac{k}{k} = p_M. \tag{7.83}$$

Thus, in this model, the Minkowksi form of the momentum includes the Abraham force on the medium and, while the Abraham form of the momentum might be more fundamental [229], the 'auxiliary' Minkowksi form represents a useful simplification. This simplification is evident in the derivations earlier for the Doppler and Cerenkov effects.

It is interesting that Garrison and Chiao [233] find good agreement with the Jones–Leslie experiment [234]—evidently one of the most careful experiments on field momentum in a dielectric—if the 'canonical' photon momentum (7.70) is assumed. The Abraham and Minkowksi forms, which both involve the group velocity v_g, yielded predictions differing by many standard deviations from the experimental data. The subject of field momentum is a complicated and very interesting one but further discussion here would take us too far afield.

7.6   Discussion

Let us now summarize and discuss a bit further the most important points thus far of this chapter. The main one, of course, is that negative values of ϵ and μ imply a negative refractive index. Negative values of ϵ or μ are well known to occur but the frequencies at which negative values of ϵ are possible are generally very different from those at which negative values of μ are possible. In fact, there are no known naturally occurring materials for which ϵ and μ are simultaneously negative—simultaneously negative ϵ and μ are realized in artificial *metamaterials*.

A negative refractive index requires consideration of dispersion for consistency, e.g. for the electromagnetic energy density to be positive.

In a metamaterial, the wavevector k of a plane wave is opposite in direction to the Poynting vector S. The Doppler effect is reversed in a metamaterial and a spontaneously emitting atom will experience a recoil momentum in the same direction as the Poynting vector of the emitted light. Similarly, if light in a metamaterial is incident on a reflecting interface, it will impart to the reflector a linear momentum in the direction *opposite* to the Poynting vector. This was noted by Veselago [29]:

> A monochromatic wave in a left-handed medium can be regarded as a stream of photons, each having a momentum $p = \hbar k$, with the vector k directed toward the source of radiation, not away from it as is the case in a right-handed medium. Therefore a beam of light propagated in a left-handed medium and incident on a reflecting body imparts to it a momentum $p = N\hbar k$ (N is the number of incident photons) directed toward the source of the radiation ... Owing to this the light pressure characteristic for ordinary (right-handed) substances is replaced in left-handed substances by a light tension or attraction.

Such strange consequences of a negative index are perfectly consistent with the conservation of energy and linear momentum.

As a consequence of Snell's law, light incident from an ordinary medium and transmitted by an NIM will refract to the same side of the normal as the incident light, i.e. the angle of transmission is negative. This implies the possibility of a planar lens made from a negative-index material, as discussed in the following section.

If absorption (or amplification) is negligible, v_g in an NIM is a positive number [equation (7.20)] but the *direction* of the group velocity vector, which is the same as the direction of the Poynting vector [equation (7.52)], is opposite to that of the wavevector k. k, in this case, points towards the source while the group velocity points away from the source. In other words, the phase velocity is opposite in direction to the direction of energy flow. When the group velocity is opposite in direction to the phase velocity, it is said to be negative: Veselago [29] remarked that the term 'left-handed substance' is equivalent in this context to the term 'substance with negative group velocity.'

In chapter 2, we discussed experiments in which v_g is negative. In those experiments, the wavevector and phase velocity are essentially the same for the incident and transmitted fields. When light is incident from a medium with positive index to a passive medium with negative index (or *vice versa*), however, the wavevector and phase velocity are reversed and the angle of transmission is negative, while the scalar v_g remains positive.

As already noted, the possibility that phase and group velocities could be in opposite directions was noted a century ago by Lamb [220] and Schuster [221]. It should also be noted that Pocklington [235] around the same time considered an example in which a displacement produced instantaneously at a point on a mechanical chain could propagate with a group velocity pointing away from that point but with a phase velocity pointing *towards* it, a result Pocklington considered 'most remarkable'.

It should be borne in mind that the characterization of a medium by a refractive index presumes that there are many scattering elements, e.g. atoms, within a wavelength: this was briefly discussed near the end of section 2.1. As discussed in the following chapter, the basic elements of metamaterials are typically metallic structures configured to produce negative ϵ and μ in a narrow spectral range and the spacing of these structures must be small compared with a wavelength in order to characterize the 'medium' by a refractive index meaningfully.

7.7 Fresnel formulas and the planar lens

How, if at all, are the laws of reflection and refraction modified when we allow for the possibility of negative refractive indices? Consider a plane monochromatic wave

$$E_i = E_{0i} e^{-i(\omega t - k_i \cdot r)} \tag{7.84}$$

incident on a planar interface $z = 0$ of two media characterized by the permittivities ϵ_i, ϵ_t and the permeabilities μ_i, μ_t (figure 7.2). For the reflected and transmitted fields, we write similarly

$$E_r = E_{0r} e^{-i(\omega t - k_r \cdot r)} \tag{7.85}$$

and

$$E_t = E_{0t} e^{-i(\omega t - k_t \cdot r)}. \tag{7.86}$$

The Maxwell equation $\nabla \times E = -\partial B / \partial t$ implies that the corresponding B fields are

$$B_i = \frac{1}{\omega} k_i \times E_i \qquad B_r = \frac{1}{\omega} k_r \times E_r \qquad B_t = \frac{1}{\omega} k_t \times E_t \tag{7.87}$$

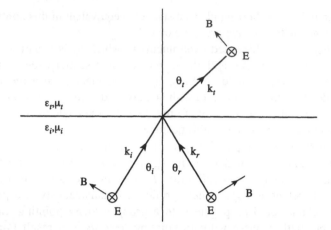

Figure 7.2. Wavevectors for incident (k_i), transmitted (k_t) and reflected (k_r) plane waves at a planar interface at $z = 0$ between two media characterized by ϵ_i, μ_i and ϵ_t, μ_t. The angles of incidence, transmission, and reflection are θ_i, θ_t, and θ_r, respectively.

and similar expressions for the fields D and H follow from the constitutive relations. Here

$$k_j^2 = \omega^2 \epsilon_j \mu_j \qquad (j = i, r, t) \qquad (7.88)$$

and, as discussed in section 7.2, we should take the *negative* square root if ϵ_j and μ_j are *both* negative.

In order to satisfy the boundary conditions for the fields at all points in the plane $z = 0$, we must have[11]

$$(k_i \cdot r)_{z=0} = (k_r \cdot r)_{z=0} = (k_t \cdot r)_{z=0}. \qquad (7.89)$$

Thus (figure 7.2), $k_i \sin \theta_i = k_r \sin \theta_r$ or, since $k_i = k_r$, $\theta_i = \theta_r$: the angles of incidence and reflection are the same regardless of the signs of the refractive indices. Similarly $k_i \sin \theta_i = k_t \sin \theta_t$ or

$$n_i \sin \theta_i = n_t \sin \theta_t \qquad (7.90)$$

so that Snell's law also holds regardless of the signs of the indices.

The derivation of the Fresnel formulas when the refractive index might be negative also proceeds along standard lines using equations (7.84)–(7.87) and the boundary conditions that the tangential components of E and H, and the normal components of D and B, are continuous. Thus, for E_{0i} perpendicular to the plane of incidence, we obtain

$$\frac{E_{0t}}{E_{0i}} = \frac{2(n_i/\mu_i) \cos \theta_i}{(n_i/\mu_i) \cos \theta_i + (n_t/\mu_t)\sqrt{1 - (n_i^2/n_t^2) \sin^2 \theta_i}} \qquad (7.91)$$

[11] The assumption that the frequency is the same on both sides of the interface is justified similarly by the fact that the boundary conditions at $z = 0$ must also be satisfied at all *times* [14].

when the angle of incidence θ_i is less than the critical angle

$$\theta_c = \sin^{-1} |n_t/n_i|. \tag{7.92}$$

For either positive- or negative-index media, the ratios n_i/μ_i and n_t/μ_t are positive and equation (7.91) holds in either case. Note, though, that a form like [14]

$$\frac{E_{0t}}{E_{0i}} = \frac{2n_i \cos \theta_i}{n_i \cos \theta_i + (\mu_i/\mu_t)\sqrt{n_t^2 - n_i^2 \sin^2 \theta_i}} \tag{7.93}$$

for the transmission coefficient follows from (7.91) only if n_t is positive. But if we replace n_i, n_t, ϵ_i, ϵ_t, μ_i and μ_t in (7.93) by their absolute values, the formula is valid regardless of whether we have positive- or negative-index media. For the reflection coefficient, similarly, we obtain

$$\frac{E_{0r}}{E_{0i}} = \frac{(n_i/\mu_i) \cos \theta_i - (n_t/\mu_t)\sqrt{1 - (n_i^2/n_t^2) \sin^2 \theta_i}}{(n_i/\mu_i) \cos \theta_i + (n_t/\mu_t)\sqrt{1 - (n_i^2/n_t^2) \sin^2 \theta_i}} \tag{7.94}$$

when E_{0i} is perpendicular to the plane of incidence. The form [14]

$$\frac{E_{0r}}{E_{0i}} = \frac{n_i \cos \theta_i - (\mu_i/\mu_t)\sqrt{n_t^2 - n_i^2 \sin^2 \theta_i}}{n_i \cos \theta_i + (\mu_i/\mu_t)\sqrt{n_t^2 - (n_i^2) \sin^2 \theta_i}} \tag{7.95}$$

which assumes $n_t > 0$, is seen to be applicable provided we again replace n_i, n_t, ϵ_i, ϵ_t, μ_i, and μ_t by their absolute values. More generally, *the standard Fresnel formulas for positive-index media are applicable to negative- as well as positive-index media for $\theta_i < \theta_c$, for any polarization of the field, provided we replace all n, ϵ, and μ in these formulas by their absolute values.*

The situation is a bit more complicated when the angle of incidence exceeds the critical angle for total internal reflection. In this case, we must replace

$$\cos \theta_t = \sqrt{1 - (n_i^2/n_t^2) \sin^2 \theta_i} \tag{7.96}$$

by

$$\pm i\sqrt{(n_i^2/n_t^2) \sin^2 \theta_i - 1}. \tag{7.97}$$

The choice of sign is dictated by the requirement that the transmitted field, which varies with z as $\exp[i(\omega z/c)n_t \cos \theta_t]$, should not diverge as $z \to \infty$. Thus, for $\theta_i > \theta_c$ at an interface with $n_t > 0$, we choose the $+$ sign in (7.97) and then, from (7.94), we obtain the amplitude reflection coefficient [236]

$$r = \frac{\cos \theta_i - i\alpha}{\cos \theta_i + i\alpha} \tag{7.98}$$

$$\alpha = \left| \frac{\mu_i}{\mu_t} \right| \sqrt{\sin^2 \theta_i - n_t^2/n_i^2}. \tag{7.99}$$

For $\theta_i > \theta_c$ at an interface in which $n_t < 0$, however, we must choose the $-$ sign in (7.97) so that the field does not diverge. Then, instead of (7.98), we obtain

$$r = \frac{\cos\theta_i + i\alpha}{\cos\theta_i - i\alpha} = |r|e^{i\phi_r}. \tag{7.100}$$

It is interesting that, since the Goos–Hänchen shift depends on ϕ_r, it will be in opposite directions when $n_t > 0$ and $n_t < 0$ [236, 237].

Consider the case in which light is incident from a positive-index medium onto a negative-index medium such that $\epsilon_t = -\epsilon_i < 0$ and $\mu_t = -\mu_i < 0$. Then, $n_t = -n_i$ and equation (7.94) gives $E_{0r}/E_{0i} = 0$. The same is true if E_{0i} is parallel to the plane of incidence. In other words, when $\epsilon_t = -\epsilon_i$ and $\mu_t = -\mu_i$, there is no reflected light, no matter what the angle of incidence or the field polarization.

It is very interesting to consider a *planar* negative-index material of width d in vacuum [29]. Suppose that the NIM has $\epsilon = -\epsilon_0$ and $\mu = -\mu_0$, so that there is no reflected light, and that a point source is at a distance $\ell < d$ from the NIM (figure 7.3) . At the front surface, the rays from the point source are refracted to the same side of the normal to the interface as the incident rays and, since the refractive index of the NIM is -1, the angle of refraction is just the negative of the angle of incidence for each ray. Since $\ell < d$, the rays are, therefore, focused to a point at a distance ℓ inside the NIM. The rays from this focal point are then refracted at the back surface of the NIM, and again the angle of refraction is just the negative of the angle of incidence, so that there is a second focal point, in vacuum, at a distance $d - \ell$ from the back surface. The net effect is that, according to geometrical optics, a point source in front of the planar NIM is focused to a point behind it. In other words, an NIM can be used to make a *planar* lens as well as convex- and concave-shaped lenses[12].

Such a focusing element is not a lens in the conventional sense, as it does not focus rays from infinity to a point. Note also that the size of the image of an extended object will equal the size of the object, i.e. there is no magnification. And unlike a conventional lens, which acts to equalize the optical pathlengths (refractive index integrated over distance) of incoming parallel rays, the lens of figure 7.3 focuses rays by *undoing* (because $n < 0$) the phases they accrue in traversing the same distance through vacuum. Finally, a conventional lens, no matter how good the glass or the surfaces, will introduce aberrations that degrade the image quality even for monochromatic light, whereas the lens of figure 7.3 is aberration-free.

---

[12] Convex and concave lenses made from an NIM will have the opposite effect of conventional lenses: a convex lens causes rays to diverge and a concave lens causes them to converge. This is another consequence of rays being refracted to the same side of the normal to the lens surface as the incident rays.

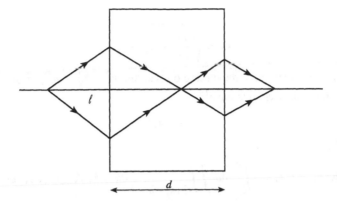

Figure 7.3. A planar slab of negative-index material of width d will produce two focal points, one inside the material and the other outside, when a point source is placed a distance $\ell < d$ from it. With $\epsilon = -\epsilon_0$ and $\mu = -\mu_0$, there are no reflected waves when the lens is in vacuum and a real image appears at a distance $2d$ from the object.

7.8 Evanescent waves

It turns out that something more subtle than 'light bending the wrong way' happens in an NIM *slab*: the behaviour of evanescent waves is very different. This leads to one of the most intriguing prospects for NIMs, namely the possibility of lenses with subwavelength resolution (section 7.9). In this section, we will briefly review some aspects of evanescent waves.

Suppose we expand the electric field in the half-space $z > 0$ in terms of spatial and temporal frequency components as

$$E(x, y, z, t) = \int_0^\infty d\omega \int_\infty^\infty dk_x \int_\infty^\infty dk_y \, A(k_x, k_y, \omega, z) e^{-i(\omega t - k_x x - k_y y)}$$

(7.101)

where, as usual, the real part of the right-hand side is implicit. If any sources of radiation are present only in the half-space $z < 0$, we have

$$\nabla^2 E - \frac{1}{c^2} \frac{\partial^2 E}{\partial t^2} = 0$$

(7.102)

for $z > 0$ and, therefore,

$$\frac{\partial^2 A}{\partial z^2} + \left(\frac{\omega^2}{c^2} - k_x^2 - k_y^2 \right) A(k_x, k_y, \omega, z) = 0$$

(7.103)

which has solutions of the form

$$A(k_x, k_y, \omega, z) \propto e^{\pm i k_z z}$$

(7.104)

where

$$k_z = \sqrt{\frac{\omega^2}{c^2} - k_x^2 - k_y^2}. \tag{7.105}$$

k_z can be real or imaginary, depending on whether $k_x^2 + k_y^2 < \omega^2/c^2$ or $k_x^2 + k_y^2 > \omega^2/c^2$. In the latter case, we have[13]

$$k_z = i|k_z|. \tag{7.106}$$

Then [238]

$$\begin{aligned}
E(x, y, z, t) &= \int_0^\infty d\omega \iint_{k_x^2+k_y^2<\omega^2/c^2} dk_x\, dk_y\, A_{\text{ord}}(k_x, k_y, \omega) \\
&\quad \times e^{i(k_x x + k_y y + k_z z - \omega t)} \\
&\quad + \int_0^\infty d\omega \iint_{k_x^2+k_y^2>\omega^2/c^2} dk_x\, dk_y\, A_{\text{ev}}(k_x, k_y, \omega) \\
&\quad \times e^{i(k_x x + k_y y - \omega t)} e^{-|k_z|z}.
\end{aligned} \tag{7.107}$$

The first term on the right-hand side is a superposition of ordinary (homogeneous[14]) plane waves propagating in the positive z direction[15]. The second term defines a superposition of *evanescent* wave components that *decay exponentially* with z.

Evanescent waves have not received much attention in the standard texts on optics or electromagnetism[16] but it is clear from the derivation of (7.107) that they are required for completeness. They are present whenever the field in some plane ($z = 0$) has spatial frequency components k_x, k_y such that $k_x^2 + k_y^2 > \omega^2/c^2$, i.e. whenever the field in this plane varies on a scale comparable to the wavelength or smaller. The best known example where evanescent waves occur is total internal reflection when a beam of light is incident from glass, for example, onto a glass–air interface at an angle greater than the critical angle for total internal reflection. On the glass side $k_x^2 + k_y^2 < n^2\omega^2/c^2$, corresponding to ordinary propagation, and the variations in the field occur on a scale larger than a wavelength. On the air side, however, $k_x^2 + k_y^2 > \omega^2/c^2$ and the spatial variations of the field are on a scale smaller than a wavelength. An evanescent field, therefore, appears on the air side and its intensity decays with distance z as $\exp(-az)$, where

$$a = \frac{4\pi n_a}{\lambda} \sqrt{\frac{n_g^2}{n_a^2} \sin^2 \theta_i - 1} \tag{7.108}$$

[13] We do not consider solutions with $k_z = -i|k_z|$ because they grow exponentially with z and are, therefore, unphysical. Similarly $k_z = 0$ is excluded because it would imply from equation (7.103) a solution that grows linearly with z.

[14] 'Homogeneous' refers to the fact that the planes of constant amplitude and phase are identical, whereas they are different for the evanescent or 'inhomogeneous' plane waves.

[15] We exclude plane waves propagating in the negative z direction in the half-space $z > 0$ because, by assumption, any sources of the field are in the $z < 0$ half-space.

[16] A serious treatment of evanescent waves may be found in Clemmow [239].

n_a and n_g are the refractive indices of air and glass, respectively, θ_i is the angle of incidence, and λ is the vacuum wavelength. If a second piece of glass is brought near the first piece, leaving an air gap between them, the internal reflection is 'frustrated' and light can propagate into the second glass with a transmission coefficient ~ 1 if the width of the air gap is small compared with the wavelength. A nice demonstration of frustrated total internal reflection is described by Zhu *et al* [46], who also discuss some of the long history of the subject.

The expansion (7.107) shows that the evanescent waves do not undergo any variation of phase in the z direction. This is related to the fact that evanescent waves—and their quantum-mechanical analog of tunnelling into a classically forbidden region [46]—can exhibit 'superluminal' behaviour, as discussed in chapter 2. The only phase variation of the evanescent waves is along their propagation direction, parallel to the interface.

An important characteristic of evanescent waves is that they do not transport energy in the direction in which they decay. Consider, for example, the monochromatic evanescent wave with electric field

$$E = x E_0 \cos(ky - \omega t)e^{-Kz}. \tag{7.109}$$

The Poynting vector $S = E \times H$ is

$$S = \frac{E_0^2}{\omega\mu}[ky\cos^2(ky - \omega t) - Kz\cos(ky - \omega t)\sin(ky - \omega t)]e^{-2Kz} \tag{7.110}$$

and the component of S in the z direction has a cycle average of zero.

Because of their exponential decay with distance, evanescent waves are said to belong to the *near field* of the object, as opposed to the 'far field' associated with the homogeneous plane-wave components. Near-field optics (e.g. a scanning tunnelling microscope) involves distances close enough to an object that the evanescent waves are captured, giving information about the object that is not available in the far field [240].

Quantization of evanescent waves has been carried out by Carniglia and Mandel [238]. In their approach, each triplet of incident, reflected, and transmitted waves is regarded as a single mode for the purpose of quantization. Experiments [241] have demonstrated that evanescent photons are emitted and absorbed as expected from this theory, so that a single photon can be associated with a homogeneous wave on one side of the interface and an evanescent wave on the other.

Expansions in evanescent waves of the field from a uniformly moving charge have been usefully employed to describe Cerenkov, Smith–Purcell and related effects [242].

7.8.1 Limit to resolution with a conventional lens

The field radiated or reflected by a two-dimensional object can be expanded in a Fourier series with spatial frequency components k_x, k_y such that

$$k^2 = k_x^2 + k_y^2 + k_z^2 \qquad (7.111)$$

with $k = \omega/c$ in vacuum. Large values of k_x and k_y correspond to small spatial features, so that for good image resolution it is necessary to access high spatial frequencies. Equation (7.111) suggests that the maximum possible resolution Δ is achieved when $k_x^2 + k_y^2 = k_{max}^2 = k^2$:

$$\Delta \sim \frac{2\pi}{k_{max}} = \frac{2\pi c}{\omega} = \lambda. \qquad (7.112)$$

Implicit in this argument is the assumption that $k_{max} = k$, i.e. that $k_z^2 \geq 0$. In other words, the limit (7.112) assumes that the accessible field from the object has no evanescent components. Since the evanescent components of the field decay with distance z from the object as

$$e^{-|k_z|z} = \exp\left(-\sqrt{k_x^2 + k_y^2 - \omega^2/c^2}\,z\right) \qquad (7.113)$$

the largest spatial frequency components decay most rapidly with z. In particular, spatial frequencies corresponding to spatial features on a scale of λ on the object do not appear in the field at distances greater than about λ from the object. For this reason, the maximum resolution possible with a conventional lens is given by (7.112), regardless of how good the glass is, or how large the aperture is. Features on the scale of the wavelength or smaller cannot be resolved by an ordinary lens because an ordinary lens does not capture the evanescent components of the object field.

7.9 The 'perfect' lens

The work of Veselago and others on negative refraction did not attract much interest until the publication of a theoretical paper by Pendry [243] entitled 'Negative Refraction Makes a Perfect Lens', which was stimulated by the first experimental observations of a negative refractive index by Smith *et al* [244,245]. General and authoritative overviews of work in this field in the few years following these publications are given by Pendry [246] and Pendry and Smith [247].

Pendry [243] began with the observation that the resolution of a conventional lens is limited by the fact that evanescent waves do not contribute to the image [equation (7.112)]. He then showed that, for the planar negative-index lens of figure 7.3, the evanescent waves *grow* with distance into the lens and are *transmitted*, so that all spatial frequency components can, in principle, contribute to the image.

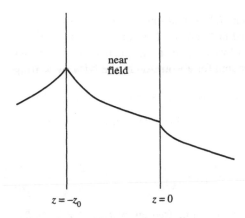

near
field

$z = -z_0$ $z = 0$

Figure 7.4. An evanescent wave incident on a negative-index half-space.

7.9.1 Evanescent wave incident on an NIM half-space

In section 7.7, we considered the transmission and reflection of a *homogeneous* plane wave incident from a positive-index medium onto either a positive- or a negative-index medium. Let us consider now the case where an evanescent wave from a source at $z = -z_0 < 0$ is incident from vacuum onto an NIM (figure 7.4). For polarization perpendicular to the plane of incidence (yz), we write the incident field as[17]

$$E = E_0 x e^{i(k_y y + k_z z)} e^{-i\omega t} \tag{7.114}$$

for $z > -z_0$, where

$$k_z = i\sqrt{k_y^2 - \omega^2/c^2} \qquad (k_y^2 > \omega^2/c^2). \tag{7.115}$$

Similarly, the transmitted field is written as

$$E' = E_0' x e^{i(k_y' y + k_z' z)} e^{-i\omega t} \tag{7.116}$$

$$k_z' = i\sqrt{k_y^2 - \epsilon\mu\omega^2} \qquad (k_y^2 > \epsilon\mu\omega^2). \tag{7.117}$$

The corresponding expression for the reflected field is written using double primes, as in Jackson [14]. Using the expressions for D, D', D'', etc that follow from Maxwell's equations, and enforcing the boundary conditions at $z = 0$, we obtain straightforwardly the amplitude transmission and reflection coefficients r and t, respectively, at the interface ($z = 0$) between vacuum and the NIM:

$$t = \frac{2\mu k_z}{\mu k_z + \mu_0 k_z'} \qquad r = \frac{\mu k_z - \mu_0 k_z'}{\mu k_z + \mu_0 k_z'}. \tag{7.118}$$

[17] Similar results for the NIM half-space and slab are obtained when the polarization is parallel to the plane of incidence. See section 8.6.

It appears from (7.118) that t and r diverge as $\epsilon \to -\epsilon_0$ and $\mu \to -\mu_0$, i.e. when the half-space in figure 7.4 has refractive index $n' = -1$. However, as in many situations in physics, a divergence is avoided when dissipation is taken into account. Let us account for absorption in the NIM by writing

$$\epsilon = -\epsilon_0(1 - i\delta) \qquad \mu = -\mu_0(1 - i\delta) \qquad (\delta > 0). \qquad (7.119)$$

Then the denominator in (7.118) does not go to zero. For small δ [$\epsilon\mu \cong \epsilon_0\mu_0(1 - 2i\delta)$],

$$t \cong \frac{2i}{\delta}\left(1 - \frac{\omega^2}{k_y^2 c^2}\right) \qquad r \cong \frac{2i}{\delta}\left(1 - \frac{\omega^2}{k_y^2 c^2}\right) \qquad (7.120)$$

and, in the limit of large spatial frequency k_y, $t \cong r \cong 2i/\delta$.

7.9.2 Evanescent wave incident on an NIM slab

In the case of an evanescent wave incident on a slab, we also require the transmission and reflection coefficients t' and r' for the $z = d$ interface in figure 7.5. t' and r' can be written immediately by simply interchanging $\mu_0 k_z'$ and μk_z in equations (7.118):

$$t' = \frac{2\mu_0 k_z'}{\mu k_z + \mu_0 k_z'} \qquad r' = \frac{\mu_0 k_z' - \mu k_z}{\mu k_z + \mu_0 k_z'}. \qquad (7.121)$$

For small δ,

$$t' \cong \frac{-2i}{\delta}\left(1 - \frac{\omega^2}{k_y^2 c^2}\right) \qquad r' \cong \frac{-2i}{\delta}\left(1 - \frac{\omega^2}{k_y^2 c^2}\right). \qquad (7.122)$$

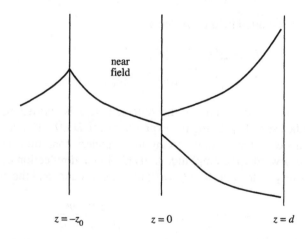

$$z = -z_0 \qquad\qquad z = 0 \qquad\qquad z = d$$

Figure 7.5. An evanescent wave incident on a negative-index slab of thickness d.

The transmission coefficient t_S for the slab is obtained in the usual fashion by adding up the contributions associated with all the possible reflections within the slab:

$$
\begin{aligned}
t_S &= tt'\mathrm{e}^{\mathrm{i}k_z'd} + tr'r't'\mathrm{e}^{3\mathrm{i}k_z'd} + tr'^4t'\mathrm{e}^{5\mathrm{i}k_z'd} + \cdots \\
&= \frac{tt'\mathrm{e}^{\mathrm{i}k_z'd}}{1 - r'^2\mathrm{e}^{2\mathrm{i}k_z'd}} \\
&= \frac{(4/\delta^2)(1 - \omega^2/k_y^2 c^2)^2 \exp\left(-\sqrt{k_y^2 - \omega^2/c^2}d\right)}{1 + (4/\delta^2)(1 - \omega^2/k_y^2 c^2)^2 \exp\left(-2\sqrt{k_y^2 - \omega^2/c^2}d\right)}.
\end{aligned}
\tag{7.123}
$$

This is a remarkable result: an evanescent wave incident on a negative-index material with $\epsilon = -\epsilon_0$ and $\mu = -\mu_0$ results in an *amplified* field:

$$
t_S = \exp(\sqrt{k_y^2 - \omega^2/c^2}d)
\tag{7.124}
$$

in the limit $\delta \to 0$ of no absorption. Let us note that, after some algebra, we can write t_S in a form that shows explicitly that it contains exponentially growing as well as exponentially decaying parts [248]:

$$
\begin{aligned}
t_S = \Bigg(&\left[\frac{1}{2} + \frac{1}{4}\left(\frac{\mu_0 K_z'}{\mu K_z} + \frac{\mu K_z}{\mu_0 K_z'}\right)\right]\mathrm{e}^{K_z'd} \\
&+ \left[\frac{1}{2} - \frac{1}{4}\left(\frac{\mu_0 K_z'}{\mu K_z} + \frac{\mu K_z}{\mu_0 K_z'}\right)\right]\mathrm{e}^{-K_z'd} \Bigg)^{-1}
\end{aligned}
\tag{7.125}
$$

where

$$
K_z = \sqrt{k_y^2 - \omega^2/c^2} \qquad K_z' = \sqrt{k_y^2 - \epsilon\mu\omega^2}.
\tag{7.126}
$$

Normally the exponentially decaying part of t_S will dominate. But, for $\epsilon \to -\epsilon_0$ and $\mu \to -\mu_0$, *the exponentially decaying part vanishes* and only the exponentially growing part remains: $t_S \to \exp(K_z d)$. This shows again the very special significance of $\epsilon = -\epsilon_0$ and $\mu = -\mu_0$ for superlensing.

In order for this amplification to occur, the absorption must be small enough that the second term in the denominator in the third line of equation (7.123) is large compared with unity. For sufficiently large k_y^2, this requires

$$
\frac{\delta}{2} \ll \exp\left(-\sqrt{k_y^2 - \omega^2/c^2}d\right).
\tag{7.127}
$$

For a given value of δ, therefore, there should be amplification up to a spatial frequency of magnitude

$$
|k_y| \sim \frac{1}{d}|\ln\delta|.
\tag{7.128}
$$

These results imply that the planar NIM lens of figure 7.3 is 'perfect' in the limit of zero absorption: there are no reflections off it (since $\epsilon = -\epsilon_0$ and

$\mu = -\mu_0$), it is aberration-free, and it transmits evanescent waves as well as homogeneous plane waves. The spatial resolution associated with a wavelength at which the absorption is characterized by δ is

$$\Delta \sim \frac{2\pi d}{|\ln \delta|}. \tag{7.129}$$

Thus, the negative-index lens can beat the resolution limit $\Delta = \lambda$ [equation (7.112)] if

$$\delta < e^{-2\pi d/\lambda}. \tag{7.130}$$

Absorption can, therefore, be a big hurdle in the way of superlensing [249] but it is not a *fundamental* limitation [250].

Aside from practical matters like a finite lens aperture, the superlensing possible with an NIM is limited primarily by absorption: NIMs designed for subwavelength resolution must have very small losses. The loss will, of course, be wavelength-dependent, as will be the smallest achievable resolution. And since NIMs are necessarily dispersive and typically narrowband (chapter 8), subwavelength resolution will be possible only within a narrow spectral range.

7.9.3   Surface modes

The amplification of the incoming evanescent wave that occurs in an NIM slab with $\epsilon = -\epsilon_0$ and $\mu = -\mu_0$ [equation (7.124)] does not arise, of course, from any gain (e.g. population inversion) in the material. As discussed in section 7.8, the Poynting vector of an evanescent wave has a vanishing cycle average in the direction of exponential decay (or growth): no energy is transported, on average, in that direction by the evanescent wave. As in the treatment by Carniglia and Mandel [238], the triplet of incoming, reflected, and transmitted waves can be regarded as a single field mode which can be normalized and, if desired, quantized. The exponential growth does not represent amplification of field energy at the expense of the (passive!) medium in which it propagates. In order to further appreciate how special is the situation when $\mu = -\mu_0$ and $\epsilon = -\epsilon_0$, we will now review briefly the concept of a *surface mode* localized at an interface between two dielectrics [251].

We consider first a dielectric characterized by ϵ and μ and occupying the half-space $z > 0$, the left half-space ($z < 0$) being vacuum. A surface mode is, by definition, localized at the interface $z = 0$, i.e. the fields decay exponentially with distance on *both* sides of the interface. We, therefore, write, for polarization perpendicular to the (yz) plane of incidence,

$$E = A\mathbf{x}e^{i(k_y y - \omega t)}e^{K_z z} \qquad (z < 0) \tag{7.131}$$

$$E = B\mathbf{x}e^{i(k_y y - \omega t)}e^{-K'_z z} \qquad (z > 0) \tag{7.132}$$

and, likewise, using $\nabla \times \mathbf{E} = -\mu \partial \mathbf{H}/\partial t$,

$$\mathbf{H} = -\frac{i}{\mu_0 \omega}(K_z \mathbf{y} - ik_y \mathbf{z})Ae^{i(k_y y - \omega t)}e^{K_z z} \qquad (z < 0) \tag{7.133}$$

$$\boldsymbol{H} = -\frac{i}{\mu\omega}(-K'_z\boldsymbol{y} - ik_y\boldsymbol{z})Be^{i(k_y y - \omega t)}e^{-K'_z z} \qquad (z > 0) \qquad (7.134)$$

where K_z and K'_z are defined by (7.126). The boundary condition that the tangential component of $\boldsymbol{E}$ is continuous implies that $A = B$, while the condition that the tangential component of $\boldsymbol{H}$ is continuous then yields the dispersion relation [252]

$$\frac{K'_z}{\mu(\omega)} + \frac{K_z}{\mu_0} = 0 \qquad (7.135)$$

which must be satisfied if a surface mode with polarization perpendicular to the plane of incidence is to exist. We obtain similarly the condition [252]

$$\frac{K'_z}{\epsilon(\omega)} + \frac{K_z}{\epsilon_0} = 0 \qquad (7.136)$$

for the existence of a surface mode with polarization parallel to the plane of incidence.

K_z and K'_z in the dispersion relations (7.135) and (7.136) for surface modes are both real and positive. Therefore, since $\epsilon\mu > 0$, these dispersion relations can only be satisfied at frequencies ω for which both $\mu(\omega)$ and $\epsilon(\omega)$ are negative. In other words, *surface modes can only be supported at a vacuum–dielectric interface if the dielectric has negative refractive index.* In this case, surface modes—or *surface polaritons*—are possible in the absence of any externally applied field, provided, of course, that the dispersion relations are satisfied.

Next consider the possibility that surface modes can exist at *both* interfaces ($z = 0$ and $z = d$) between vacuum and a dielectric slab of thickness d. The calculation proceeds as in the case of the dielectric half-space, and the imposition of the boundary conditions leads easily to the dispersion relations [253]

$$\frac{K'_z}{K_z}\tanh\left(\frac{1}{2}K'_z d\right) = -\frac{\mu(\omega)}{\mu_0} \qquad (7.137)$$

$$\frac{K'_z}{K_z}\coth\left(\frac{1}{2}K'_z d\right) = -\frac{\mu(\omega)}{\mu_0} \qquad (7.138)$$

for polarization perpendicular to the yz plane and

$$\frac{K'_z}{K_z}\tanh\left(\frac{1}{2}K'_z d\right) = -\frac{\epsilon(\omega)}{\epsilon_0} \qquad (7.139)$$

$$\frac{K'_z}{K_z}\coth\left(\frac{1}{2}K'_z d\right) = -\frac{\epsilon(\omega)}{\epsilon_0} \qquad (7.140)$$

for polarization parallel to the yz plane.

These dispersion relations determine what surface modes can be resonantly excited: for imaging by a planar NIM lens, they are deleterious in that they will skew the spatial frequency distribution in favour of the resonant modes. However, when $\epsilon(\omega) = -\epsilon_0$ and $\mu(\omega) = -\mu_0$ these relations can only be satisfied for $k_y = \infty$. In fact, *it is only when* $\epsilon(\omega) = -\epsilon_0$ *and* $\mu(\omega) = -\mu_0$ *that no surface modes are resonantly excited for any finite value of the spatial frequency* k_y [248]. Here we have another example of the special significance and advantage of $\epsilon/\epsilon_0 = \mu/\mu_0 = -1$.

7.10   Elaborations

There are now hundreds of papers on the subject of negative refractive index and the possibility of superlensing. As discussed in the following chapter, there are compelling demonstrations of 'metamaterials' with negative refractive index. The special case $\epsilon = -\epsilon_0$ and $\mu = -\mu_0$ of interest for superlensing will likely be more difficult to realize: the main challenge in this case is to design a metamaterial with very low absorption [250] since, as the preceding discussion indicates, even a small degree of absorption can substantially limit the resolution that is theoretically possible with a planar NIM lens.

Since the object has been assumed to be at a distance $\ell < d$ from the lens (figure 7.3), $\exp(-2\pi d/\lambda) < \exp(-2\pi \ell/\lambda)$ and the condition (7.130) requires that

$$\delta < e^{-2\pi \ell/\lambda} \qquad (7.141)$$

or that the absorption coefficient[18]

$$a < \frac{4\pi}{\lambda}e^{-2\pi \ell/\lambda}. \qquad (7.142)$$

Thus, if the absorption is not sufficiently small, it may be necessary to have the object very close to the lens surface [254]. Then it is not clear what major advantages an NIM lens might offer compared with other near-field microscopies.

Even in the absence of absorption, the real parts of ϵ/ϵ_0 and μ/μ_0 must not deviate far from -1 if perfect lensing is to be achieved. Merlin [255] has obtained an analytical solution for the resolution scale $\Delta = 2\pi/|k_y|_{max}$ [cf equations (7.112) and (7.128)] under the assumption that the refractive index $n = -\sqrt{1-\sigma}, |\sigma| \ll 1$:

$$\Delta = \frac{-2\pi d}{\ln|\sigma/2|} \qquad (7.143)$$

which is consistent with a heuristic argument by Smith *et al* [248] as well as an analysis by Gómez-Santos [256]. This shows the crucial importance for perfect lensing of having n very nearly equal to -1.

Ziolkowksi and Heyman [222] have carried out a numerical study of wave propagation in an NIM. They conclude that the perfect lens effect is only possible when $\epsilon/\epsilon_0 = \mu/\mu_0 = -1$ *and* when there is no absorption or dispersion; otherwise a monochromatic wave 'is channeled into beams rather than being

---

[18] Note that $n_R^2 - n_I^2 + 2in_R n_I = \epsilon\mu/\epsilon_0\mu_0 = 1 - 2i\delta$, so that $a = 2\omega n_I/c = 4\pi\delta/\lambda$ for $n_R = -1$.

focused and, hence, the Pendry perfect lens effect is not realizable with any realistic metamaterial'.

It is evident that the question of the practicality of perfect lensing will only be decided definitively by experiment.

7.11 No fundamental limit to resolution

Claims of subwavelength resolution occasionally elicit scepticism on the grounds that diffraction sets a fundamental resolution limit $\Delta = \lambda$. This, of course, is false.

In 1928, Synge [257] published a paper in which he suggested a way to obtain subwavelength resolution. Synge's proposal was very simple, the key idea being to use a microscope aperture smaller than a wavelength. Such an aperture in an opaque screen is illuminated and the light from it is incident on a specimen placed very close to the screen, so that the light does not diffract as it propagates to the specimen and can be used to form an image that is not limited by diffraction. Technological difficulties made Synge's proposal impracticable at the time but subwavelength resolution using a subwavelength aperture was demonstrated with microwaves in 1972 [258] and with optical wavelengths in the mid-1980s [259]. These near-field techniques demonstrate that the resolution limit $\Delta = \lambda$ of a conventional lens is certainly *not* a fundamental limit. The superlensing based on a negative-index lens also demonstrates this point, albeit mainly theoretically thus far[19].

That the diffraction limit is not a fundamentally unbreakable one can also be seen from a purely formal standpoint. The Fourier transform of a finite two-dimensional source is an analytic function of the spatial frequencies k_x, k_y and, if this function is given in any finite region of the $k_x k_y$ plane, it can be determined over the entire plane. In particular, there is no upper limit to the spatial frequencies at which this function can, in principle, be determined and, therefore, no limit, in principle, to resolution.

7.12 Summary

There is nothing to prevent the refractive index from being negative, although negative-index materials (NIMs) are only known to occur in specially designed metamaterials. Propagation of light in an NIM provides an example of what we have called left-handed light, in which the Poynting vector $E \times H$ points in the direction opposite to the wavevector k and the phase and group velocities are, therefore, in opposite directions.

Causality requires that the refractive index be negative when *both* the electric permittivity ϵ and the magnetic permeability μ are negative. Negative index can

[19] Experimental evidence for subwavelength resolution using negative refraction is described in the following chapter.

be paradoxical (e.g. the energy density can be negative) unless dispersion is taken into account. Even in the absence of paradoxes, a negative refractive index leads to such unfamiliar consequences as reversed Doppler and Cerenkov effects.

One of the most intriguing potential applications of NIMs is in imaging: a *planar* NIM with $\epsilon = -\epsilon_0$ and $\mu = -\mu_0$ can function as a 'perfect' lens in which there are no reflections off the lens, there are no aberrations, and *all spatial frequencies* including those associated with evanescent waves contribute to the image, so that subwavelength resolution is, in principle, possible. The condition $\epsilon = -\epsilon_0$ and $\mu = -\mu_0$ also prevents surface modes from being resonantly excited, so that their contribution to the image is not overly weighted compared with the spatial frequency distribution of the field from the source being imaged. The principal challenge in actually realizing superlensing with an NIM is to make ϵ/ϵ_0 and μ/μ_0 very nearly equal to -1 *and* to have the absorption very small at a frequency for which this occurs. Otherwise, the absorption limits the range of spatial frequencies that can contribute to the image. As discussed in the following chapter, some consequences of negative refraction have already been demonstrated experimentally.

Chapter 8

Metamaterials for left-handed light

As noted in the preceding chapter, there are no naturally occurring materials known to have simultaneously negative electric permittivity ϵ and magnetic permeability μ. Negative values of ϵ and μ tend to occur over narrow frequency ranges around very different frequencies. However, *metamaterials* having simultaneously negative values of ϵ and μ at particular wavelengths have been fabricated and shown to exhibit the predicted negative refractive index. In this chapter, we describe some of the theoretical and experimental work on materials exhibiting a negative refractive index.

We begin by considering the type of metamaterial that was first used to produce a negative index of refraction. This material consists of a periodic thin-wire array that gives rise to an effective permittivity that is negative and a periodic array of 'split ring resonators' giving a negative effective permeability. Combining the two periodic arrangements in a composite structure has been shown to produce a negative index of refraction for microwaves. Metamaterials consisting of periodic microwave transmission line structures are considered next and experiments demonstrating the negative-index, focusing and evanescent-wave properties of such structures are described. Finally we describe other materials, such as photonic crystals, that have been used to demonstrate negative refraction and some of its consequences.

8.1 Negative permittivity

Recall that the simplest model for a plasma yields the permittivity

$$\epsilon(\omega)/\epsilon_0 = 1 - \frac{\omega_p^2}{\omega^2} \qquad (8.1)$$

where the plasma frequency ω_p is given by

$$\omega_p^2 = \frac{Ne^2}{m\epsilon_0} \qquad (8.2)$$

with N, e, and m the electron density, charge, and mass, respectively. To get a negative ϵ, all we have to do is choose the field frequency ω to be below the plasma frequency. Or, for a given field frequency, we can choose a metal with a plasma frequency larger than ω. The plasma frequency for the electron gas in a metal is typically in the ultraviolet: for copper, for example, $\lambda_p = 2\pi c/\omega_p = 115$ nm. Visible radiation does not penetrate very far into such a medium, the only allowed modes being evanescent.

Things are not quite so simple. Dissipation changes the dielectric function (8.1) to

$$\epsilon(\omega)/\epsilon_0 = 1 - \frac{\omega_p^2}{\omega(\omega + i\gamma)} \tag{8.3}$$

where the damping rate γ is typically significant at infrared and lower frequencies; and, for copper, $\gamma \cong 4 \times 10^{13}$ rad s^{-1}. At frequencies $\omega \ll \gamma$, the imaginary part of $\epsilon(\omega)$ is dominant. Instead of surface modes associated with collective electron oscillations (surface plasmons), 'Life becomes rather dull again' [260]. At low frequencies, the attenuation coefficient is

$$a_{abs}(\omega) = \frac{2\omega}{c} n_I(\omega) \cong \frac{\omega_p}{c}\sqrt{\frac{2\omega}{\gamma}} \tag{8.4}$$

and the field does not penetrate significantly into the metal. At high frequencies ($\omega \gg \omega_p$), there is transmission but, in this case, $\epsilon(\omega) > 0$ and is not of interest for the purpose of realizing a negative refractive index.

Prior to the interest in negative-index materials (NIMs), Sievenpiper *et al* [261] demonstrated that a three-dimensional copper wire mesh with a periodic diamond-like structure can act as photonic bandgap 'crystal' for microwaves. The wires were 1 cm long and had a 1.25-mm square cross section and the lattice spacing was 2.33 cm. A range of forbidden transmission frequencies was observed to extend from zero frequency up to a cutoff frequency of 6.5 GHz corresponding to half the frequency $c/(2.33$ cm$)$. The cutoff frequency acted, in effect, as a plasma frequency. Microwave transmission measurements indicated that the wire mesh structure supports a longitudinal plane-wave mode as well as two transverse modes. Penetration into the structure was demonstrated with the introduction of defects in the form of cut wires.

Equation (8.4) suggests another way to increase the penetration of the field: if the plasma frequency could be made much smaller, the attenuation coefficient would decrease proportionally. Pendry *et al* [260, 262] showed that the plasma frequency could be substantially reduced by using wires much thinner than those of Sievenpiper *et al*.

Thin wire structures can have much smaller plasma frequencies than the individual wires because the effective electron number density is smaller than that in the wires and because the effective electron mass is larger than m. Consider the structure shown in figure 8.1. If N is the average free electron density in

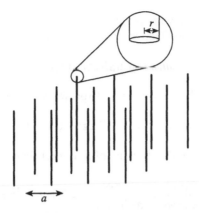

Figure 8.1. Infinitely long wires arranged to define a square lattice in a plane normal to the wires. The wires have radius r and the lattice spacing is a. From [262], with permission.

the wires, then the average electron density of this structure is just N times the fraction of space occupied by the wires. That is, the structure has an effective average electron density [260]

$$N_{\text{eff}} = N \frac{\pi r^2}{a^2}. \tag{8.5}$$

For wires of radius $r = 1 \ \mu m$ and spacing $a = 5$ mm,

$$N_{\text{eff}} = 1.3 \times 10^{-7} N = 1.1 \times 10^{22} \ \text{m}^{-3}. \tag{8.6}$$

The effect of a force acting on an electron will depend not only on the electron mass m but also on the self-inductance of the structure, which effectively increases the electron mass in its response to a force. In the case of a single wire carrying a current I, the magnetic field at a distance R from the wire has the magnitude

$$H(R) = \frac{I}{2\pi R} \tag{8.7}$$

and can be obtained from the vector potential pointing along the z-axis with magnitude

$$A(R) = \frac{\mu_0 I}{2\pi} \ln(R_0/R) \tag{8.8}$$

where R_0 is any constant length. As discussed later, the vector potential at a point on the surface of a wire for the structure shown in figure 8.1 is approximately [262]

$$A(r) = \frac{\mu_0 I}{2\pi} \ln(a/r) = -\frac{\mu_0}{2\pi}(Nev)(\pi r^2) \ln(a/r) \tag{8.9}$$

where v is the average electron velocity.

An electron on the wire picks up a contribution $-eA(r)$ to its kinetic momentum[1], implying an effective electron mass $m_{\text{eff}} = -eA(r)/v$:

$$m_{\text{eff}} = \tfrac{1}{2}\mu_0 Ne^2 r^2 \ln(a/r). \tag{8.10}$$

For copper wires of radius $r = 1\ \mu$m and spacing $a = 5$ mm,

$$m_{\text{eff}} = 1.2 \times 10^{-26}\ \text{kg} = 1.3 \times 10^4\ \text{m}. \tag{8.11}$$

The plasma frequency for the structure of figure 8.1 then becomes

$$\omega_{\text{p}}^2 = \frac{N_{\text{eff}} e^2}{m_{\text{eff}} \epsilon_0} = \frac{2\pi c^2}{a^2 \ln(a/r)} = 5.1 \times 10^{10}\ \text{rad s}^{-1} \tag{8.12}$$

or $\nu_{\text{p}} = \omega_{\text{p}}/2\pi = 8.2$ GHz for our example. Note that this result depends only on the wire radii and their spacing. The plasma frequency in our example with copper wires is reduced by a factor $\sim 10^6$ compared with ω_{p} for a single wire.

The $[\ln(a/r)]^{-1}$ dependence of ω_{p}^2 shows the advantage of using thin wires. The large reduction in the plasma frequency implies a large increase in the penetration depth of the field into the structure [cf equation (8.4)]. Moreover, if the wires were not thin and $\ln(a/r) \sim 1$, then the wavelength corresponding to the plasma frequency is of the order of the lattice constant a:

$$\lambda_{\text{p}} = 2\pi c\omega_{\text{p}} \sim a\sqrt{2\pi} \tag{8.13}$$

implying that diffraction effects would come into play at wavelengths comparable to λ_{p}. For very thin wires, by contrast, λ_{p} is much larger than the lattice constant and the continuous medium approximation is a sensible one.

It remains to justify the expression (8.9) for the vector potential for the wire mesh shown in figure 8.1. For this purpose, Pendry *et al* [262] consider a longitudinal plasmonic excitation

$$D = zD_0 e^{-\text{i}(\omega t - kz)} \tag{8.14}$$

of the type inferred by Sievenpiper *et al* [261] in their experiments. We want to solve the Maxwell equation

$$\nabla \times H = J + \frac{\partial D}{\partial t} \tag{8.15}$$

for H, given that J is confined to the thin wires while, in the long-wavelength approximation, D is uniform over the xy plane. The geometry suggests dividing the xy plane into squares of side a, each wire at the centre of a square. Given the form (8.7) of the magnetic field of a single wire, it is convenient to approximate each square by a circle of radius R_{c} chosen so that the circle has the same area

---

[1] Recall that $F = eE = -e\partial A/\partial t$ implies a change $-eA$ in the kinetic momentum mv.

as the square ($R_c = a/\sqrt{\pi}$). To approximate the behaviour of the field in the squares, it is assumed that, within a given circle, there is a current density

$$\boldsymbol{J} = z\frac{I}{\pi R}\delta(R) \qquad \left(2\pi \int_0^\infty dRRJ(R) = I\right) \tag{8.16}$$

and a uniformly distributed $\boldsymbol{D}$; that the other circles make no contribution to the field inside the given circle; and that the field vanishes on the circumference. In the circle approximation, (8.15) takes the form

$$\frac{1}{R}\frac{\partial}{\partial R}(RH) = \frac{I}{\pi R}\delta(R) + K \tag{8.17}$$

where K is constant. Choosing K so that $H(R_c) = 0$, we obtain for the field inside a circle

$$H(R) = \frac{I}{2\pi R}\left(1 - \frac{R^2}{R_c^2}\right) \tag{8.18}$$

where R is the radial distance from the wire. The vector potential of magnitude $A(R)$, pointing in the z direction and satisfying $\boldsymbol{H} = \mu_0^{-1}\nabla \times \boldsymbol{A}$ ($H(R) = -\mu_0^{-1}\partial A/\partial R$), can be taken to be

$$A(R) = \frac{\mu_0 I}{2\pi}\left[\ln(R_c/R) + \frac{R^2 - R_c^2}{2R_c^2}\right] \tag{8.19}$$

for $R < R_c$ and $A(R) = 0$ for $R \geq R_c$. With this choice, $A(R)$ vanishes at the wires centred at all the other circles and the mutual inductance (involving $\boldsymbol{J}_i \cdot \boldsymbol{A}_j$ between wires i and j) is, in effect, accounted for. For very thin wires, the logarithm in (8.19) is dominant and

$$A(r) \cong \frac{\mu_0 I}{2\pi}\ln(a/\sqrt{\pi}r) \cong \frac{\mu_0 I}{2\pi}\ln(a/r) \tag{8.20}$$

which is (8.9).

Pendry *et al* [262] also reported the results of numerical solutions of the Maxwell equations that are in close agreement with this model under the assumption that the plasma frequency is 8.2 GHz.

Experiments on the structure shown in figure 8.2 supported the theory [262]. The most important conclusion from the experiments was that the structure shown in figure 8.2 exhibited a sharp transmission cutoff at about 9 GHz. Below this frequency, the transmission decreases exponentially with distance ($\epsilon < 0$) but increases rapidly towards unity as the frequency is increased above 9 GHz. In other words, *the wire mesh structure behaved in its transmission characteristics like a low-density plasma* with a plasma frequency of 9 GHz.

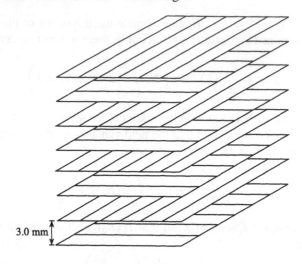

3.0 mm

Figure 8.2. Wire mesh structure with $\epsilon < 0$ at frequencies below the plasma frequency. The gold-plated wires are nominally 20 μm in diameter and laid in 120 mm × 120 mm polystyrene sheets with a 5 mm spacing between parallel wires. From [262], with permission.

8.2 Negative permeability

Whereas the plasma dispersion function (8.1) suggests ways to realize a negative ϵ, the realization of a negative μ is a more challenging task, as familiar materials do not exhibit this property. The key ideas leading to materials with $\mu < 0$ were provided by Pendry *et al* [263] and this section is devoted to a discussion of these ideas. We will focus our attention on the type of negative-μ material that was utilized shortly after publication of the theoretical work.

As in the case of the thin-wire meshes having an effective $\epsilon < 0$, the structures proposed by Pendry *et al* consist of periodically arranged conductors with a lattice spacing small compared with the wavelength of radiation of interest. Then the structure behaves, in effect, as a continuous medium characterized by a magnetic permeability (as well as an electric permittivity).

Consider first an array of parallel conducting cylinders of radius r arranged so that the distance between the centres of nearest-neighbour cylinders is $2a$. A magnetic field of amplitude H_0 and frequency ω ($2\pi c/\omega \gg a$), H_0 parallel to the cylinder axes, is applied. Inside a given cylinder, the magnetic field has magnitude

$$H = H_0 + j - \frac{\pi r^2}{a^2} j \tag{8.21}$$

where j is the current per unit length circulating around the circumference. The third term on the right-hand side arises from the 'depolarizing' effect of the sources at the ends of the cylinders and is uniform over the unit cell of the periodic array [263]. The electromotive force depends on the resistance σ of the cylinder surface as well as the rate of change of the magnetic flux:

$$\text{emf} = -2\pi r\sigma j - \mu_0 \frac{\partial}{\partial t}\left(\pi r^2 \left[H_0 + j - \frac{\pi r^2}{a^2} j\right]\right)$$

$$= -2\pi r\sigma j + i\omega\mu_0\pi r^2 \left(H_0 + j - \frac{\pi r^2}{a^2} j\right) \qquad (8.22)$$

and the loop condition that the net emf vanishes around the circumference of a cylinder gives the relation between the current and the applied field:

$$j = -\frac{H_0}{[1 - \pi r^2/a^2] + 2i\sigma/\mu_0\omega r^2}. \qquad (8.23)$$

This current determines the magnetic field H and can, therefore, be used to determine the permeability. But the effective permeability of the 'medium' is defined in terms of the *average B* and *H* fields and these averages must be carefully defined. Based on the integral form of the Maxwell curl equations,

$$\oint_C H \cdot dr = \frac{\partial}{\partial t}\int_S D \cdot dS \qquad (8.24)$$

$$\oint_C E \cdot dr = -\frac{\partial}{\partial t}\int_S B \cdot dS \qquad (8.25)$$

Pendry *et al* find it convenient (see later) to define the components of H_{avg} as averages along the axes of a unit cell, e.g.

$$(H_{\text{avg}})_x = \frac{1}{a}\int_{r=(0,0,0)}^{r=(a,0,0)} H \cdot dr \qquad (8.26)$$

whereas the components of B_{avg} are defined as averages over the surface of a unit cell, e.g.

$$(B_{\text{avg}})_x = \frac{1}{a^2}\int_{S_x} B \cdot dS \qquad (8.27)$$

where S_x is the surface defined by the y-, z-axes of the unit cell. The x component of the effective permeability is

$$(\mu_{\text{eff}})_x = (B_{\text{avg}})_x/(H_{\text{avg}})_x \qquad (8.28)$$

and, likewise, for the y and z components.

The average of H over a line element lying just outside a cylinder is

$$H_{\text{avg}} = H_0 - \frac{\pi r^2}{a^2} j$$

$$= H_0 \frac{1 + 2i\sigma/\mu_0\omega r}{[1 - \pi r^2/a^2] + 2i\sigma/\mu_0\omega r} \qquad (8.29)$$

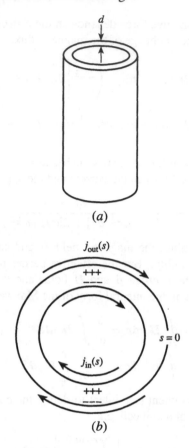

(a)

(b)

Figure 8.3. (*a*) Split ring cylindrical structure in which gaps prevent currents around a ring. However, the capacitance between the sheets separated by *d* allows current to flow when a magnetic field is applied along the cylinder axis, as indicated in (*b*). From [262], with permission.

where we have used (8.23), while the average of *B* over a unit cell is just $\mu_0 H_0$. Thus,

$$\frac{\mu}{\mu_0} \equiv \frac{\mu_{\text{eff}}}{\mu_0} = \frac{B_{\text{avg}}}{H_{\text{avg}}} = 1 - \frac{\pi r^2/a^2}{1 + 2i\sigma/\mu_0\omega r}. \tag{8.30}$$

For this 'medium', μ cannot be negative. However, the introduction of internal capacitance to the cylinders leads to a resonance feature that does allow a negative permeability. Capacitance is introduced, for example, in the 'split ring' configuration shown in figure 8.3. The effect of the capacitance introduced by the split rings is indicated roughly by adding the emf

$$\frac{-q}{C} = \frac{-2\pi i r j}{\pi r^2 \omega C} \tag{8.31}$$

to (8.22), where C is the capacitance per unit area between the sheets. This leads to the replacement of (8.30) by

$$\frac{\mu}{\mu_0} = 1 - \frac{\pi r^2/a^2}{1 + 2i\sigma/\mu_0\omega r - 2/\pi\mu_0\omega^2 C r^3}. \tag{8.32}$$

Pendry *et al* [262] report the following result of detailed calculations:

$$\frac{\mu}{\mu_0} = 1 - \frac{\pi r^2/a^2}{1 + 2i\sigma/\mu_0\omega r - 3/\pi^2\mu_0\omega^2 C r^3} \tag{8.33}$$

with $C = \epsilon_0/d = 1/\mu_0 d c_0^2$, where d is the separation between the split rings (figure 8.3) and c_0 ($= 1/\sqrt{\epsilon_0\mu_0}$) is the speed of light in vacuum. Thus,

$$\frac{\mu}{\mu_0} = 1 - \frac{\pi r^2/a^2}{1 + 2i\sigma/\mu_0\omega r - 3dc_0^2/\pi^2\omega^2 r^3}$$

$$\equiv 1 - \frac{F_1\omega^2}{\omega^2 - \omega_1^2 + i\omega\Gamma} \tag{8.34}$$

with

$$F_1 = \pi r^2/a^2 \tag{8.35}$$

$$\omega_1 = \sqrt{\frac{3d^2c_0^2}{\pi^2 r^3}} \tag{8.36}$$

$$\Gamma = 2\sigma/\mu_0 r. \tag{8.37}$$

The form (8.34) of the permeability now contains a resonance frequency ω_1, *a consequence of the capacitance of the split rings*, that allows the real part of μ to be negative, depending on the applied field frequency ω. μ is negative for frequencies ω such that

$$\omega_1 < \omega < \omega_{mp} \tag{8.38}$$

where $\omega_{mp} = \omega_1/\sqrt{1 - F_1}$ is the 'magnetic plasma frequency' [262], above which $\mu > 0$.

Note that, in addition to introducing a capacitance and a resonance frequency as a consequence of having a gap between the two rings, the split ring can have a resonance at wavelengths much larger than the ring diameter, i.e. there is no requirement that an integral number of wavelengths fit in the ring.

It has been assumed that the magnetic field is parallel to the axes of the cylinders. In order to have an isotropic negative-permeability structure, Pendry *et al* [262] suggested the following approach, based again on the idea of split rings with capacitance. Sheets, each with an imprinted square array of split rings shown in figure 8.4(*a*), are attached to inert solid blocks of thickness a and the blocks are then stacked along the z-axis to give columns of split rings (figure 8.4(*b*)).

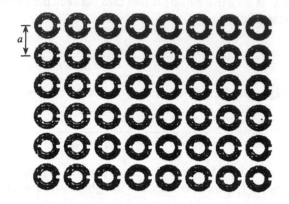

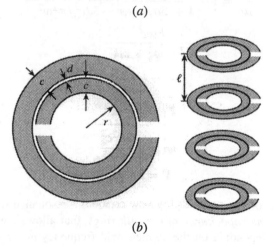

Figure 8.4. (*a*) Split ring resonators in a square array with lattice spacing *a*. (*b*) Stacked split ring resonators, with the dimensions *r*, *c*, and *d* defined. From [262], with permission.

The unit cell for the resulting structure is shown at the left in figure 8.5. Then the resulting structure is cut into a series of slabs of thickness *a* by cutting in the *yz* plane, and split rings are printed onto the surface of each slab. The slabs are then stacked along the *x*-axis to give a structure with the unit cell shown in the central cell of figure 8.5. In the final step, a third set of slabs is made by cutting in the *xz* plane, printing split rings on the slabs, and then restacking them. The unit cell for the resulting structure with a cubic symmetry is shown on the right-hand side in figure 8.5.

Pendry *et al* [262] calculate for the three-dimensional structure, assuming

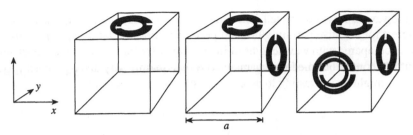

Figure 8.5. Unit cells resulting from the successive steps in making a three-dimensional negative permeability material in the way suggested by Pendry *et al.* From [262], with permission.

$r \approx a$, an approximate effective permeability [compare to (8.33)]

$$\mu = 1 - \frac{\pi r^2/a^2}{1 + 2\ell\sigma_1/\mu_0\omega r - 3\ell/\pi^2\mu_0\omega^2 C_1 r^3}$$
$$\equiv 1 - \frac{F\omega^2}{\omega^2 - \omega_0^2 + i\omega\Gamma} \tag{8.39}$$

where σ_1 is the resistance per unit length around a ring circumference and $C_1 = (\epsilon_0/\pi)\ln(2c/d)$ is the capacitance per unit length due to the gaps in the split ring resonators. In this case, the resonance frequency ω_0 is defined by

$$\omega_0^2 = \frac{3\ell c_0^2}{\pi r^3 \ln(2c/d)}. \tag{8.40}$$

Assuming $c = 1$ mm, $d = 0.1$ mm, $\ell = 2$ mm, and $r = 2$ mm, $\omega_0 = 2\pi \times 13.5$ GHz. Together with the numerical estimates for negative ϵ given in the preceding section, therefore, these estimates suggest that $\epsilon < 0$ and $\mu < 0$ might be realized simultaneously over a common frequency range by combining thin wire and split ring structures. As discussed in section 8.3, this surmise has been borne out by experiment.

8.2.1 Artificial dielectrics

The idea of using periodic arrays of conductors to obtain effective permittivities and permeabilities is not new. In 1948, Kock [264] proposed using such 'artificial dielectrics' as microwave lenses and constructed working examples of such structures[2]. An advantage of artificial dielectrics is that the volume of actual material is small compared with the volume of the 'medium' and, consequently, a lens built from an artificial dielectric could be lighter and less bulky than the large

---

[2] Chapter 12 of the book by Collin [265] is devoted to artificial dielectrics and contains references to some of the early theoretical and experimental work in this area.

microwave lenses constructed from solid dielectrics. As in the periodic arrays of interest for negative permittivities and permeabilities, the lattice spacing must be small compared with wavelengths of interest in order for the artificial dielectric to mimic a continuous medium characterized by a permittivity and a permeability.

It might be interesting to note that Kock stated the formula

$$\mu/\mu_0 = 1 - 2\pi N a^3 \qquad (8.41)$$

for the permeability of an array of N conducting spheres of radius a but did not comment on the possibility that μ could be negative.

8.3   Realization of negative refractive index

The preceding two sections suggest that a composite structure consisting of periodically arranged thin wires and split ring resonators (SRRs) can have a negative permittivity as well as a negative permeability, i.e, a negative index of refraction. The thin wires produce a negative effective ϵ, while the SRRs produce a negative effective μ. Assuming that the lattice spacing is small compared with a wavelength, the composite structure should be characterized by the approximate dispersion relation

$$k^2 = n^2(\omega)\omega^2/c^2 = \frac{\omega^2}{c^2}\left(1 - \frac{\omega_p^2}{\omega^2}\right)\left(1 - \frac{F\omega^2}{\omega^2 - \omega_0^2}\right)$$
$$\cong \frac{\omega^2 - \omega_p^2}{c^2}\frac{\omega^2 - \omega_b^2}{\omega^2 - \omega_0^2} \qquad (8.42)$$

where $\omega_b = \omega_0/\sqrt{1-F}$. There should be a passband for allowed propagation ($k^2 > 0$) defined by

$$\omega_0 < \omega < \omega_b. \qquad (8.43)$$

This was first demonstrated in the microwave experiments of Smith *et al* [244]. In their first reported experiments, they used copper split ring resonators of the type shown in figure 8.4(*b*) with $r = 1.5$ mm, $c = 0.8$ mm, and $d = 0.2$ mm. There were 17 rows of SRRs with a lattice spacing of 8 mm, eight elements deep along the direction of propagation: such a structure could theoretically exhibit a negative refractive index for a single polarization and propagation direction. Microwave transmission measurements were made with H polarized normally to the planes of the SRRs and E along parallel wires interspersed between the SRRs and parallel to the SRR planes, so that a unit cell consisted of an SRR and a wire post. Without the wires, the SRR structure exhibited a gap between propagation passbands below and above a resonance frequency band around 4.8 GHz, as shown in the full curve of figure 8.6. With wires alone, there was no propagation mode. These results are consistent with a negative ϵ when there are no SRRs and a negative μ in the gap region when there are no wires. With both the SRRs

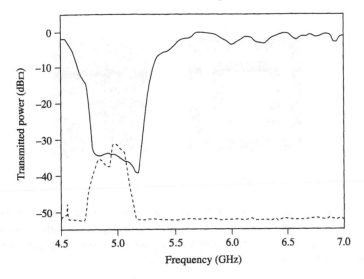

Figure 8.6. Measured transmission for an array of SRRs (full curve), showing a gap in the propagation around 4.8 GHz, consistent with a negative permeability in the gap region. When wires are uniformly interspersed between the SRRs, there is propagation in the region where there had been a gap, consistent with μ and ϵ both being negative in that frequency band. From [244], with permission.

and the wires in place, a passband was observed in the region where both ϵ and μ are negative. The results shown in figure 8.6 provided the first experimental evidence that *a composite structure with simultaneously negative ϵ and μ allows propagating electromagnetic modes.*

In a second set of experiments [266], a two-dimensional metamaterial was fabricated with square SRRs such that a unit cell consisted of six copper SRRs and two copper wire strips on thin fibreglass boards (figure 8.7). The isotropy of this structure was demonstrated and, as in [244], numerical solutions of Maxwell's equations were used to obtain dispersion curves. Figure 8.8 shows a dispersion curve and the passband that results when both the SRRs and the wires are in place. Note that $d\omega/dk < 0$ in the passband, i.e. the group velocity is negative, as is necessary for the propagating waves to be left-handed.

In a paper that triggered much additional interest in the area of left-handed light, Shelby *et al* [245] used this metamaterial to provide the first experimental evidence of a negative refractive index when ϵ and μ are both negative. Figure 8.9 shows the basic idea of the experiment. The detector was rotated in steps of 1.5 degrees and the data shown in figure 8.10 were obtained when teflon and the metamaterial were alternatively used as the prism material. Because of the corrugation at the refracting surface of the metamaterial, the angular transmission pattern has modulations: to wash out these modulations, the data for

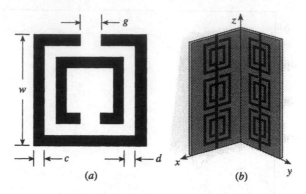

Figure 8.7. (*a*) Split ring resonator of Shelby *et al* [266]: $w = 2.62$ mm, $g = 0.46$ mm, $d = 0.30$ mm, and $c = 0.25$ mm. (*b*) Unit cell with orthogonal fibreglass boards and a lattice constant of 5 mm. The wire strips are 1 cm long and on the opposite sides of the boards from the SRRs. From [266], with permission.

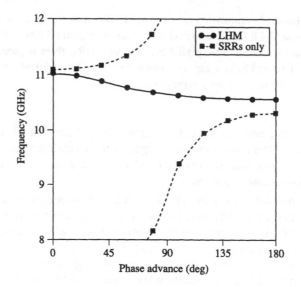

Figure 8.8. Dispersion curve computed by Shelby *et al*. The phase advance is proportional to the wavenumber k. Note that the group velocity $d\omega/dk < 0$ in the passband. From [266], with permission.

the metamaterial are presented as an average over eight different sample positions. The data for teflon are consistent with a refractive index of 1.4 ± 0.1 at 10.5 GHz. For the metamaterial, however, Snell's law implies a *negative* refractive index of -2.7 ± 0.1.

Further experiments using the SRR and thin-wire metamaterial lent

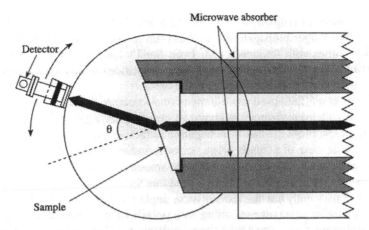

Figure 8.9. Schematic diagram of the experiment by Shelby *et al* [245]. The sample is a two-dimensional metamaterial (see text) in the shape of a prism. The refraction angle θ as indicated is positive, as would be the case for a positive refractive index. From [245], with permission.

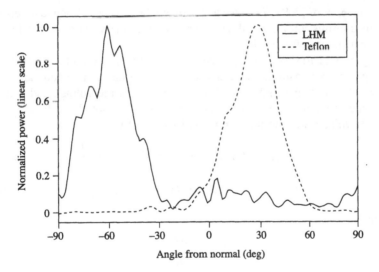

Figure 8.10. Transmitted power *versus* angle θ in the experiment indicated in figure 8.9. With teflon as the prism sample, the angle of refraction is positive, whereas for the metamaterial, it is negative. The width of the beam is due to diffraction and the angular detection sensitivity. From [245], with permission.

additional support to the conclusion that the refractive index is negative. Parazzoli *et al* [267] used a one-dimensional structure (SRRs and thin wires lying on parallel planes) and found that their structure *in free space* exhibits a negative refractive index at frequencies between 12.6 and 13.2 GHz. Their experiments also involved larger propagation distances—up to 66 cm. Their data show that the index varies with frequency and, from Snell's law, has the value -1.05 at 12.6 GHz. They also carried out numerical simulations which were found to be in good agreement with their experimental data.

Houck *et al* [268] used a two-dimensional structure similar to that of Shelby *et al* to provide even stronger evidence of negative refraction. They emphasize that a strict test of Snell's law with negative refractive index requires at least two prisms: in the case of a single prism, an observation of a negative transmission angle could conceivably be due to different amounts of attenuation on the longer and shorter sides of the prism. They verified that Snell's law with a negative index accounts consistently for the transmission angle for two different metamaterial prisms. Using an input antenna acting as a point source in front of a planar slab, and scanning the transmitted field, these authors also observed what appeared to be some degree of the theoretically predicted focusing [29].

8.4   Transmission line metamaterials

Eleftheriades *et al* [269–272] have generalized the concept of negative refractive index to transmission lines and have designed microwave circuits exhibiting an effective negative refractive index.

Figure 8.11(a) shows an equivalent circuit representation for a transmission line (e.g. a coaxial cable). Let $\mathcal{L}$ and $\mathcal{C}$ be the the inductance and capacitance, respectively, per unit length, and $V(x, t)$ and $I(x, t)$ the voltage and current at time t and at point x along the transmission line. Denote by ΔV and ΔI the change in voltage and current over a small section Δx:

$$V(x, t) + \Delta V(x, t) = V(x, t) - \mathcal{L}\Delta x \frac{\partial I}{\partial t} \tag{8.44}$$

$$I(x, t) + \Delta I(x, t) = I(x, t) - \mathcal{C}\Delta x \frac{\partial V}{\partial t}. \tag{8.45}$$

Letting $\Delta x \to 0$,

$$\frac{\partial V}{\partial x} + \mathcal{L}\frac{\partial I}{\partial t} = 0 \tag{8.46}$$

$$\frac{\partial I}{\partial x} + \mathcal{C}\frac{\partial V}{\partial t} = 0 \tag{8.47}$$

or[3]

$$\frac{\partial^2 V}{\partial x^2} - \mathcal{L}\mathcal{C}\frac{\partial^2 V}{\partial t^2} = 0 \tag{8.48}$$

---

[3] More generally, when we include a resistance $\mathcal{R}$ and a conductance $\mathcal{G}$ per unit length, we obtain the *telegrapher's equation*, $\partial^2 V/\partial x^2 = \mathcal{R}\mathcal{G}V + (\mathcal{L}\mathcal{G} + \mathcal{R}\mathcal{C})\partial V/\partial t + \mathcal{L}\mathcal{C}\partial^2 V/\partial t^2$. See Jordan and Balmain [273] for the circuit representation of a transmission line when resistance and conductance are included.

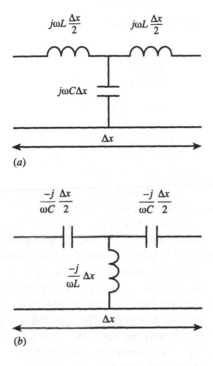

Figure 8.11. (*a*) An equivalent circuit, consisting of a sequence of inductors in series and capacitors in parallel, representing a transmission line. (*b*) The dual transmission line representation in which the inductors and capacitors are reversed. The time variations are indicated according to the electrical engineering usage in which exp(jωt) is used instead of the exp($-i\omega t$) used in the text. From [269], with permission.

$$\frac{\partial^2 I}{\partial x^2} - \mathcal{LC}\frac{\partial^2 I}{\partial t^2} = 0 \tag{8.49}$$

implying the propagation velocity $v_{\mathrm{p}} = 1/\sqrt{\mathcal{LC}}$.

Consider a sinusoidal oscillation such that $\partial V/\partial t = -i\omega V$ and $\partial I/\partial t = -i\omega I$. Then (8.46) and (8.47) become

$$\frac{\partial V}{\partial x} - i\omega \mathcal{L} I = 0 \tag{8.50}$$

$$\frac{\partial I}{\partial x} - i\omega C V = 0. \tag{8.51}$$

These equations have the same form as the Maxwell curl equations

$$\frac{\partial E_x}{\partial x} - i\omega\mu H_z = 0 \tag{8.52}$$

$$\frac{\partial H_z}{\partial x} - i\omega\epsilon E_y = 0 \tag{8.53}$$

for a transverse electromagnetic wave propagating in the x direction.   The correspondence

$$\mu \leftrightarrow \mathcal{L}$$

$$\epsilon \leftrightarrow \mathcal{C}$$

$$v_{\mathrm{p}} = \frac{1}{\sqrt{\epsilon\mu}} \leftrightarrow \frac{1}{\sqrt{\mathcal{L}\mathcal{C}}} \qquad (8.54)$$

establishes a quantitative correspondence between the transmission line equations (8.50) and (8.51) and the equations for transverse electromagnetic wave propagation in a homogeneous medium [273].

Thus, the propagation of voltage and current along a transmission line is analogous to the propagation of a plane wave in a homogeneous medium. The analogy applies not only to the propagation equations but also to the boundary conditions: the tangential components E_x and H_z are continuous at the boundary between two dielectric media, while V and I are continuous at a junction in a transmission line. In other words, there is a one-to-one correspondence, based on (8.54), between the propagation of a plane electromagnetic wave along a sequence of homogeneous dielectric sections and the propagation of current and voltage along a transmission line with junctions. The solution of a propagation problem in one case can be applied directly to obtain the solution in the other. The analogy extends straightforwardly to two-dimensional structures.

Eleftheriades *et al* have exploited this analogy in the case that the equivalent transmission line corresponds to a *negative*-index medium.   For the 'dual' transmission line indicated in figure 8.11, with d the length of a unit cell, equations (8.50) and (8.51) are replaced by

$$\frac{\partial V}{\partial x} + \frac{i}{\omega \mathcal{C} d} I = 0 \qquad (8.55)$$

$$\frac{\partial I}{\partial x} + \frac{i}{\omega \mathcal{L} d} V = 0 \qquad (8.56)$$

in the continuous-medium approximation ($2\pi c/\omega \gg d$).   These equations correspond to the plane-wave equations (8.52) and (8.53) when we make the substitutions

$$\mu \leftrightarrow -\frac{1}{\omega^2 \mathcal{C} d}$$

$$\epsilon \leftrightarrow -\frac{1}{\omega^2 \mathcal{L} d} \qquad (8.57)$$

or, in other words, the propagation of voltage and current in the dual transmission line corresponds to plane-wave propagation with $\epsilon < 0$ and $\mu < 0$. The fact that $\epsilon, \mu < 0$ implies a refractive index $n < 0$ in the case of a dielectric medium suggests that the propagation constant β in the equation

$$\frac{\partial^2 V}{\partial x^2} + \beta^2 V = 0 \qquad (8.58)$$

that follows from (8.55) and (8.56) should be taken to be negative:

$$\beta = -\frac{1}{\omega d \sqrt{\mathcal{L}\mathcal{C}}} \tag{8.59}$$

so that the phase velocity

$$v_\mathrm{p} = \frac{\omega}{\beta} = -\omega^2 d \sqrt{\mathcal{L}\mathcal{C}} \tag{8.60}$$

while the group velocity

$$v_\mathrm{g} = \left(\frac{d\beta}{d\omega}\right)^{-1} = \omega^2 d \sqrt{\mathcal{L}\mathcal{C}}. \tag{8.61}$$

The phase and group velocities defined in this way are opposite. As in the case of a negative-index dielectric medium, the phase velocity is negative and the group velocity, which is in the direction of energy flow, is positive.

Equations (8.57) suggest that a host transmission line loaded with lumped capacitors in series and inductors in parallel can be described by an effective permittivity and an effective permeability defined by [269]

$$\epsilon_\mathrm{eff} = \epsilon - \frac{1}{\omega^2 \mathcal{L}d}$$
$$\mu_\mathrm{eff} = \mu - \frac{1}{\omega^2 \mathcal{C}d} \tag{8.62}$$

where ϵ, μ are the effective material parameters of the unperturbed transmission line. Evidently ϵ_eff and μ_eff can both be negative if the frequency ω is low enough and the spacing d of interconnecting transmission lines is short enough.

The synopsis just given is only a superficial summary of the transmission line approach to the realization of a negative effective refractive index. Using this approach of loading a host transmission line with lumped series capacitors and shunt inductors, Eleftheriades *et al* [269–272] have performed a number of microwave experiments that confirm the theoretical prediction that ϵ_eff and μ_eff can both be negative and that, therefore, the effective refractive index n_eff of the perturbed transmission line is negative. In one experiment [269], it was demonstrated that $n_\mathrm{eff} < 0$ leads to backward-travelling waves associated with opposite phase and group velocities. In particular, it was shown that a guiding structure with $n_\mathrm{eff} < 0$ could emit a *backward* cone of radiation into free space, analogous to the reversal of Cerenkov radiation predicted when the refractive index in which a charged particle moves is negative. This is a direct consequence of opposite phase and group velocities and the continuity of the wavevectors at the interface of positive- and negative-index media. Grbic and Eleftheriades [269] note that, while there are other radiating structures known to have phase and group velocities of opposite sign, theirs is evidently 'the first to operate in the

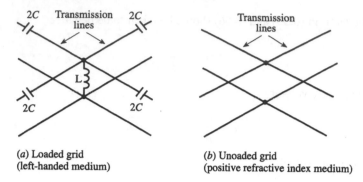

(a) Loaded grid
(left-handed medium)

(b) Unoaded grid
(positive refractive index medium)

Figure 8.12. Transmission line unit cell for (a) a loaded tranmsission line having an effective refractive index that is negative and (b) an unloaded transmission line with an effective index that is positive. From [272], with permission.

long-wavelength regime and demonstrate backward-wave radiation in its lowest passband of operation'. Previous backward-wave sources they cite radiate in higher-order spatial harmonics.

Grbic and Eleftheriades [272] have demonstrated a transmission-line lens that produces images narrower than the diffraction limit. For this purpose, it is necessary to have the analog of a negative-index dielectric slab in which it is possible to have a growing exponential wave (section 7.9.2). Figure 8.12 shows schematically the unit cell for a loaded (dual) transmission line (negative index) as well as for an unloaded transmission line (positive index) and figure 8.13 shows the actual configuration used to make an isotropic and effectively negative-index planar lens. The loaded and unloaded grids were impedance matched so that the lens had an effective index of −1 at 1 GHz. The half-power beam width of the image was measured to be 0.21λ compared with the theoretical diffraction-limited width of 0.36λ: even greater resolution ought to be possible if losses can be reduced. *The electric field measurement also showed the amplified evanescent wave behaviour in the 'slab'*, as shown by the data in figure 8.14.

It is noteworthy that these transmission line metamaterials do not involve a plasma-like resonance as in the wire-and-SSR structures and, consequently, they can have an effective negative index over a large bandwidth with small dispersion. They can also be virtually lossless. Liu *et al* [274] have suggested that these structures can be used to design novel antennas which can scan all directions in space from backfire to endfire as the frequency is varied.

8.5 Negative refraction in photonic crystals

Propagation of light in photonic bandgap structures, or *photonic crystals*, exhibits properties similar to those of electrons in solids, e.g. dispersion curves with forbidden gaps [275]. They typically involve a periodic array of holes in

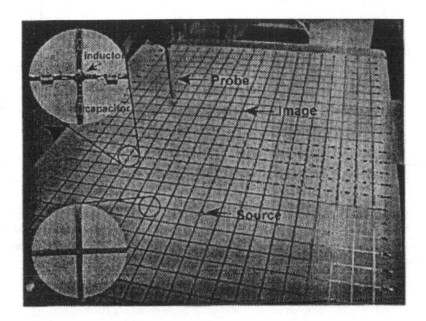

Figure 8.13. Implementation of a negative-index planar lens using the loaded and unloaded transmission line unit cells indicated in figure 8.12. The planar lens consists of five columns of loaded transmission lines printed on a grounded microwave substrate, shown between the source and image points which are located symmetrically at 2.5 cells (0.135λ) from the lens. The area external to the loaded lines consists of unloaded lines. The source is excited by a coaxial cable and the vertical electric field is measured 0.8 mm above the surface with a probe as shown. From [272], with permission.

a dielectric medium and are of interest, for instance, for the suppression of spontaneous emission. As in the case of electron energy bands in solids, the propagation of light in a photonic crystal is determined by the (photonic) band structure. Notomi [276], following experimental observations of negative refraction and large beam steering in a photonic crystal [277], showed, among other things, that under certain circumstances the dispersion relation for a photonic crystal can imply a negative effective refractive index[4].

Luo *et al* [278,279] have shown theoretically that a photonic band is possible such that the refractive index (and the group velocity) is positive while the effective 'photon mass' $\partial^2 \omega / \partial k_i \partial k_j$ is such that negative refraction—*without a negative refractive index*—is possible for all incident angles. This conclusion

---

[4] Notomi [276] discusses this and other 'anomalous light propagation phenomena', apparently independently of the work on negative index of refraction.

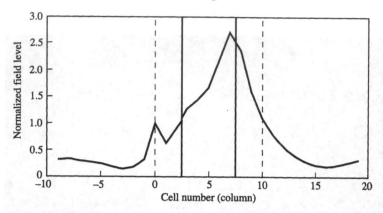

Figure 8.14. Measured vertical electric field along the source–image line in figure 8.13. The vertical broken lines indicate the source and image planes, while the vertical full lines indicate the interfaces between the positive- and negative-index 'media'. From [272], with permission.

relies heavily on numerical computations of constant-frequency surfaces in k space, which give the k vectors for allowed propagation modes at a given frequency in the photonic crystal. The group velocity $\nabla_k \omega$ is orthogonal to these surfaces and in the direction of increasing ω. Based on such computations, Luo *et al* conclude that negative refraction can occur even though the phase and group velocities are both positive, i.e. even though the effective refractive index is positive. The fact that a photonic crystal can exhibit negative refraction at all angles is, of course, important for superlensing in order for all diverging rays from a point source to be focused.

Cubukcu *et al* [281] demonstrated microwave (13.10–15.44 GHz) negative refraction and subwavelength resolution in a two-dimensional photonic crystal slab and found good agreement between their experimental data and numerical simulations. Their photonic crystal consisted of a square array of dielectric rods (diameter 3.15 mm, length 15 cm, lattice constant 4.79 mm) embedded in a dielectric with $\epsilon = 9.61$. The refractive index inferred from their measurements was -1.94 compared with the theoretical value of -2.06. The full-width-at-half-maximum width of the focused beam was found to be 0.21λ. They also observed subwavelength resolution of two incoherent point sources separated by $\lambda/3$.

Independently, Parimi *et al* [282] observed *wide-angle* subwavelength microwave (9.0–9.4 GHz) focusing using a flat photonic crystal lens. They also demonstrated explicitly that their flat lens has no optical axis as in a conventional lens.

8.6 Remarks

The number of publications on left-handed light, negative refraction, and metamaterials has grown very rapidly in the past few years, exceeding 200 in 2003; and it appears that this growth is likely to continue at an even faster pace. In this and the preceding chapter, we have attempted to provide an introduction to the basic theory and to describe some of the early seminal experiments. We conclude this chapter by touching on a few further developments in the field.

In his 'perfect lens' paper, Pendry [243] suggested that, in the electrostatic limit ($\omega \to 0$), it might be possible to obtain superlensing with a thin film for which only ϵ is negative. For polarization *parallel* to the plane of incidence, the reflection and transmission coefficients (7.118) and (7.121) are replaced by corresponding expressions with ϵ and μ interchanged and we obtain for the transmission coefficient in the limit $\omega \to 0$

$$t_P = \frac{4\epsilon \exp(\mathrm{i}k_z d)}{(\epsilon + 1)^2 - (\epsilon - 1)^2 \exp(2\mathrm{i}k_z d)} \tag{8.63}$$

with $k_z = \mathrm{i}\sqrt{k_x^2 + k_y^2}$. For $\epsilon \to -1$, this becomes

$$t_P = \exp(-\mathrm{i}k_z d) = \exp(\sqrt{k_x^2 + k_y^2}d). \tag{8.64}$$

Thus, it should be possible, in the limit of large wavelength compared with size scales of interest, to achieve superlensing using a thin planar film *without requiring μ to be negative*. Pendry suggested that this could be done with a thin layer of a metal like silver, for which the permittivity is well described by the idealized plasma form and is negative below the plasma frequency. Experiments by Fang *et al* [283] using evanescent waves produced by surface roughness scattering indeed showed that evanescent wave amplification by factors ~ 30 occurred for film thicknesses up to about 50 nm, beyond which absorption became dominant.

At the time of writing, all of the experimental demonstrations of negative refraction and the focusing properties of metamaterials have been in the microwave region. Podolsky *et al* [284] have numerically modelled a metamaterial that would appear to be applicable in the near-infrared and visible. The unit cell of their metamaterial consists of two parallel nanowires with radii small compared with the wavelength and with length comparable to the wavelength. The spacing between the wires is small compared with their length. Numerical modelling reported by Podolskiy *et al* indicates that both the permittivity and the permeability can be negative.

Ziolkowski [285] has presented results of a detailed numerical analysis of the transmission by a negative-index slab of both pulsed and continuous-wave Gaussian beams and he describes some potential applications based on these results.

There are many potential applications of negative refraction and, obviously, the field is in a very early stage of development. The question remains as to whether its great promise will be realized. This question will only be answered after further research and, in particular, after further experiments and fabrication of metamaterials.

Bibliography

[1] Maxwell J C 1954 *A Treatise on Electricity and Magnetism* (New York: Dover) p ix [republication of 3rd edition published by Clarendon Press in 1891]

[2] Longair M S 1984 *Theoretical Concepts in Physics* (Cambridge: Cambridge University Press)

[3] Wroblewski A 1985 *Am. J. Phys.* **53** 620

[4] Bialynicka-Birula Z 1996 *SPIE Proc.* **2729** 8

[5] Harman P M *The Scientific Letters and Papers of James Clerk Maxwell* (Cambridge: Cambridge University Press) p 685

[6] Whittaker E 1951 *A History of the Theories of Aether and Electricity* (London: Thomas Nelson and Sons) p 254

[7] Bates H E 1988 *Am. J. Phys.* **56** 682

[8] Milonni P W 1994 *The Quantum Vacuum. An Introduction to Quantum Electrodynamics* (San Diego, CA: Academic)

[9] Born M and Wolf E 1999 *Principles of Optics* 7th edn (Cambridge: Cambridge University Press)

[10] Fearn H, James D F V and Milonni P W 1996 *Am. J. Phys.* **64** 986

[11] Ladenburg R and Kopfermann H 1928 *Nature (Paris)* **122** 438

[12] Wood R W 1904 *Phil. Mag.* **8** 293

[13] Toll J S 1956 *Phys. Rev.* **104** 1760

[14] Jackson J D 1975 *Classical Electrodynamics* 2nd edn (New York: Wiley)

[15] Nussenzveig H M 1972 *Causality and Dispersion Relations* (New York: Academic)

[16] Naus H and Ubachs W 2000 *Opt. Lett.* **25** 347

[17] Hamilton W R 1839 *Proc. R. Irish Acad.* **1** 341

[18] Russell J S 1844 *Brit. Assoc. Rep.* 311

[19] Lord Rayleigh (J W Strutt) 1881 *Nature* **24** 382 and **25** 52

[20] Feynman R P *et al* 1964 *The Feynman Lectures on Physics* vol I (Reading, MA: Addison-Wesley) pp 48–7

[21] Oughstun K E and Sherman G C 1994 *Electromagnetic Pulse Propagation in Causal Dielectrics* (Berlin: Springer)

[22] Panofsky W K H and Phillips M 1962 *Classical Electricity and Magnetism* (Reading, MA: Addison-Wesley)

[23] Shankland R S 1964 *Am. J. Phys.* **32** 16

[24] Einstein A 1952 *Relativity* (New York: Crown Publishers) p 40

[25] Hannay J 1976 Cambridge University Hamilton Prize Essay (unpublished)

[26] Berry M V, Chambers R G, Large M D, Upstill C and Walmsley J C 1980 *Eur. J. Phys.* **1** 154

[27] Cook R J, Fearn H and Milonni P W 1995 *Am. J. Phys.* **63** 705

[28] Veselago V G 1967 *Sov. Phys. Solid State* **8** 2854

[29] Veselago V G 1968 *Sov. Phys.–Usp.* **10** 509

[30] Mandelstam L I 1972 *Lectures on Optics, Relativity and Quantum Mechanics* (Moscow: Nauka)

[31] Mandelstam L I 1945 *Sov. Phys.–JETP* **15** 475

[32] Brillouin L 1960 *Wave Propagation and Group Velocity* (New York: Academic)

[33] Bolda E L, Chiao R Y and Garrison J C 1993 *Phys. Rev.* A**48** 3890

[34] Chiao R Y 1993 *Phys. Rev.* A **48** 34

[35] Basov N G *et al* 1966 *Sov. Phys.–JETP* **23** 16

[36] Icsevgi A and Lamb W E Jr 1969 *Phys. Rev.* **185** 517

[37] Garrett C G B and McCumber D E 1970 *Phys. Rev.* A **1** 305

[38] Crisp M D 1971 *Phys. Rev.* A **4** 2104

[39] Chiao R Y, Kozhekin A E and Kurizki G 1996 *Phys. Rev. Lett.* **77** 1254

[40] Faxvog F R *et al* 1970 *Appl. Phys. Lett.* **17** 192

[41] Chu S and Wong S 1982 *Phys. Rev. Lett.* **48** 738

[42] Chiao R Y 1996 *Amazing Light: A Volume Dedicated to Charles Hard Townes on His 80th Birthday* ed R Y Chiao (Berlin: Springer) p 91

[43] Chiao R Y and Steinberg A M 1997 *Progress in Optics 37* ed E Wolf (Amsterdam: Elsevier) p 345

[44] Chiao R Y and Steinberg A M 1998 *Phys. Scr.* T **76** 61

[45] MacColl L A 1932 *Phys. Rev.* **40** 621

[46] Zhu S *et al* 1986 *Am. J. Phys.* **54** 601

[47] Milonni P W, Fearn H and Zeilinger A 1996 *Phys. Rev.* A **53** 4556

[48] Hong C K, Ou Z Y and Mandel L 1987 *Phys. Rev. Lett.* **59** 2044

[49] Wang L J, Kuzmich A and Dogariu A 2000 *Nature* **406** 277

[50] Steinberg A M and Chiao R Y 1994 *Phys. Rev.* A **49** 2071

[51] Nimtz G and Heitmann W 1997 *Prog. Quantum Electron.* **21** 81

[52] Martin Th and Landauer R 1992 *Phys. Rev.* A **45** 2611

[53] Nimtz G 1999 *Eur. Phys. J.* B **7** 523

[54] Diener G 1996 *Phys. Lett.* A **223** 327

[55] Peatross J and Ware M 2001 *J. Opt. Soc. Am.* A **18** 1719

[56] Durnin J, Miceli J J and Eberly J H 1987 *Phys. Rev. Lett.* **58** 1499

[57] Milonni P W and Eberly J H 1988 *Lasers* (New York: Wiley)

[58] Saari P and Reivelt K 1997 *Phys. Rev. Lett.* **79** 4135

[59] Mugnai D *et al* 2000 *Phys. Rev. Lett.* **84** 4830

[60] Bigelow N P and Hagen C R 2001 *Phys. Rev. Lett.* **87** 059401

[61] Ringermacher H and Mead L R 2001 *Phys. Rev. Lett.* **87** 059402

[62] Landau L D and Lifshitz E M 1975 *Electrodynamics of Continuous Media* (Oxford: Pergamon) section 61

[63] Loudon R 1970 *J. Phys. A: Math. Gen.* **3** 233

[64] Schultz-DuBois E O 1969 *Proc. IEEE* **57** 1748

[65] Diener G 1997 *Phys. Lett.* A **235** 118

[66] Peatross J, Glasgow S A and Ware M 2000 *Phys. Rev. Lett.* **84** 2370

[67] Stratton J S 1941 *Electromagnetic Theory* (New York: McGraw-Hill)

[68] Kikuchi S 1930 *Z. Phys.* **60** 558

[69] Fermi E 1932 *Rev. Mod. Phys.* **4** 87

[70] Louisell W H 1964 *Radiation and Noise in Quantum Electronics* (New York: McGraw-Hill)

[71] Ferretti B and Peierls R E 1947 *Nature London* **160** 531

[72] Ferretti B 1968 *Old and New Problems in Elementary Particles* ed G Puppi (New York: Academic)

[73] Hamilton J 1949 *Proc. R. Soc.* A **62** 12

[74] Heitler W and Ma S T 1949 *Proc. R. Irish Acad.* A **52** 109

[75] Fierz M 1950 *Helv. Phys. Acta* **23** 731

[76] Milonni P W and Knight P L 1974 *Phys. Rev.* A **10** 1096

[77] Milonni P W and Knight P L 1975 *Phys. Rev.* A **11** 1090

[78] Shirokov M I 1978 *Sov. Phys.–Usp.* **21** 345

[79] Hegerfeldt 1994 G C *Phys. Rev. Lett.* **72** 596

[80] Maddox J 1994 *Nature* **367** 509

[81] Milonni P W 1994 *Nature* **372** 325

[82] De Haan M 1985 *Physica* A **132** 375

[83] Compagno G, Passante R and Persico F 1990 *J. Mod. Opt.* **37** 1377

[84] Power E A 1993 *Physics and Probability. Essays in Honor of E T Jaynes* ed W T Grandy Jr and P W Milonni (Cambridge: Cambridge University Press)

[85] Milonni P W, James D F V and Fearn H 1995 *Phys. Rev.* A **52** 1525

[86] Berman P R and Dubetsky B 1997 *Phys. Rev.* A **55** 4060

[87] Stephen M J 1964 *J. Chem. Phys.* **40** 669

[88] Milonni P W and Knight P L 1976 *Am. J. Phys.* **44** 741

[89] Glauber R J 1963 *Phys. Rev.* **130** 2529
      Glauber R J 1963 *Phys. Rev.* **131** 2766

[90] Glauber R J 1965 *Quantum Optics and Electronics* ed C DeWitt, A Blandin and C Cohen-Tannoudji (New York: Gordon and Breach)

[91] Mandel L and Wolf E 1965 *Rev. Mod. Phys.* **37** 231

[92] Loudon R 1983 *The Quantum Theory of Light* (London: Oxford University Press)

[93] Mandel L and Wolf E 1995 *Optical Coherence and Quantum Optics* (Cambridge: Cambridge University Press)

[94] Scully M O and Zubairy M S *Quantum Optics* (Cambridge: Cambridge University Press)

[95] Bykov V P and Tatarskii V I 1989 *Phys. Lett.* A **136** 77

[96] Bykov V P and Tatarskii V I 1990 *Phys. Lett.* A **144** 491

[97] Milonni P W, Furuya K and Chiao R Y 2001 *Opt. Express* **8** 59

[98] Berman P R and Milonni P W 2004 *Phys. Rev. Lett.* **92** 053601

[99] Bohr N and Rosenfeld L 1950 *Phys. Rev.* **78** 794

[100] Einstein A 1947 Letter to Max Born, 3 March 1947. Reprinted translation in *The Born–Einstein Letters* (New York: Walker and Company)

[101] Peres A 1995 *Quantum Theory: Concepts and Methods* (Dordrecht: Kluwer)

[102] Herbert N 1982 *Found. Phys.* **12** 1171

[103] Milonni P W and Hardies M L 1982 *Phys. Lett.* A **92** 321

[104] Wooters W K and Zurek W H 1982 *Nature* **299** 802

[105] Dieks D 1982 *Phys. Lett.* A **92** 371

[106] Mandel L 1983 *Nature* **304** 188

[107] Glauber R J 1986 *Ann. N. Y. Acad. Sci.* **480** 336
[108] Gisin N 1998 *Phys. Lett.* A **242** 1
[109] Bužek V and Hillery M 1996 *Phys. Rev.* A **54** 1844
[110] Lamas-Linares A *et al* 2002 *Science* **296** 712
[111] Kwiat P G *et al* 1995 *Phys. Rev. Lett.* **75** 4337
[112] Bennett C H *et al* 1993 *Phys. Rev. Lett.* **70** 1895
[113] Furusawa A *et al* 1998 *Science* **282** 706
[114] Garuccio A 1997 *The Present Status of the Quantum Theory of Light* ed S Jeffers *et al* (Dordrecht: Kluwer Academic Publishers)
[115] Boyd R W *et al* 1987 *Opt. Lett.* **12** 42
[116] Furuya K *et al* 1999 *Phys. Lett.* A **251** 294
[117] Gaeta A L and Boyd R W 1988 *Phys. Rev. Lett.* **60** 2618
[118] Milonni P W, Bochove E J and Cook R J 1989 *Phys. Rev.* A **40** 4100
[119] Agarwal G S 1987 *J. Opt. Soc. Am.* B **4** 1806
[120] Peres A 1999 Private communication
[121] Haroche S and Kleppner D 1989 *Physics Today* **42** (January) 24
[122] Berman P R (ed) 1994 *Cavity Quantum Electrodynamics* (San Diego, CA: Academic)
[123] Fearn H, Cook R J and Milonni P W 1995 *Phys. Rev. Lett.* **74** 1327
[124] Cook R J and Milonni P W 1987 *Phys. Rev.* A **35** 5081
[125] Branning D, Kwiat P and Migdall A 2003 *Quantum Communication, Measurement and Computing* ed J H Shapiro and O Hirota (Paramus, NJ: Rinton)
[126] Herzog T J *et al* 1994 *Phys. Rev. Lett.* **72** 629
[127] Kauranen M *et al* 1998 *Phys. Rev. Lett.* **80** 952
[128] Einstein A, Podolsky B and Rosen N 1935 *Phys. Rev.* **47** 777
[129] Einstein A 1954 *Ideas and Opinions* (New York: Crown Publishers)
[130] Bell J S 1981 *J. Physique Coll.* C **2** (supplément 3) C2-41
[131] Feynman R P 1971 *Lectures on Gravitation* (Pasadena, CA: California Institute of Technology) pp 15–16
[132] Bohm D 1952 *Phys. Rev.* **85** 169
[133] Bell J S 1966 *Rev. Mod. Phys.* **38** 447
[134] Steinberg A M, Kwiat P G and Chiao R Y 1996 *Atomic, Molecular, & Optical Physics Handbook* ed G W F Drake (New York: American Institute of Physics)
[135] Jammer M 1974 *The Philosophy of Quantum Mechanics* (New York: Wiley)
[136] Centini M *et al* 2003 *Phys. Rev.* E **68** 016602
[137] Stenner M D, Gauthier D J and Neifeld M A 2003 *Nature* **425** 695
[138] Stenner M D and Gauthier D J 2003 *Phys. Rev.* A **67** 063801
[139] Aharonov Y, Reznik B and Stern A 1998 *Phys. Rev. Lett.* **81** 2190
[140] Segev B *et al* 2000 *Phys. Rev.* A **62** 022114
[141] Glauber R and Haake F 1978 *Phys. Lett.* A **68** 29
[142] Polder D, Schuurmans M F H and Vrehen Q H F 1979 *Phys. Rev.* A **19** 1192
[143] Maki J J *et al* 1989 *Phys. Rev.* A **40** 5135
[144] Burnham D C and Chiao R Y 1969 *Phys. Rev.* **188** 667
[145] Kuzmich A *et al* 2001 *Phys. Rev. Lett.* **86** 3925
[146] Desurvire E 1994 *Erbium-Doped Fiber Amplifiers: Principles and Applications* (New York: Wiley) ch 2
[147] Caves C M 1982 *Phys. Rev.* D **26** 1817
[148] Dogariu A, Kuzmich A and Wang L J 2001 *Phys. Rev.* A **63** 053806

[149] Yamamoto Y 1980 *IEEE J. Quantum Electron.* **16** 1073
[150] Wang L-G *et al* 2002 *Europhys. Lett.* **60** 834
[151] Wynne K 2002 *Opt. Commun.* **209** 85
[152] Dogariu A *et al* (eds) 2003 *IEEE J. Selected Topics Quantum Electron.* **9** no 1
[153] Segard B and Macke B 1985 *Phys. Lett.* A **109** 213
[154] Tanaka H *et al* 2003 *Phys. Rev.* A **68**, 053801
[155] Macke B and Segard B 2003 *Eur. Phys. J.* D **23** 125
[156] Longhi S *et al* 2003 *IEEE J. Selected Topics Quantum Electron.* **9** 4
[157] Mojahedi M *et al* 2003 *IEEE J. Selected Topics Quantum Electron.* **9** 30
[158] Mitchell M W and Chiao R Y 1998 *Am. J. Phys.* **66** 14
[159] Kitano M, Nakanishi T and Sugiyama K 2003 *IEEE J. Selected Topics Quantum Electron.* **9** 43
[160] Büttiker M and Thomas H 1998 *Ann. Phys. (Leipzig)* **7** 602
[161] Winful H G 2003 *Phys. Rev. Lett.* **90** 023901
[162] Casperson L and Yariv A 1971 *Phys. Rev. Lett.* **26** 293
[163] Grischkowsky D 1973 *Phys. Rev.* A **7** 2096
[164] McCall S L and Hahn E L 1969 *Phys. Rev.* A **183** 457
[165] Slusher R E and Gibbs H M 1972 *Phys. Rev.* A **5** 1634
[166] Arimondo E 1996 *Progress in Optics* vol 35, ed E Wolf (Amsterdam: Elsevier) p 257
[167] Alzetta G *et al* 1976 *Nuovo Cimento* B **36** 5
[168] Harris S E 1997 *Phys. Today* **50** 36
[169] Harris S E, Field J E and Kasapi A 1992 *Phys. Rev.* A **46** R29
[170] Lukin M D *et al* 1997 *Phys. Rev. Lett.* **79** 2959
[171] Fleischhauer M and Lukin M D 2002 *Phys. Rev.* A **65** 022314
[172] Boller K-J, Imamoglu A and Harris S E 1991 *Phys. Rev. Lett.* **66** 2593
[173] Boyd R W and Gauthier D J 2002 *Progress in Optics* vol 43, ed E Wolf (Amsterdam: Elsevier) p 497
[174] Oreg J, Hioe F T and Eberly J H 1984 *Phys. Rev.* A **29** 690
[175] Grobe R, Hioe F T and Eberly J H 1994 *Phys. Rev. Lett.* **73** 3183
[176] Fleischhauer M and Manka A S 1996 *Phys. Rev.* A **54** 794
[177] Kasapi *et al* 1995 *Phys. Rev. Lett.* **74** 2447
[178] Harris S E and Hau L V 1999 *Phys. Rev. Lett.* **82** 4611
[179] Hau L V *et al* 1999 *Nature* **297** 594
[180] Kash M M *et al* 1999 *Phys. Rev. Lett.* **82** 5229
[181] Agrawal G P 1995 *Nonlinear Fiber Optics* 2nd edn (San Diego, CA: Academic)
[182] Boyd R W 2003 *Nonlinear Optics* 2nd edn (San Diego, CA: Academic)
[183] Ham B, Hemmer P and Shahriar M 1997 *Opt. Commun.* **144** 227
[184] Turukhin A V *et al* 2002 *Phys. Rev. Lett.* **88** 023602
[185] Bigelow M S, Lepeshkin N N and Boyd R W 2003 *Phys. Rev. Lett.* **90** 113903
[186] Bigelow M S, Lepeshkin N N and Boyd R W 2003 *Science* **301** 200
[187] Schwartz S E and Tan T Y 1967 *Appl. Phys. Lett.* **10** 4
[188] Bloembergen N and Shen Y-R 1964 *Phys. Rev.* **133** A37
[189] Mollow B R 1972 *Phys. Rev.* A **5** 2217
[190] Boyd R W *et al* 1981 *Phys. Rev.* A **24** 411
[191] Hillman L W *et al* 1983 *Opt. Commun.* **45** 416
[192] Agarwal G S and Dey T N 2004 *Phys. Rev. Lett.* **92** 203901
[193] Malcuit M S *et al* 1984 *J. Opt. Soc. Am.* B **1** 73

[194] Tewari S P and Agarwal G S 1986 *Phys. Rev. Lett.* **56** 1811
[195] Xiao M, Li Y-Q, Jin S-Z and Gea-Banacloche J 1995 *Phys. Rev. Lett.* **74** 666
[196] Budker D *et al* 1999 *Phys. Rev. Lett.* **83** 1767
[197] Agarwal G S, Dey T N and Menon S 2001 *Phys. Rev.* A **64** 053809
[198] Talukder Md A I, Amagishi Y and Tomita M 2001 *Phys. Rev. Lett.* **86** 3546
[199] Kim K *et al* 2003 *Phys. Rev.* A **68** 103810
[200] Bayindir M and Ozbay E 2000 *Phys. Rev.* B **62** R2247
[201] Fleischhauer M and Lukin M D 2000 *Phys. Rev. Lett.* **84** 5094
[202] Fleischhauer M, Yelin S F and Lukin M D 2000 *Opt. Commun.* **179** 395
[203] Lukin M D, Yelin S F and Fleischhauer M 2000 *Phys. Rev. Lett.* **84** 4232
[204] Mazets I E and Matisov B G 1996 *JETP Lett.* **64** 515
[205] Zibrov A S *et al* 2002 *Phys. Rev. Lett.* **88** 103601
[206] Liu C *et al* 2001 *Nature* **409** 490
[207] Phillips D F *et al* 2001 *Phys. Rev. Lett.* **86** 783
[208] Abella I D, Kurnit N A and Hartmann S R 1966 *Phys. Rev.* **141** 391
[209] Hahn E L 1950 *Phys. Rev.* **80** 580
[210] Allen L and Eberly J H 1975 *Optical Resonance and Two-Level Atoms* (New York: Wiley)
[211] Leung K P, Mossberg T W and Hartmann S R 1982 *Opt. Commun.* **43** 145
[212] Kimble H J 1998 *Phys. Scr.* **76** 127
[213] Bergmann K, Theuer H and Shore B W 1998 *Rev. Mod. Phys.* **70** 1003
[214] Cirac J I *et al* 1997 *Phys. Rev. Lett.* **78** 3221
[215] Kocharovskaya O, Rostovtsev Y and Scully M O 2001 *Phys. Rev. Lett.* **86** 628
[216] Leonhardt U and Piwnicki P 2001 *J. Mod. Opt.* **48** 977
[217] Bullough R K and Gibbs H M 2004 *J. Mod. Opt.* **51** 255
[218] Bullough R K *et al* 1976 *Opt. Commun.* **18** 200
[219] Reference [62] section 60
[220] Lamb H 1904 *Proc. Lond. Math. Soc.* **1** 473
[221] Schuster A 1904 *An Introduction to the Theory of Optics* (London: Edward Arnold) p 313
[222] Ziolkowski R W and Heyman E 2001 *Phys. Rev.* E **64** 056625
[223] Reference [62] section 64
[224] Schuurmans F J P *et al* 1998 *Phys. Rev. Lett.* **80** 5077
[225] Milonni P W and Maclay G J 2003 *Opt. Commun.* **228** 161
[226] Yariv A 1975 *Quantum Electronics* 2nd edn (New York: Wiley)
[227] Milonni P W 1995 *J. Mod. Opt.* **42** 1991
[228] Pafomov V E 1959 *Sov. Phys.–JETP* **36** 1321
[229] Ginzburg V L 1979 *Theoretical Physics and Astrophysics* (Oxford: Pergamon) ch 7
[230] Agranovich V M and Ginzburg V L 1966 *Spatial Dispersion in Crystal Optics and the Theory of Excitons* (New York: Wiley)
[231] Brevik I 1979 *Phys. Rep.* **52** 133
[232] Loudon R 2002 *J. Mod. Opt.* **49** 821
[233] Garrison J C and Chiao R Y 2004 *Preprint*
[234] Jones R V and Leslie B 1978 *Proc. R. Soc. Lond.* A **360** 347
[235] Pocklington H C 1905 *Nature* **71** 607
[236] Berman P R 2002 *Phys. Rev.* E **66** 067603
[237] Lakhtakia A 2003 *Electromagnetics* **23** 71
[238] Carniglia C K and Mandel L 1971 *Phys. Rev.* D **3** 280

[239] Clemmow P C 1966 *The Plane Wave Spectrum Representation of Electromagnetic Fields* (New York: Pergamon)
[240] de Fornel F 2002 *Evanescent Waves from Newtonian Optics to Atomic Optics* (Berlin: Springer)
[241] Carniglia C K, Mandel L and Drexhage K H 1972 *J. Opt. Soc. Am.* **62** 479
[242] Toraldo di Francia G 1960 *Nuovo Cimento* **16** 61
[243] Pendry J B 2000 *Phys. Rev. Lett.* **85** 3966
[244] Smith D R *et al* 2000 *Phys. Rev. Lett.* **84** 4184
[245] Shelby R A, Smith D R and Schultz S 2001 *Science* **292** 77
[246] Pendry J B 2004 *Contemp. Phys.* **45** 191
[247] Pendry J B and Smith D R 2004 *Physics Today* **57** 37
[248] Smith D R *et al* 2003 *Appl. Phys. Lett.* **82** 1506
[249] Garcia N and Nieto-Vesperinas M 2002 *Phys. Rev. Lett.* **88** 207403
[250] Pendry J B 2003 *Phys. Rev. Lett.* **91** 099701
[251] Sernelius B E 2001 *Surface Modes in Physics* (Berlin: Wiley–VCH)
[252] Ruppin R 2000 *Phys. Lett.* A **277** 61
[253] Ruppin R 2001 *J. Phys.: Condens. Matter* **13** 1811
[254] Nieto-Vesperinas M 2004 *J. Opt. Soc. Am.* A **21** 491
[255] Merlin R 2004 *Appl. Phys. Lett.* **84** 1290
[256] Gómez-Santos G 2003 *Phys. Rev. Lett.* **90** 077401
[257] Synge E H 1928 *Phil. Mag.* **6** 356
[258] Ash E A and Nicholls G 1972 *Nature* **237** 510
[259] Pohl D W, Denk W and Lanz M 1984 *Appl. Phys. Lett.* **44** 651
[260] Pendry J B *et al* 1996 *Phys. Rev. Lett.* **76** 4773
[261] Sievenpiper D F, Sickmiller M E and Yablonovitch E 1996 *Phys. Rev. Lett.* **76** 2480
[262] Pendry J B *et al* 1998 *J. Phys.: Condens. Matter* **10** 4785
[263] Pendry J B *et al* 1999 *IEEE Trans. Microwave Theory Tech.* **47** 2075
[264] Kock W E 1948 *Bell System Tech. J.* **27** 58
[265] Collin R E 1960 *Field Theory of Guided Waves* (New York: McGraw-Hill)
[266] Shelby R A *et al* 2001 *Appl. Phys. Lett.* **78** 489
[267] Parazzoli C G *et al* 2003 *Phys. Rev. Lett.* **90** 107401
[268] Houck A A, Brock J B and Chuang I L 2003 *Phys. Rev. Lett.* **90** 137401
[269] Grbic A and Eleftheriades G V 2002 *J. Appl. Phys.* **92** 5930
[270] Eleftheriades G V, Iyer A K and Kremer P C 2002 *IEEE Trans. Microwave Theory Tech.* **50** 2701
[271] Iyer A K, Kremer P C and Eleftheriades G V 2003 *Opt. Express* **11** 696
[272] Grbic A and Eleftheriades G V 2004 *Phys. Rev. Lett.* **92** 117403
[273] Jordan E C and Balmain K G 1968 *Electromagnetic Waves and Radiating Systems* 2nd edn (Englewood Cliffs, NJ: Prentice-Hall)
[274] Liu L, Caloz C and Itoh T 2002 *Electron. Lett.* **38** 1414
[275] Yablonovitch E 1987 *Phys. Rev. Lett.* **58** 2059
[276] Notomi M 2000 *Phys. Rev.* B **62** 10696
[277] Kosaka H *et al* 1998 *Phys. Rev.* B **58** R10096
[278] Luo C *et al* 2002 *Phys. Rev.* B **65** 201104
[279] Luo C *et al* 2003 *Opt. Express* **11** 746
[280] Povinelli M L *et al* 2003 *Appl. Phys. Lett.* **82** 1069
[281] Cubukcu E *et al* 2003 *Phys. Rev. Lett.* **91** 207401
[282] Parimi P V *et al* 2003 *Nature* **426** 404

[283] Fang N *et al* 2003 *Opt. Express* **11** 682
[284] Podolskiy V A, Sarychev A K and Shalaev V M 2003 *Opt. Express* **11** 735
[285] Ziolkowski R W 2003 *Opt. Express* **11** 662

Index

spontaneous emission, 89, 90, 95, 111, 113, 124, 125, 136, 143, 177
spontaneous emission noise, 111, 112, 128
spontaneous emission rate, 87
spontaneous parametric down-conversion, 97
spontaneous parametric down-conversion, inhibited, 97
spontaneous radiation noise, 94
spooky action at a distance, 59
stimulated emission, 75, 89, 90, 120
stopped light, 164, 165, 169, 172, 175–179
subwavelength resolution, 206, 209, 210
superfluorescence, 113, 115, 116, 118, 122, 125
superfluorescent radiation, 116
superlensing, 205, 206, 208–210
superluminal communication, 49, 89, 92, 94
superluminal group velocity, 30, 35, 47, 108–111, 122–125, 131
superluminal propagation, observability, 111
superluminal signal velocity, 17
superposition state, 89
surface modes, 206–208, 210, 212
surface plasmons, 212

tachyon, 111
telegrapher's equation, 226
teleportation, 91
thin wire mesh, transmission characteristics, 215
thin wire structures, 213
thin wires, 214
Thomas–Reiche–Kuhn sum rule, 16
threshold detector level, 109
total internal reflection, 197, 200, 201
transit-time broadening, 140, 150
transmission line, 226, 228
transmission line metamaterials, 226, 228–230
transparency window, 143, 170, 177, 178
tunnelling, 39, 40

ultracold gas, 146, 148, 152
ultraslow group velocity, 172, 173, 176

vacuum field, 80
vacuum field fluctuations, 118, 123
vacuum fluctuations, 80
velocity of light, 3
virtual transitions, 80, 81

Weisskopf–Wigner approximation, 65
wire mesh, 212

Milton Keynes UK
Ingram Content Group UK Ltd.
UKHW040106071024
449327UK00019B/859